Cavitation and the Centrifugal Pump

Cavitation and the Centrifugal Pump

A Guide for Pump Users

Edward Grist

TAYLOR & FRANCIS
ALERE FLAMMAM
Founded 1798

USA	Publishing Office:	TAYLOR & FRANCIS
		325 Chestnut Street
		Philadelphia, PA 19106
		Tel: (215) 625-8900
		Fax: (215) 625-2940
	Distribution Center:	TAYLOR & FRANCIS
		47 Runaway Road, Suite "G"
		Levittown, PA 19057-4700
		Tel: (215) 269-0400
		Fax: (215) 269-0363
UK		TAYLOR & FRANCIS
		1 Gunpowder Square
		London EC4A 3DE
		Tel: +44 171 583 0490
		Fax: +44 171 583 0581

CAVITATION AND THE CENTRIFUGAL PUMP: A Guide for Pump Users

1 2 3 4 5 6 7 8 9 0

Edited by Edward A. Cilurso and Jean Anderson. Cover design by Joseph Dieter. Printed by Edwards Brothers, Ann Arbor, MI, 1998.

A CIP catalog record for this book is available from the British Library.
   The paper in this publication meets the requirements of the ANSI Standard Z39.48-1984 (Permanence of Paper).

Library of Congress Cataloging-in-Publication Data
        Grist, Edward.
                Cavitation and the centrifugal pump: a guide for pump users /
                Edward Grist.
                p.   cm.
        Includes bibliographical references and index.
        ISBN 1-56032-591-7 (alk. paper)
        1.   Centrifugal pumps – Cavitation – Handbooks, manuals, etc.
        I. Title
        TJ919.G75   1998   98-17168
        621.6'7–dc21   CIP

ISBN 1-56032-591-7 (case)

CONTENTS

PREFACE

Cavitation in a centrifugal pump that limits performance unacceptably or compromises safe operation is evidence of an engineering failure. Numerous options are available that avoid such undesirable consequences. Diligence in appreciating the risks inherent in design proposals and taking appropriate action to reduce them to an acceptable level is necessary.

From the plant failures I have seen and remedial works I have had to undertake, both in the United Kingdom and world wide, it is clear to me that there is a need, particularly by pump users, to acquire a comprehensive insight into the industrially significant problems that can arise from the occurrence of cavitation in a centrifugal pump. This book is aimed at meeting this need.

The fundamentals that must be fully understood to provide a basis for developing commercially viable guidelines are detailed in Chapters 2, 3, 4 and 5. These are essential to subsequent chapters where the means of dealing with the risks arising from cavitation are enumerated. The academic content is reduced to that essential to understanding the logic of the arguments put forward; the inclusion of unnecessary diversions has been avoided.

The acquisition of knowledge by those working in the engineering industry, including pump designers, pump system designers and pump users, has always rested heavily on regular reviews of past experience. Evidence from the real world can mercilessly denounce an unjustifiable hypothesis just as it can give added confidence that a chosen path has viability. My earliest papers on the subject of cavitation erosion erred, necessarily, on the side of caution. A further twenty years of evaluating operational records leads me now to reduce some of the margins by which confidence in securing erosion-free operation can be given. The evidence now available makes it extremely unlikely that further reductions in these margins will take place.

My guidelines on acceptable operating conditions, both by net positive suction head (NPSH) margins and by minimum flowrates, can, by their very nature, only be general. Their applicability to a particular pumping system needs to be assessed carefully. I have described the evidence and logic by which I have arrived at these guidelines to assist those whose task it is to bear the heavy responsibility of making such an assessment. Information is presented in a format that addresses the needs of pump users. A method is described by which likely commercial choice and the boundaries of risk can be evaluated before a pump supplier is approached. The same logic helps deal with problems arising from cavitating pumps in service.

I do not profess to have covered in this book every aspect of cavitation in centrifugal pumps. Those of significance for the majority of pump applications are included. My hope is that by providing a greater understanding my book will significantly reduce the incidence of unacceptable performance caused by cavitation in centrifugal pumps.

E. Grist
Congleton, England
May 1998

ACKNOWLEDGEMENTS

Illustrations and data from existing publications and other sources have proved invaluable in compiling my presentation. All the work of individuals is referenced at its point of inclusion. I gratefully acknowledge the permission to publish material in this book given by the following organisations:

American Society of Mechanical Engineers, New York, NY, USA
British Pump Manufacturers' Association, Birmingham, UK
Hopkinsons Ltd., Huddersfield, UK
Ingersoll – Dresser Pump Company, Liberty Corner, NJ, USA
Institution of Mechanical Engineers, London, UK
    and their publishers MEP, Bury St. Edmunds, UK
Masoneilan Division, Dresser Produits Industriels, Paris, France
National Engineering Laboratory, Glasgow, UK
Newcastle Discovery Museum, Newcastle, UK
Rütschi Pumpen AG, Brugg, Switzerland
Science Museum/Science & Society Picture Library, London, UK
Sulzer Pumps Ltd., Winterthur, Switzerland
Valtek Engineering Ltd., Pershore, UK
Weir Pumps Ltd., Glasgow, UK
Western Power Corporation, Perth, Australia
Westinghouse Electric Corporation, Pittsburgh, PA, USA
Yarway Division, Keystone Valve (UK) Ltd., Glasgow, UK

I also acknowledge the help in securing photographs and other data given by my peers in the UK pump and valve industries particularly John Frew, Angus Grant, Ron Palgrave, Les Statham and Ashley Crossland. Additionally, for encouraging my efforts to produce this book, I thank C. A. Eubanks of the Westinghouse Electric Corporation and Andrew Lichtarowicz of Nottingham University. Finally I wish to place on record my thanks to the three directors with engineering responsibility who, over a number of years, encouraged me to obtain the experience necessary to write the book. They are H. R. Wilshaw of Holden & Brooke Ltd., Donald McLellan of Crane Ltd. and Tom Leith of Weir Pumps Ltd.

NOMENCLATURE

A	flow area between impeller blades (total)
B	a power in the relationship $NPSH \propto n^B$
b	impeller flow passage width at outlet
CLA	impeller surface roughness (centre line average height)
C_P	specific heat of liquid
D	impeller outer diameter
D_p	inlet pipe diameter
f	functional dependence
g	gravitational constant
H	head
H_t	head preventing ebullition
H_1	inlet total head
H_2	outlet total head
H_T	head available to suppress ebullition
ΔH	generated head
ΔH_G	ΔH at duty point ("guarantee" conditions)
ΔH_P	plant aggregate ΔH per pumping application
δH	head amplitude of a cavitation surge
h	enthalpy
h_F	total variable ("frictional") head loss (equation 4.7)
h_{F1}	component of h_F up to the pump inlet
h_{F2}	component of h_F after the pump outlet
h_{sys}	total system head loss (equation 4.5)
h_S	total static head loss (equation 4.6)
h_{S1}	component of h_S up to the pump inlet
h_{S2}	component of h_S after the pump outlet
K	thermal diffusivity $= \dfrac{k_1}{\rho_L C_p}$
k_1	thermal conductivity of liquid
k_2	a constant for a particular system (equation 4.7)
k_3	a constant for a particular pump (equation 5.8)
k_4	a Thoma/specific speed constant (equation 15.3)
L	thermal diffusion length
l	flow path length along impeller blade from inlet to outlet diameter
l_c	cavitation zone length
$l_{c'}$	distance along impeller blade at which r_{max} occurs
NPSH(I)	NPSH at visually observable cavitation inception

NPSH	Net positive suction head (ISO 3555-1977 definition)
NPSH(A)	NPSH available at the pump inlet
NPSH(B)	NPSH at generated head breakdown conditions
NPSH(R)	NPSH required to prevent unacceptable cavitation
ΔNPSH	NPSH reduction resulting from inlet pipe transients
δNPSH	pressure producing cavity growth at breakdown NPSH(I) - NPSH(B)
NPSH(xpc)	NPSH at x% head-drop where x can have any value
NPSH(3pc)	NPSH at 3% head-drop
NPSH(xmm)	NPSH at xmm cavity length where x can have any value
NPSH(4mm)	NPSH at 4mm cavity length
NPSH(spl)	NPSH at unspecified sound pressure level
n	shaft rotational speed
n_S	specific speed (defined by equation 3.4)
n_{SS}	suction specific speed (defined by equation 3.11)
P_A	pump input power
P_{AG}	P_A at duty point ("guarantee" conditions)
P_{AO}	P_A at zero flowrate
P_1	pressure driving cavity growth and collapse in a stationary liquid
ΔP	generated pressure
Δp	pressure decay in inlet vessel
δp	pressure amplitude of a cavitation surge
p	pressure
p_b	barometric pressure
p_d	pressure produced by duct constraints
p_g	pressure generated in an impeller by centrifugal effects
p_{gc}	pressure generated by the end of the cavitating zone
p_{gl}	pressure generated in non-cavitating zone of the impeller
p_n	pressure producing cavity growth
p_{st}	restraining pressure resulting from surface tension
p_{th}	pressure equivalent of thermodynamic property effects
p_{vap}	vapour pressure
p_{V1}	pressure in inlet vessel
p_{V2}	pressure in outlet vessel
p_x	cavity internal pressure
p_1	pressure at pump inlet
p_2	pressure at pump outlet
p_3	pressure beyond an outlet non-return valve
Q	pump volumetric flowrate
Q_1	Q at the pump inlet
Q_2	Q at the pump outlet
Q_{bep}	Q at best efficiency point
Q_G	Q at duty point ("guarantee" conditions)

Q_H	maximum Q in operating range
Q_L	minimum Q in operating range
Q_{lo}	leak-off rated flowrate
Q_M	pump mass flowrate
Q_m	inlet vessel mass flowrate
Q_P	plant aggregate Q per pump application
Q_{tr}	lower limit of transitional flowrate band during leak-off operation
Q_V	minimum continuous flowrate to prevent vapour locking
Q_{WC}	Q at worst case conditions
q	dryness fraction (Chapter 8 only)
q	heat content
Δq	change in heat content
q_{fo}	forward flowrate during leak-off operation
q_{lo}	leak-off flowrate during leak-off operation
q_{vap}	flowrate of vapour by volume
R	cavitation protection ratio (defined by equation 4.2)
R(3pc)	**R** based upon 3% head-drop data (equation 4.3)
R(4mm)	**R** based upon 4mm cavity length data (equation 4.4)
R_O	Universal gas constant
r	cavity radius
r_o	radius of cavity in equilibrium
r_{max}	maximum cavity radius
S	specific volume
T	temperature of liquid (at impeller blade entry where applicable)
T_o	temperature of liquid in pump inlet pipe
T_r	cavity wall temperature
t	time
t_{fo}	time for flowrate to change from zero to Q_{lo}
t_{lo}	time for flowrate to change Q_{lo} to Q_{tr}
t_p	transit time from inlet vessel to pump inlet ($t_p = 0$ when $t = 0$)
t_x	time increment (Chapter 6)
V_C	volume of liquid contained in pump casing
V_L	specific volume of liquid
V_v	specific volume of vapour
v_1	mean velocity of liquid at pump inlet
v_2	mean velocity of liquid at pump outlet
W	vessel liquid mass
ΔW	change in liquid mass
Y	area ratio (Fig. 3.6)
z	number of impeller stages
z_C	cavitation site density (cavities/unit time)
z_1	static head - gauge centreline to reference plane (Fig. 2.5)
z_2	static head - gauge centreline to reference plane (Fig. 2.5)

β   the group $\left[\dfrac{3K}{\pi}\right]\dfrac{p_{vap}{}^2\,\lambda^2}{k_1\,R_O\,T^3}$

λ   latent heat of vaporisation
φ   rate of heat input to fluid from pump power losses
η   pump efficiency (%)
η_{bep} η at best efficiency flowrate (%)
φ   pipichum value (defined by equation 10.4)
ν   kinematic viscosity
π   3.14
ρ_L liquid density
σ   Thoma cavitation parameter (defined by equation 15.1)
σ_T surface tension
τ   churned mass multiplier (defined by equation 10.5)

Special subscripts (Chapter 6)
in   inflow
out   outflow
(p)   relating to pump inlet
(v)   relating to the inlet vessel
(t)   at time t

Chapter 1

Cavitation — An Unacceptable Phenomenon

1.1 Cavitation Defined

Cavitation is defined as the occurrence of vapour-filled cavities in a liquid.

The definition describes all conditions in a centrifugal pump from the growth and collapse of small bubbles in the pumped liquid to the occurrence of large amounts of vapour that are sufficient to vapour lock a pump in a predominantly liquid-filled system.

1.2 Unacceptability Defined

A cavitating centrifugal pump can be found unacceptable in a number of ways. Those of industrial significance can be categorised as follows.

Category	Unacceptable Feature
1	Hydraulic Performance Loss
2	Cavitation Surging — Hydrodynamically Induced
3	Cavitation Surging — Thermodynamically Induced
4	Cavitation Erosion

The potentially serious consequences that can arise when a centrifugal pump cavitates are loss of economic performance, damage to expensive equipment, and, in the extreme, loss of life. Such consequences are unacceptable.

The ways in which unacceptable behaviour can arise are presented in outline in this chapter. This highlights the need to understand the basic pump terminology, cavitation performance characteristics, and the limited amount of engineering fluid mechanics presented in Chapters 2, 3, 4 and 5.

Later chapters expand on the categories of unacceptability. They give a more detailed description of the underlying mechanisms. They also provide an engineering methodology that identifies risk and describes how such risk can be minimised.

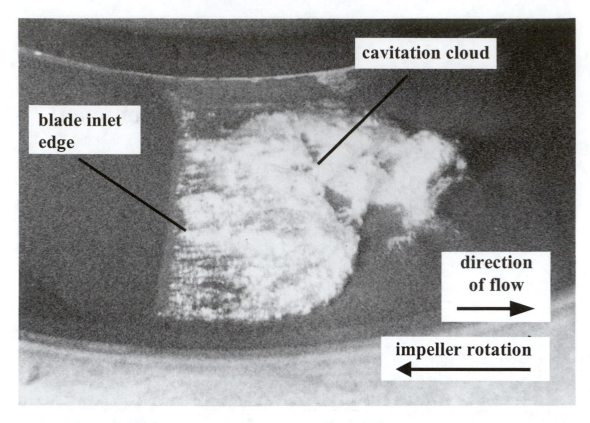

Fig. 1.1   Cavitation on a centrifugal pump impeller blade

Courtesy of Weir Pumps Ltd.

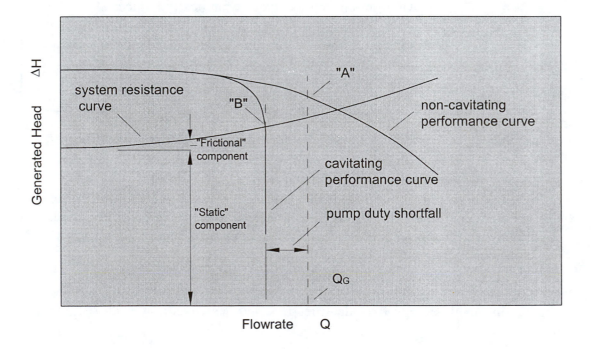

Fig. 1.2   Pump performance and system resistance curves

1.3 The Categories of Unacceptability — A Brief Résumé

1.3.1 *Category 1. Hydraulic performance loss*

The presence of cavitation in a centrifugal pump impeller, such as shown in Fig. 1.1, can give rise to a loss of generated head. Whilst a modest amount of cavitation may have no measurable effect, a large volume of cavities can result in the generated head being reduced considerably.

Two conditions are of industrial significance:

1. Failure to meet the required duty thereby resulting in long-term economic loss.
2. A catastrophic reduction in head caused by large volumes of cavitation, which give rise to vapour locking of the pump and lead to an immediate cessation of pumping. This is sometimes called vapour binding. Mechanical damage to the rotating element of the pump nearly always occurs as a consequence.

Pump duty shortfall. In a centrifugal pump the local pressure drops experienced by the liquid flowing within its passages determine where cavitation occurs. This is usually close to the inlet tips of the impeller blades. Cavitation can be suppressed by raising the pressure in the region of its occurrence. In a near-stationary liquid a pressure a little above vapour pressure will suffice. In a cavitating centrifugal pump the effect of continuing to increase the pressure in the pipe immediately prior to the pump inlet is to reduce and eventually eliminate the presence of cavitation.

Centrifugal pump performance is usually expressed as a curve showing generated head ΔH plotted against volumetric flowrate Q as shown in Fig. 1.2. Typically a pump is tested with an excess of inlet pipe pressure to ensure cavitation does not occur; the non-cavitating pump performance curve shown is obtained. Insufficient provision of inlet pipe pressure to prevent cavitation leads to a reduction in achievable generated head at higher flowrates; the cavitating performance curve shown is then obtained.

Every pump has to overcome a system resistance which comprises fixed ("static") and variable ("frictional") components. The pump operates at the intersection of the system resistance curve and the pump performance curve appropriate to the particular inlet pressure conditions. The pump duty, point "A", is where the pump is intended to operate. A test under non-cavitating conditions would confirm its acceptability for this task since the system resistance curve can be raised by increasing the "frictional" component, typically by the partial closure of an outlet control valve. A cavitating pump would operate at point "B" instead of the required point "A" resulting in a substantial shortfall in flowrate. Where pump flowrate is directly linked to plant output this can represent an unacceptable economic loss. Where conditions allow cavitation to persist indefinitely the cumulative loss can be considerable.

Chapter 4 describes how to quantify the inlet conditions controlling cavitation in a centrifugal pump at the pump duty and over the pump operating range. Chapter 6 explore how to ensure the pump duty requirement is met.

Vapour locking. As flow through a pump is reduced an impeller becomes more and more inefficient. Much of the energy expended in the associated erratic turbulent internal flows goes into heating the liquid. Should throughflow become very small or

(a) Excessive casing clearance wear impeller

(b) Neck ring seizure onto

Fig. 1.3 Pump damage after a vapour lock

Courtesy of Western Power Corporation

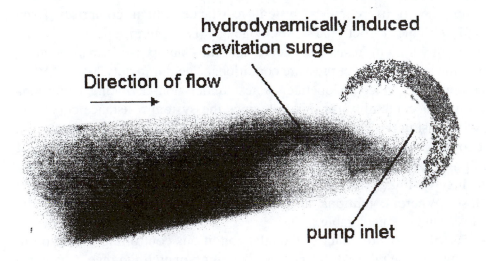

Fig. 1.4 Hydrodynamically induced cavitation surging

Courtesy of Ingersoll-Dresser Pump Company

stop altogether then practically all the power absorbed by the impeller goes into producing heat. Continuous operation in this mode ultimately leads to vaporisation of the liquid within the pump. In low-energy pumps (e.g. domestic central heating pumps) it can take many hours to produce sufficient cavitation for a full vapour lock. In high-energy pumps (e.g. large power station feedwater pumps) it can take less than one second.

The consequences of vapour locking are unacceptable. Many pumps, especially multistage pumps, incorporate internal clearances such as neck rings, interstage bushes, or balance drums, which act as pumped-liquid lubricated bearings. These clearances support the radial loads on the pump rotor, which originate principally from the unbalanced hydraulic pressure contour surrounding it. Without internal support the resulting rotor deflections lead to contact between rotating and stationary components and seizure ensues. Typically this appears as shown in Fig. 1.3.

Some pump designs include mechanical seals. These often depend upon the pumped liquid being present to provide a lubricating film between rotating and stationary components. Rapid mechanical failure can ensue if this supply of liquid is removed by vapour locking.

Centrifugal pumps can be vapour locked when NPSH(A), a measure of conditions suppressing cavitation defined later, falls to a very low value. This can be the consequence of a malfunction of the low-flow protection system. It can also be the result of a plant-induced operating transient which significantly reduces the pressure in the inlet vessel that, as a consequence, reduces the pressure in the pump inlet pipe.

Chapter 6 explores the problems and the available solutions.

1.3.2 *Category 2. Cavitation surging — Hydrodynamically induced*

At part-load flowrates, typically below about 50% of best efficiency flowrate, the angular mismatch between incoming liquid flowpaths and impeller blades is significant. This is usually more so at the leading edge of blades near the outer diameter of the impeller eye. In many impellers a recirculation develops in the low-pressure region behind each blade making it possible for flow to pass back into the inlet pipe at its outer diameter. This recirculation becomes progressively stronger as flowrate is reduced. Where the pressure drop behind the impeller blade is compounded by a sufficiently low inlet pressure cavitation occurs.

Cavitation is normally suppressed in milliseconds as it enters the strongly increasing pressure gradient associated with flows along a centrifugal pump impeller blade at near best efficiency conditions. Where backflow recirculation into the inlet pipe occurs the cavities find themselves in the relatively much weaker pressure gradient of the inlet pipe. Their growth and collapse take longer, and this allows them to be propagated impressive distances back down the inlet pipe. At very low flowrates and with a straight inlet pipework configuration this spiralling outer core of cavitating liquid can travel several pipe diameters against the general direction of flow before returning to the impeller. An example of this is shown in Fig. 1.4.

The production of vapour on the impeller blade surface is disrupted by the returning recirculation flow of liquid-borne cavities. The whole process becomes unstable. A periodic surging motion is set up that is predominantly dependent upon the recycle time of the recirculating liquid. Typically this surge has a frequency of 2 Hz to 10 Hz.

Throughout the region where low-pressure gradients exist in the inlet pipe such cavitation surging can persist. This can travel through to the outlet side of low generated-head types of impellers such as inducers and those for axial flow machines. In centrifugal pump impellers where flow no longer enters the high-pressure region in the normal direction of flow the growth and collapse of large volumes of cavitation occurs.

Experience shows the intensity of cavitation surging to vary considerably for different pump-impeller and inlet-pipework configurations. Some combinations give rise to a leisurely "sloshing" motion. Others give rise to a violent piston-like effect that is accompanied by a powerful "chuffing" noise — very reminiscent of a reciprocating steam engine exhaust. The resulting pressure pulsations often produce severe pipework vibrations. The wild variations in axial thrust and torsional loading on the pump rotor frequently lead to thrust bearing failure and occasionally to shaft breakage.

It is important to note that this form of cavitation surging is (a) independent of the plant design beyond the run of inlet piping immediately preceding the pump and (b) is more easily propagated where impeller and inlet piping are conducive to axially symmetric motions. An end-suction pump with a long straight inlet pipe is a common example of the latter.

Chapter 7 explores the problem and the available solution.

1.3.3 *Category 3.   Cavitation surging — Thermodynamically induced*

The consequences of pressure containment failure are potentially so horrific it is essential to appreciate all the underlying pump related mechanisms that could give rise to it. A high-energy centrifugal pump, such as a power station feedwater pump, has the capability of initiating a catastrophic failure of the pump inlet pipe. Such a breach creates large changes in momentum. The forces that then arise can be more than sufficient to rip pipework from supporting structures. More serious consequences can occur when the liquid in the inlet pipe is at a temperature much greater than that necessary to keep it as a liquid on release into the environment. The explosive escape of the pumped liquid as it vaporises can seriously injure individuals caught by the scalding blast.

The damage to the power station feedwater pump shown in Fig. 1.5 resulted from a catastrophic failure of the pump inlet pipe. The explosive escape of hot condensate caused a number of fatalities. In this particular instance the cause was attributed to flow-accelerated corrosion in the inlet pipework. The example is used to focus clearly on the risk which might be present should pipework over-pressurisation occur.

Although catastrophic failures associated with centrifugal pumps are rare it is necessary to study the features inherent in each centrifugal pump and plant design combination to ensure that essential reliability and redundancy is identified and an appropriate protection system is provided. Safe operating conditions for high-energy pumps are crucially dependent upon plant design. The provision of low-flow protection, always essential for such pumps, takes on greater importance in particular plant design arrangements. It is necessary to understand why this is. It is also necessary

Fig. 1.5   Catastrophic failure of pump inlet pipework

to understand the potentially serious consequences arising from the maloperation or the operational failure of such low-flow protection.

High-energy centrifugal pumps that operate at zero or very low flowrate will experience vapour locking. The process of vapour locking can take place in less than one second. The amount of vapour produced in this time will approximate to the volume of the impeller plus that of the liquid passages within the pump and inlet pipework where liquid is churned by the impeller. In a typical boiler feedwater pump this might be 0.5 m^3. The sudden need to accommodate the additional vapour volume needs to be addressed. In inlet pipework systems where the pumped liquid is drawn from a vessel which has a free surface the increase in pressure in the pump casing may be small, particularly if the inlet pipe length is very short. Some types of shaft sealing systems may also provide relief by allowing escape of liquid. The pipework and pump casing also grow slightly to provide added volume as high pressures increase stresses in pipework materials toward their elastic limit.

The pressure generated in the inlet pipe of a pump that is preceded by a non-return valve poses a potentially more serious risk. In a system where the pump discharges liquid into a high-pressure system through a non-return valve the structural integrity of the pressure boundary may be challenged. This occurs when pump speed is reduced below that which is necessary to provide a pressure sufficient to open the outlet non-return valve. Failure to provide adequate relief between upstream and downstream non-return valves means that the pressure rapidly increases as the vapour being generated by the pump tries to expand in the near incompressible trapped liquid. Two outcomes are possible:

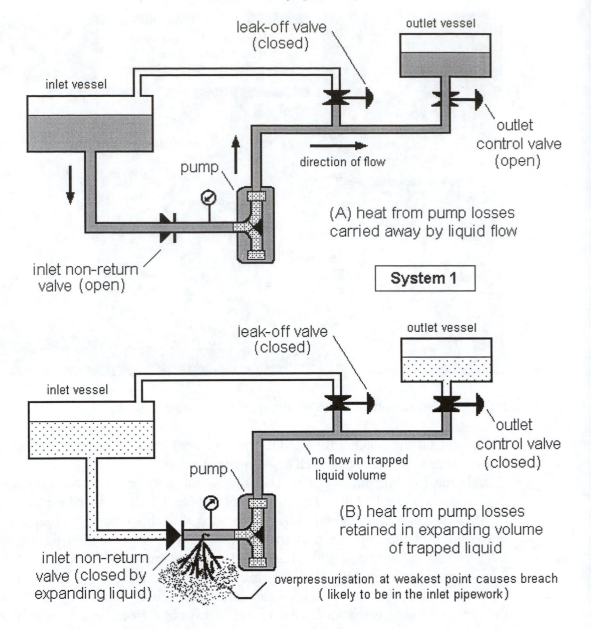

Fig. 1.6   Over-pressurisation of pump inlet pipework resulting from
thermodynamically induced cavitation surging

1. Where the inlet pipework is sufficiently strong the pressure in the pump and
the inlet pipework rises to the pressure in the pipework beyond the outlet non-
return valve.

As the cavity continues to expand this valve opens. The pump then tries to
commence normal pumping. A small amount of new liquid is drawn into the
impeller, the generated head then increases, and some of the vapour/liquid mixture
is displaced toward the outlet pipe thereby allowing the inlet pressure to fall. This
relieves the outlet pressure and allows the outlet non-return valve to close again.
The cycle then repeats itself. The pipework vibrations associated with such
thermodynamically induced surging are often frightening to witness.

2. Where the inlet pipework is not strong enough the pressure containment boundary fails catastrophically. Rupture occurs at the weakest point and this can reveal a design deficiency other than in the pump. The role of thermodynamically induced cavitation surging in providing the pressure pulse to finally breach the pressure boundary in such circumstances can become overlooked.

In a system such as that shown in Fig. 1.6 where flow through the pump is stopped — typically by the rapid closure of an outlet control valve — the outcome is one in which the mechanical integrity of the pump, pump pipework, and inlet non-return valve can be challenged. In practice this scenario occurs when the time to close the outlet control valve is much less than the time to open a flowpath through the low flow protection "leak-off" line. Mechanical damage is inevitable unless relief of the pressure produced by the expanding liquid/vapour mixture is available.

Chapter 8 explores the problem and the available solution.

1.3.4 *Category 4. Cavitation erosion*

When cavitation occurs cavitation erosion might take place — or it might not! It is important to appreciate this lack of inevitability when considering the risk of cavitation erosion and when attempting to quantify its intensity by extrapolating from the damage observed on other impellers. Equally, it is necessary to understand why such uncertainty exists.

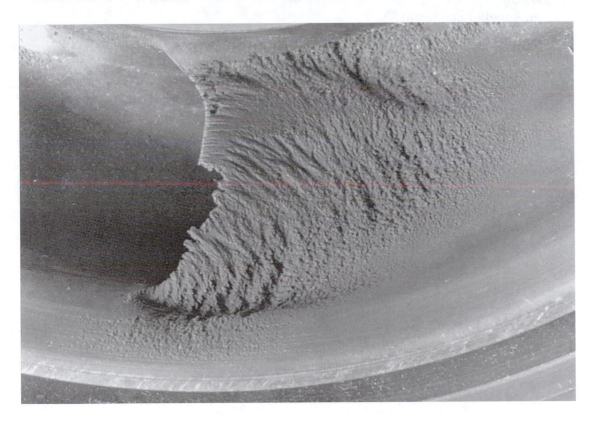

Fig. 1.7 Typical cavitation erosion damage on an impeller blade

Courtesy of Sulzer Pumps Ltd.

Cavitation erosion is caused as vapour-filled cavities enter a region of higher pressure within a pump and collapse. When this violently implosive cavitation process takes place in close proximity to a metal surface a damage mechanism that produces local pitting can result. The uncertainty surrounding both the onset of erosion and the intensity of attack when cavitation does occur in a commercially manufactured pump arises from the lack of awareness of how near the cavities get to the pump internal surfaces. Collapse of cavities within the body of liquid flowing through the pump that are distant from an internal surface may not result in cavitation erosion damage. However, as the point of low pressure that initiates cavitation is nearly always at such a surface (e.g. close to the impeller blade tip at inlet), damage is more often than not observed. A characteristic audible crackling often accompanies such a damaging cavity collapse.

The probability of damage increases where a pump is used over a range of flowrates since the changing flowpaths of cavitation within the liquid in the machine are more likely at some point to impinge upon an internal surface. Interestingly, examination of
damaged pumps shows that sometimes when cavitation erosion does occur it may be inconsequential to the safe running of the pump even after cavitation attack has persisted for thousands of hours. Conversely cavitation damage, such as that shown in Fig. 1.7, is sometimes observed after a matter of a few hours. Where cavitation erosion damage is so severe or cracks are initiated the structural failure of an impeller can result.

Chapter 9 explores the problem and the available solutions.

Chapter 2

Independent Variables and Terminology

The independent variables that are used to describe centrifugal pump performance come from a background where choice has been exercised for particular beneficial reasons. Also, custom and practice in the pump industry have led to a number of terms being used that have a specific meaning when applied to the technology. It is useful to review these independent variables and terminology.

2.1 Primary Independent Variables

Non-cavitating pump performance is, for a majority of applications, found by test to be completely defined by four independent variables:

(1)	Generated head	ΔH
(2)	Volumetric flowrate	Q
(3)	Shaft rotational speed	n
(4)	Impeller outer diameter	D

Cavitating pump performance is, for a majority of applications, found by test to be completely defined by four independent variables:

(5)	Net positive suction head	NPSH
6 (2)	Volumetric flowrate	Q
7 (3)	Shaft rotational speed	n
8 (4)	Impeller outer diameter	D

From the bracketed numbers it can be seen that five independent variables between them describe both non-cavitating and cavitating performance. These are considered in turn.

(1) *Generated head* ΔH. Often referred to as "pump total head" or more simply as "head", this is the rise in head between the inlet (suction) and outlet (discharge) branches of a pump. It is a primary variable.

The choice of generated head rather than generated pressure is significant and intentional. Centrifugal pump performance is determined by the kinetic energy of the pumped liquid as it leaves the impeller at a velocity close to its peripheral velocity. The generated head, which has a value directly proportional to this energy, is constant for a given shaft rotational speed irrespective of the liquid density. The use of this single independent variable to describe the added energy aids analysis. It also mirrors the way in which system losses — which determine the pump duty — are calculated.

In a strict analytical sense the product of generated head and gravitational constant

11

is the correct measure of pump performance. However, industrial practice is to assume the gravitational value is unchanging and to describe pump performance by head measurement alone. This has the obvious perceptual advantage — which was the original basis for the choice — of being relatable to physical heights in pumping plant design.

It must be remembered that when generated head is used in the dimensional analysis of pump performance its product with the gravitational constant has to be included.

(2) *Volumetric flowrate* Q. Often referred to simply as "flowrate", this is a primary variable. The choice of volumetric flowrate rather than mass flowrate is significant and intentional. Centrifugal pump performance is controlled by the velocity of liquid passing through the impeller outlet flow passage area so making the volumetric flowrate a constant for a given speed irrespective of the liquid density. The use of this single independent variable to describe flowrate aids analysis and also mirrors the way in which system losses — which determine the pump duty — are calculated.

(3) *Shaft rotational speed* n. Often referred to simply as "speed", this is a primary variable.

(4) *Impeller outer diameter* D. This is a primary variable. As a length measurement it characterises the dimensional proportions of a pump. It also provides the basis for performance analysis.

(5) *Net positive suction head* NPSH. Usually referred to simply by the initials NPSH, this is a primary variable. A measure of conditions suppressing cavitation is the difference between the inlet total head (which includes the velocity head in the inlet pipe) and the head equivalent of the vapour pressure of the liquid being pumped. This head difference is termed "net positive suction head".

Within an impeller the local pressure controlling cavitation is related to the interaction between rotating impeller blades and liquid flow paths. The head drop on the downstream side of a blade results from the kinetic energy loss due to the mismatch of the two. A true measure of pump performance is the product of NPSH and the gravitational constant. Industrial practice again assumes the gravitational constant is unchanging and cavitating pump performance is usually described by an NPSH measurement alone. It must be remembered that when NPSH is used in the dimensional analysis of pump performance its product with gravitational constant has to be included.

The choice of NPSH to measure cavitation has a perceptual advantage where a pump draws from a vessel containing a liquid at its boiling point and the inlet velocity head is insignificant. A large number of pump installations where cavitating problems occur match this description. In these circumstances the physical height of the liquid surface level to pump shaft centreline is a close approximation to the value of NPSH at the pump inlet.

2.2 Secondary Independent Variables

A number of other independent variables are known to influence the performance of a centrifugal pump in particular circumstances. For the majority of pump applications where cold water or liquids having properties similar to those of cold water are pumped they are insignificant. They are regarded as secondary independent variables.

The circumstances in which these secondary independent variables need to be taken into account and a methodology for dealing with the risks and uncertainties they bring to engineering applications are discussed in Chapter 3.

The secondary independent variables fall into two groups: Group 1 for those having their principal effect on non-cavitating performance and Group 2 for those having their principal effect on cavitating performance.

Group **1**

Kinematic viscosity	ν
Impeller surface roughness	CLA

Group **2** *(Thermodynamic properties affecting cavity dynamics)*

Vapour pressure	p_{vap}
Latent heat of vaporisation	λ
Specific volume of vapour	V_v
Liquid density	ρ_L
Thermal conductivity of liquid	k_l
Specific heat of liquid	C_p

2.3 Hydraulic Configuration Conventions

2.3.1 *Centrifugal pump impeller design terminology*

A typical centrifugal pump impeller is shown in Fig. 2.1. Flow enters the impeller through a single "eye" and passes to blades contained within front and back shrouds.

These blades "impel" the pumped liquid through the pump whilst at the same time moving it radially outward from the pump shaft. Liquid leaving the periphery of the impeller does so at a higher circumferential velocity. Energy is gained by the liquid as, relative to the inlet, it accelerates rotationally with the impeller blades. This energy is transformed into an increase in pressure by being constrained to slow down by the geometry of a collecting chamber.

The impeller is the head generating component within a centrifugal pump. The combined outlet geometry of the impeller and the collecting chamber surrounding it determines the value of the generated head rise and the best efficiency point flowrate. The impeller inlet geometry determines the cavitation performance characteristics of the pump.

The outlet blade tip usually finishes straight and parallel to the shaft axis as shown in Fig. 2.1. By exception, it finishes at an angle to the axis or is shaped. This latter design feature is provided on high speed pumps to reduce the pressure pulse that occurs as the blades pass close to stationary parts in the collecting chamber. This has little or no influence on cavitation performance except where, for good manufacturing reasons, it affects the design of the blade at inlet.

The blade shape at inlet is of crucial importance in preventing cavitation. Ideally the blade should be shaped so that the angle at which the blade meets the flow is the same at all points across the flowpath. This reduces the propensity of an impeller to local recirculations behind the blade at or near best efficiency flowrate. Where local pressure

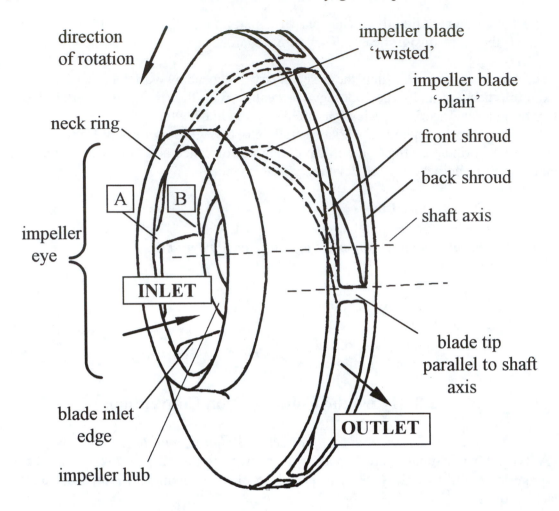

Fig. 2.1 Impeller terminology

drops are low it delays the onset of cavitation. To achieve a common flow impingement angle across the blade inlet edge there has to be a three-dimensional twist since with a uniform axial flow into the impeller eye the circumferential velocity at the neck ring (Fig. 2.1 point "A") is much greater than that at the impeller hub (Fig. 2.1 point "B"). Such twisted blades can, as shown in Figs. 1.1 and 1.7, exhibit a near flat surface when seen from the impeller eye. Cavitation and cavitation damage is thus spread uniformly across the blade. Repair of damaged areas becomes a possibility.

Manufacturing twisted blades is difficult (Ref. 2.1) and costly. For pumps where cavitation is not a problem and for those pumps where the cavitation can be easily avoided by small changes to plant design three-dimensional twisted blades are an unnecessary added expense. The use of simple-to-manufacture two-dimensional blades, usually referred to as "plain" vanes, then makes good commercial sense.

Blades that spiral radially outward in a clockwise direction, like those in Fig. 2.1, are termed right-handed impellers (because they are easier to hand-file in a bench vice by a right-handed person). Those that spiral anticlockwise are termed left-handed the impellers. This "handing" of an impeller is of crucial importance where the prime mover driving the pump is unidirectional. An example of this is a pump driven by a steam turbine that has internal nozzles that are not suited to operation in both directions.

2.3.2 *Centrifugal pump impeller design configurations*

There are three basic centrifugal pump constructional configurations. They are (i) the single-entry pump, (ii) the double-entry pump and (iii) the multistage pump. They are shown in cross-sectional view in Fig. 2.2. The illustration shows variations based on shrouded impellers. In some pumps one or both of the shrouds are omitted to give an easier-to-manufacture "open" impeller design. The multistage assembly also comes with a variety of design possibilities. Figure 2.2(c) shows a multistage pump made up of single-entry impellers. In practice a combination of single-entry and double-entry impellers is often used to maximise the benefits inherent in these particular hydraulic designs.

A multitude of variations in centrifugal pump construction has been derived from the three basic configurations. A fuller description of them is outside the scope of this book. Karassik and Carter (Ref. 2.2) give a good pictorial guide to many of the pump and pump impeller designs that have been used.

2.4 Performance Measurement Levels

Centrifugal pump performance is described at three levels. The context that data are used in makes it self-evident which level is being referred to.

2.4.1 *Level* 1. *Impeller performance* (ΔH, Q)

To be able to compare impeller performance generically it is necessary to define a common basis for all the impeller design variants. The convention adopted by the pump industry is to use a single-entry impeller as a basis for performance analysis. Performance is always quoted at best efficiency flowrate unless otherwise stated.

The use made of best efficiency based data as a measure of impeller performance, as will be seen in later chapters, demands no great accuracy. It is therefore common practice to include losses in the collecting chamber surrounding the impeller (e.g. volute or diffuser ring) in determining the generated head. For example, the head measured across the inlet/outlet branches of an end-suction pump is used. Similarly, the head across the inlet/outlet to branches divided by the number of stages is used where two or more impellers of equal diameter are mounted in series. Double-entry impellers are treated as two identical impellers mounted back-to-back.

The data for impeller performance comparisons are obtained as follows:

    1. the best efficiency flowrate Q_{bep},

    2. the generated *head per impeller stage* at $Q_{bep,}$

    3. the volumetric *flowrate per impeller eye** at Q_{bep}.

* Flowrate per impeller eye is strongly recommended. Occasionally in the technical literature the pump performance for double-entry pumps has been described using total flowrate. Some manufacturers have considered this advantageous when correlating performance of double-entry volutes. The flowrate-per-eye used here has greater hydraulic consistency and follows the method described by Karassik and Carter (Ref. 2.2), Stepanoff (Ref. 2.3) and Anderson (Ref. 2.4).

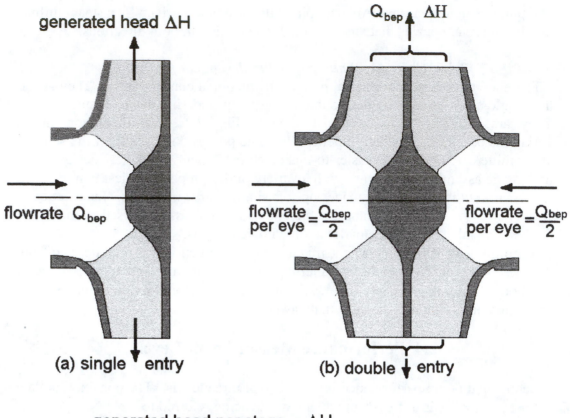

generated head ΔH

flowrate Q_{bep}

(a) single entry

Q_{bep} ▲ ΔH

$\text{flowrate} \atop \text{per eye} = \dfrac{Q_{bep}}{2}$

$\text{flowrate} \atop \text{per eye} = \dfrac{Q_{bep}}{2}$

(b) double entry

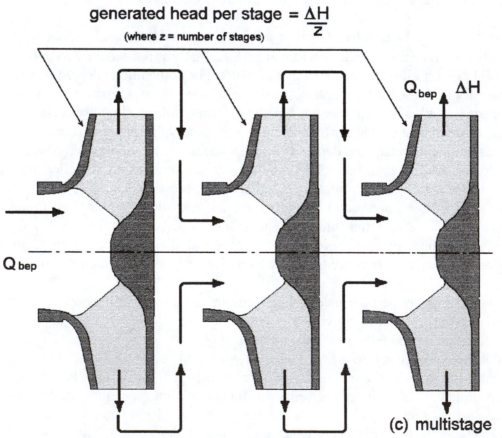

generated head per stage $= \dfrac{\Delta H}{z}$

(where z = number of stages)

Q_{bep}

Q_{bep} ▲ ΔH

(c) multistage

Fig. 2.2 Impeller configurations (cross-sectional views)

2.4.2 *Level 2. Pump performance* $(\Delta H_G, Q_G)$

This describes the "guaranteed" duty of a pump. Level 2 is, without exception, used as the basis for procurement specifications and contractual acceptance tests. The pump duty is usually close to the best efficiency point, but it seldom has this value.

When selecting a new pump to meet the guaranteed duty the numerous permutations of hydraulic configuration that provide diversity in economic choice are further amplified by the option of running off best efficiency flowrate. This strategy can be used to secure a different operational advantage (e.g. a steeply falling $\Delta H/Q$ curve) whilst still meeting the essential requirement of eliminating the risk of unacceptable cavitation.

2.4.3 *Level 3. Plant performance* $(\Delta H_p, Q_p)$

This describes the overall plant performance where a number of pumps, either in series or in parallel, are used at a particular location in aggregate to meet the requirements of the plant. Selecting combinations of new pumps from the numerous permutations of pump set configurations provides further diversity in economic choice whilst meeting the requirement of eliminating the risk of unacceptable cavitation. Pump selection is considered in some detail in Chapter 12. The basic layouts that can be used are shown in Fig. 2.3.

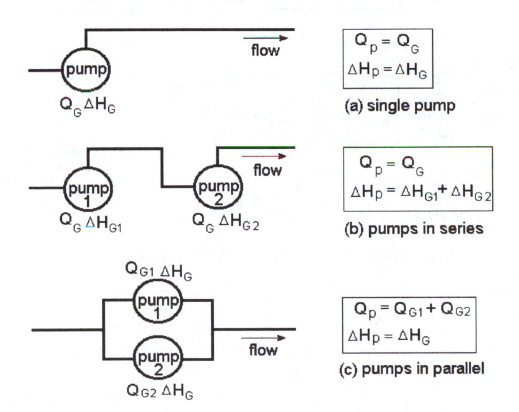

Fig. 2.3 Pump layout configurations

A number of useful options are available when pumps are arranged in series. Examples of those frequently used in power plant are shown in Fig. 2.4. The usual purpose of a series arrangement is to enable a slow running "booster" pump to provide sufficient pressure at the inlet of a high-speed pump to suppress cavitation. The method of attaining the high speed determines the configuration.

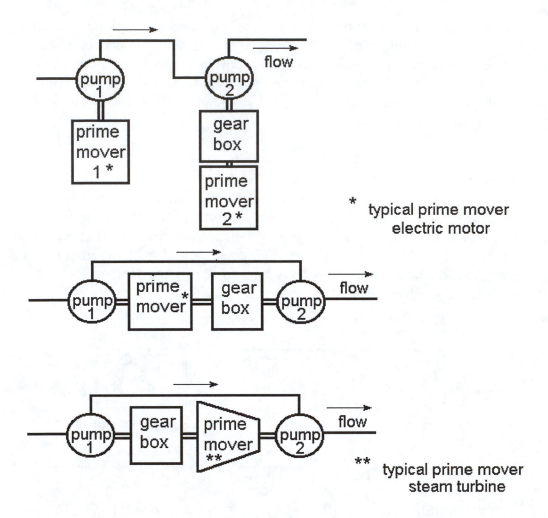

Fig. 2.4 Pumps in series — Some pump set configurations

2.5 Performance Measurement Terminology

Total head H.

The notation used is illustrated in Fig. 2.5.

1. Inlet total head $H_1 = z_1 + \dfrac{p_1}{\rho_1 g} + \dfrac{v_1^2}{2g}$ ---------------------------2.1

2. Outlet total head $H_2 = z_2 + \dfrac{p_2}{\rho_2 g} + \dfrac{v_2^2}{2g}$ ---------------------------2.2

Pressures p_1 and p_2 are "gauge" values. For cavitation performance assessments the liquid densities ρ_1 and ρ_2 are assumed to be constant and have a value ρ_L. This assumption has validity where the inlet performance of the pump for a given liquid at a given temperature is being evaluated.

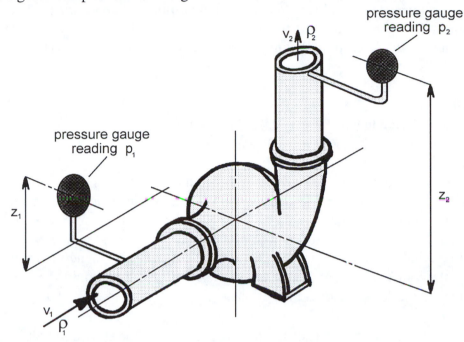

Fig. 2.5   Pictorial view of notation conventions

Generated head ΔH.

The word "generated" and the symbol ΔH are used to provide an unambiguous reminder that they represent a *change* in value.

$$\Delta H = H_2 - H_1 \qquad \text{-----------------------2.3}$$

ISO 3555-1977 (Ref. 2.5) uses "pump total head H" instead of "generated head ΔH".

Net positive suction head **NPSH.**

This is the inlet total head, plus the head corresponding to the atmospheric pressure, minus the head corresponding to the vapour pressure.

$$NPSH = H_1 + \frac{p_b}{\rho_L g} - \frac{p_{vap}}{\rho_L g} \qquad \text{------------------------2.4}$$

The definition and industrial usage of NPSH are dealt with in detail in Chapters 3 and 4. The choice of NPSH measurement reference plane is discussed in Chapter 11.

Pump input power P_A. This is the power absorbed measured at the pump coupling.

Best efficiency flowrate Q_{bep}. Cavitation performance data in pump user assessments of performance are always related to best efficiency conditions. That is, they correspond to the volumetric flowrate at which the maximum value of efficiency is attained for a centrifugal pump operating at a specified shaft rotational speed.

Pump Efficiency η.

$$\text{Pump efficiency } \eta = \frac{\rho_L\, g\, Q\, \Delta H}{P_A} \times 100\%$$
$$\text{(as a percentage)} \quad\quad\quad\quad\quad\quad \text{-------------------2.5}$$

Where a significant change in liquid density occurs — such as in a boiler feedwater pump where a considerable compression of the pumped liquid is made before discharging it at a very high pressure — the additional useful work added to the pumped liquid needs to be included where energy utilisation assessments are being made. In cavitation performance calculations where pump efficiency is used only to define the value of Q_{bep}, the error arising by assuming the liquid density does not change from inlet value ρ_L is insignificant.

Pump duty. The "duty" of a centrifugal pump is described by the specified values of generated head ΔH_G and volumetric flowrate Q_G, which the pump is required to deliver when it runs at a specified shaft rotational speed. The values ΔH_G and Q_G are used in specifications to define the contractual "guarantee" point in performance acceptance tests. Q_G is usually close to Q_{bep} but is seldom of the same value.

2.6 Metric Units

The units of measurement for pump selection and specifications are taken from the System International (SI) with reference, where appropriate, to ISO 31 — Quantities, units and symbols. SI allows choice in the use of subdivisions. Those used in the following text are listed in Table 2.1.

Table 2.1 Metric units

Variable	Notation	Units	Abbreviation
Flowrate	Q	litres per second	$m^3/s \times 10^{-3}$ see Note 2
Head	H ⎤		
Generated head	ΔH ⎦	metres	m
Pump shaft speed	n	revolutions per minute	rpm see Note 3
Impeller/pump dimensions	D	millimetres	mm

Table 2.1 continued

Variable	Notation	Units	Abbreviation
Pressure	p ⎤	bar	bar
Generated pressure	ΔP ⎦		
Liquid density	ρ_L	kilogram per cubic metre	kg/m^3

Note 1.　1 bar = 1 x 10^5 newtons per square metre.

Note 2.　The SI unit of first choice for flowrate is the cubic metre per second (m^3/s). Numerically this is only convenient for measuring extremely large flows such as those of rivers. For most pump applications pump duty is better expressed as one-thousandth of this by the litre per second (m^3/s x 10^{-3}). Unfortunately the abbreviation for "litres per second" (l/s) is too easily confused with that for "one per second" (1/s) so litres per second is written in full where the expression m^3/s x 10^{-3} is not convenient or is not appropriate. The alternative abbreviation dm^3/s is also used.

Note 3.　The practice of measuring rotational speed in rpm is widespread. Choice of this unit results from strong user requirement for consistency with the established method of measuring the rotational speed of other rotating machinery.

2.7　Useful Conversions

Mass flowrate/volumetric flowrate.　Pumped product output is usually measured by mass in the chemical process and related industries.

$$Q = \frac{Q_M}{\rho_L} \text{---} 2.6$$

Pressure/head.　Whilst pump performance and associated hydraulic calculations are normally expressed as head measurements there are occasions when the pressure at a particular point has to be determined. A pipework pressure containment integrity assessment is an example of this.

The pressure p corresponding to a value of head H is given by

$$p = \rho_L g H \text{---------------------------------------} 2.7$$

Care should always be taken to make the distinction between static head and generated head. (n.b. generated pressure $\Delta P = \rho_L g \Delta H$).

The conversion of a cold water (20°C approx.) pressure (bars) to head (metres) is frequently required. Using equation 2.7 with p = 1 bar = 1 x 10^5 N/m^2, ρ_L = 1000 kg/m^3 and g = 9.81 m/s^2 it can be seen that 1 bar of cold water pressure is equivalent to a head 10.2 m. This value appears frequently in calculations. It must be remembered however that it only applies to a liquid with a density of about 1000 kg/m^3.

References

2.1   Church, A. H., <u>Centrifugal Pumps and Blowers</u>, Wiley, New York, 1956.

2.2   Karassik, I. and Carter R., <u>Centrifugal Pumps</u>, McGraw-Hill, New York, 1960.

2.3   Stepanoff, A. J., <u>Centrifugal and Axial Flow Pumps</u>, second edition, Wiley, New York, 1957.

2.4   Anderson, H. H., <u>Centrifugal Pumps and Allied Machinery</u>, p58, fourth edition, Elsevier, Oxford, 1994.

2.5   ISO 3555 - 1977, <u>Acceptance Tests for Centrifugal, Mixed Flow and Axial Pumps</u>, 1977.

Chapter 3

Centrifugal Pump Performance Characteristics

PART 1.   NON-CAVITATING PERFORMANCE

3.1 A Typical Characteristic

Non-cavitating centrifugal pump performance characteristics provide a base from which deviations caused by cavitation can be measured. The changes of interest are almost exclusively those which produce a drop in the generated head. A typical non-cavitating characteristic is shown in Fig. 3.1.

Most pumps run at a single speed. A pump characteristic for single speed operation, as shown by the solid lines in Fig. 3.1, is normally available from the manufacturer. The essential parts of the characteristic can be, and often are, the subject of contractual acceptance tests.

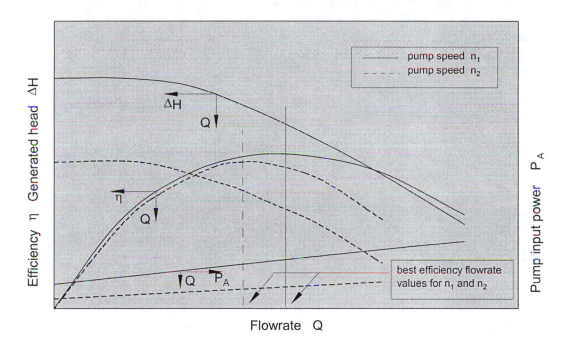

Fig. 3.1   Typical non-cavitating performance characteristics $(\Delta H/Q, P_A/Q, \eta/Q)$

3.2 Affinity Relationships and Specific Speed

It is evident from Fig. 3.1 that a change in pump speed results in a significant change in the values of generated head and flowrate at, and near, best efficiency flowrate.

23

Rules for calculating such changes are fundamental to the centrifugal pump selection process. These rules, commonly known as "affinity laws", can be derived using a simplified form of dimensional analysis.

Extending the analysis from a particular pump to a comparison of generic pump designs shows that both physical design factors and the shape of performance curves can be characterised by a single term, "specific speed".

The validity of relationships derived is limited to conditions where Q, ΔH, n and D are the only significant variables to describe pump performance. This holds true for most liquids. Pumps for liquids that fall outside this description are considered separately in section 3.4.

3.2.1 *Affinity laws*

The volumetric flowrate through a non-cavitating centrifugal pump is known, by experience, to be dependent upon (a) the product of the generated head and gravitational constant, (b) the pump speed, and (c) a characteristic dimension, the impeller diameter:

$$\text{i.e.} \quad Q = f(\Delta H\, g, n, D) \text{---3.1}$$

where f denotes functional dependence of the terms within the bracket but does not imply any particular form of dependence.

Both sides of equation 3.1 must have consistency in the fundamental units mass (M), length (L) and time (T). Dimensionally equation 3.1 can be rewritten as:

$$Q = (\Delta H\, g)^a, (n)^b, (D)^c \text{---3.2}$$

where a, b and c are numerical indices.

Expressing the independent variables in equation 3.2 in terms of M, L and T.

$$\frac{L^3}{T} = \left[\frac{L^2}{T^2}\right]^a \left[\frac{1}{T}\right]^b [L]^c$$

Equating powers of L gives $3 = 2a + c$
from which $c = 3 - 2a$

Equating powers of T gives $-1 = -2a - b$
from which $b = 1 - 2a$

Therefore equation 3.2 can be rewritten as $Q = (\Delta H\, g)^a, (n)^{1-2a}, (D)^{3-2a}$

Separating the indices gives $\dfrac{Q}{n\, D^3} = f\left[\dfrac{\Delta H\, g}{n^2\, D^2}\right]$ ----------------------3.3

The important finding from dimensional analysis is that centrifugal pump non-cavitating performance can be expressed by two independent non-dimensional groups.

These are $\dfrac{Q}{n\,D^3}$ and $\dfrac{\Delta H\,g}{n^2\,D^2}$

Equation 3.3 can be used to describe a *particular* centrifugal pump (where D = constant and g = constant):

i.e. $\dfrac{Q}{n} = f\left[\dfrac{H}{n^2}\right]$

Summarising, dimensional analysis applied to a particular pump predicts that for any given value of $\dfrac{Q}{n}$ there is only one value of $\dfrac{H}{n^2}$.

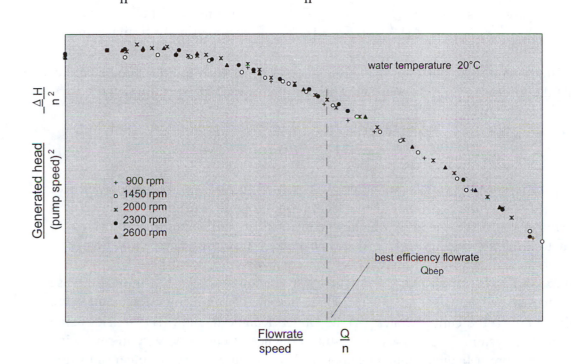

Fig. 3.2 Typical non-dimensionalised performance of a centrifugal pump

Plotting the test data used to produce Fig. 3.1 in the form shown in Fig. 3.2 results in a single curve. Additional data obtained at other speeds reinforce this finding. This confirms the validity of assumptions made in the analysis. The so-called "affinity laws" are derived from this result.

AFFINITY LAWS: Non-cavitating pumps

For a *particular* non-cavitating centrifugal pump,

(i) flowrate varies directly with the pump speed,

(ii) generated head varies directly with the square of the pump speed.

Note: As shown in Figs. 3.1 and 3.2, best efficiency flowrate varies directly with the speed.

3.2.2 Specific speed

Equation 3.3 can also be used to describe a *generic* type of centrifugal pump.

The dimension D is representative of pump geometry. Two pumps of the same generic type, that is, scale models in the sense of having similar physical geometry, can be evaluated to establish a group, not necessarily dimensionless, that represents their generic type. Consider two pumps of the same generic type but of different size subscripted 1 and 2. From equation 3.3,

$$\frac{Q_1}{n_1 D_1^3} = \frac{Q_2}{n_2 D_2^3} \quad \text{and} \quad \frac{\Delta H_1 g}{n_1^2 D_1^2} = \frac{\Delta H_2 g}{n_2^2 D_2^2}$$

$$\frac{D_1}{D_2} = \left[\frac{Q_1 n_2}{Q_2 n_1}\right]^{1/3} = \left[\frac{\Delta H_1^{1/2} n_2}{\Delta H_2^{1/2} n_1}\right]^{1/2} \quad \text{where g, the gravitational. term, is assumed constant}$$

rearranging gives

$$\frac{n_1 Q_1^{1/2}}{\Delta H_1^{3/4}} = \frac{n_2 Q_2^{1/2}}{\Delta H_2^{3/4}} = \text{a constant for all geometrically similar pumps.}$$

Summarising, dimensional analysis applied to generically similar pumps predicts that the group $\dfrac{n Q^{1/2}}{\Delta H^{3/4}}$ is constant. This constant is termed "specific speed".

$$\text{Specific speed } n_S = \frac{n Q^{1/2}}{\Delta H^{3/4}} \quad\text{------------------------3.4}$$

Specific speed is a "shape" number. It is used to classify the performance of centrifugal pumps of similar geometrical proportions. It was given its name as it corresponds to the speed at which a geometrically similar pump would have to run to generate one unit of head at one unit of flowrate. This is, of course, a mathematical curiosity of no practical consequence.

Units — Engineering usage. Practical expediency in the past prompted the use of inconsistent units for n, Q and ΔH. Different sets of inconsistent units were adopted in (i) the United States, (ii) the United Kingdom, and (iii) countries using the metric system. These practices result in the need to be very wary in reviewing past literature. Misunderstandings abound. It is essential to exercise care in using such data.

In this book specific speed is calculated using (i) pump speed measured in revolutions per minute (rpm), (ii) best efficiency flowrate measured in litres per second ($m^3/s \times 10^{-3}$), and (iii) generated head measured in metres (m).

Expressing specific speed (equation 3.4) as units gives: $\dfrac{(rpm)\,(m^3/s \times 10^{-3})^{1/2}}{(m)^{3/4}}$

The metric units chosen for specific speed are consistent throughout this book. All the data taken from other published work have been converted to these units. It is a pragmatic choice. Reviewing 1997 usage revealed that the attempts made to use a truly non-dimensional n_S and describe it as a "shape" number (a move supported for some 25 years by the author) have comprehensively failed. The initially popular use of m^3/h for flowrate also diminished rapidly, largely due to the inconvenience of relating the data to flow velocities which are always measured on a "per second" basis. The author accepts the reality of market forces and recommends others do the same.

The specific speeds of centrifugal pumps described in this book are encompassed by the range n_s = 300 to 2500.

Useful conversion factors — Specific speed

$$\frac{(\text{rpm}) \, (\text{m}^3/\text{s x } 10^{-3})^{1/2}}{(\text{m})^{3/4}} = 0.612 \; x \; \frac{(\text{rpm}) \, (\text{US gpm})^{1/2}}{(\text{feet})^{3/4}} \quad \text{----------3.5}$$

$$\frac{(\text{rpm}) \, (\text{m}^3/\text{s x } 10^{-3})^{1/2}}{(\text{m})^{3/4}} = 0.670 \; x \; \frac{(\text{rpm}) \, (\text{UK gpm})^{1/2}}{(\text{feet})^{3/4}} \quad \text{-----------3.6}$$

$$\frac{(\text{rpm}) \, (\text{m}^3/\text{s x } 10^{-3})^{1/2}}{(\text{m})^{3/4}} = 0.527 \; x \; \frac{(\text{rpm}) \, (\text{m}^3/\text{h})^{1/2}}{(\text{m})^{3/4}} \quad \text{-----------3.7}$$

1 litre per second = 1 m^3/s x 10^{-3} = 1 dm^3/s
1 Imperial gpm = 4.546 litres
1 US gpm = 3.785 litres

3.3 Performance Curves — Generic Differences

The work of Stepanoff (Ref. 3.1) in establishing that specific speed truly represents centrifugal pumps generically has been a cornerstone in the forecasting of their performance characteristics. The fact that it also characterises the internal hydraulic shapes — impeller, volute, etc., for a particular design method has also proved invaluable to pump designers. It enabled Stepanoff (Ref. 3.2) to produce a method of designing centrifugal pumps based on a highest efficiency criterion. The work of Anderson (Ref. 3.3) showed that centrifugal pumps produced using a different design method — i.e. not based on best efficiency — could provide flexibility in meeting other design objectives such as a steeply falling generated head/flowrate performance curve. As the majority of commercially manufactured pumps are designed to maximise efficiency (and thereby reduce the cost of power absorbed) it can be expected that the characteristics of most pumps will approximate to those described by Stepanoff.

An opportunity has been taken to repeat this important work for a large range of centrifugal pumps designed by the author. The results in Fig. 3.3 show the typical trends previously established. It is instructive for designers and pump users alike to repeat this using existing pump test data and to become aware how insensitive the curves are to many minor variations in geometry. These curves can be used to predict the shape of non-cavitating performance characteristics when Q and ΔH are known and the value of n is limited only by choice within defined practical constraints.

Plotting the data obtained from a great number of pump tests enables best performance norms to be established. Figure 3.4 shows what the typical well-designed pump might achieve for a given value of specific speed and flowrate.

Church (Ref. 3.4) and Anderson (Refs. 3.5 and 3.6) analysed the variations that the centrifugal pump design process produced. Anderson in Fig. 3.5 took this further in Ref. 3.7 by examining the performance of all pumps at the "optimum specific speed" line. From this work a measure of the uncertainty can be obtained.

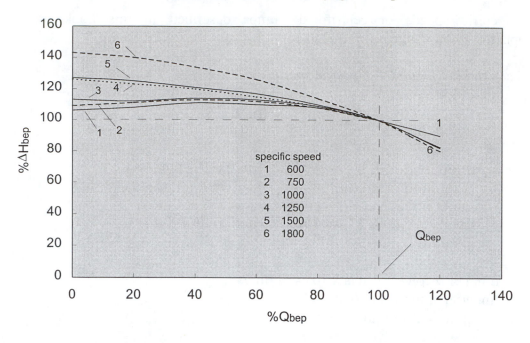

(a)   Generated head/flowrate

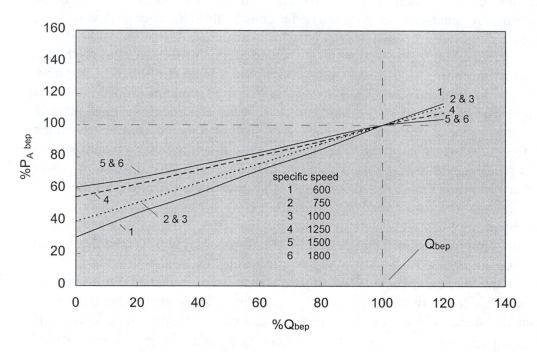

(b)   Pump input power/flowrate

Fig. 3.3 Typical performance curve shapes

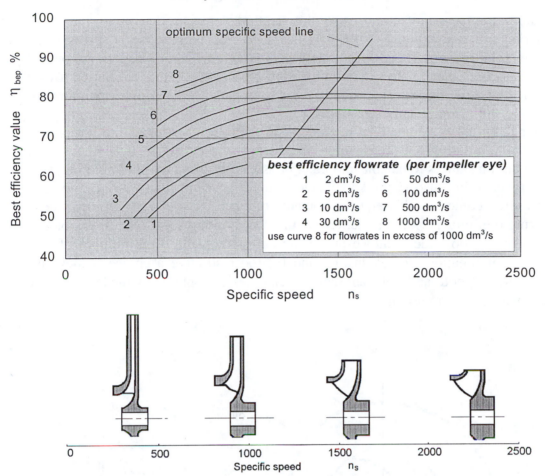

Fig. 3.4 Best efficiency value and impeller shape relationships with specific speed

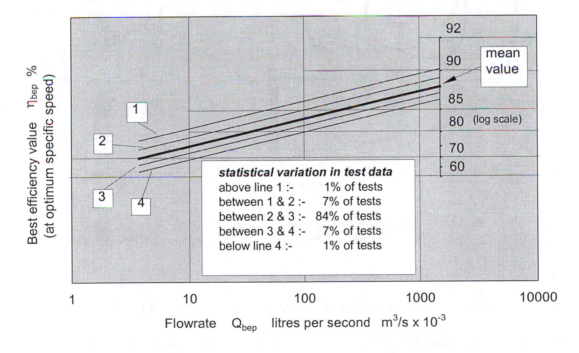

Fig. 3.5 Recorded variation of best efficiency with flowrate according to Anderson
(value at optimum specific speed)   *Private communication: Ref. 3.7*

Many pump manufacturers use this methodology to determine the likely outcome when new pumps have to be designed. For the purpose of assessing the commercial risk of failure to meet the design requirements of a new pump on test, the "in-house" data relate to a known design process (only) and are biased toward the nearest existing proven design. Such design data are, of course, manufacturer specific and commercially sensitive.

For pump users commercial pressures and the need to be exact are not present. Indeed, a variation between pumps offered must be expected. It is important in these circumstances to be able to predict or assess performance not only in terms of best practice but also in the knowledge of the likely uncertainty in achieving the desired outcome. To this end the contents of Figs. 3.3 and 3.4 are adequate for making preliminary pump selections so that the resulting specification is a balance between (a) too wide a choice producing an unacceptably high number of plant design combinations for evaluation and (b) too restrictive a choice, which risks eliminating good and relevant commercial design options. Chapter 12 gives guidance on how to do this.

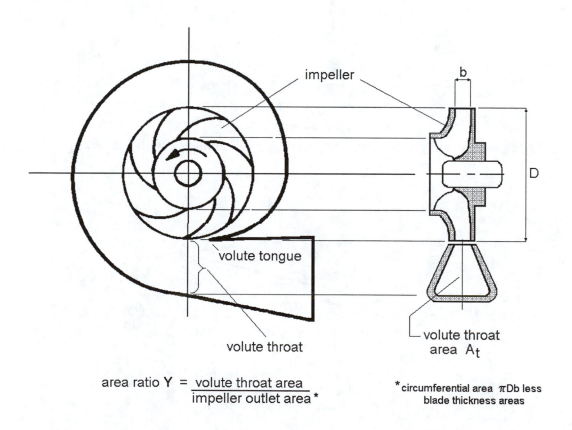

area ratio $Y = \dfrac{\text{volute throat area}}{\text{impeller outlet area}^{*}}$

* circumferential area πDb less blade thickness areas

Fig. 3.6 A definition of the area ratio for a volute-type centrifugal pump

Performance evaluation based upon a pump internal geometry that gives the highest possible pump efficiency for a given pump speed and flowrate is sound where a progressive and consistent design methodology is maintained. However, outside these design constraints there is a further design variant that has particular advantages for centrifugal impellers which are enclosed by a "volute" casing, that is, as shown in Fig. 3.6, when within a casing in which the area at the periphery of the impeller grows progressively as a spiral from a minimum at the volute tongue to a maximum at the volute throat. It is well known from tests that for a particular impeller (i.e. a constant outlet flow area at the circumference) a change in the area of the throat can bring about a significant change in the shape of the ΔH/Q curve. Often this is achieved with only a small reduction in pump efficiency. Clearly manipulation of this "area ratio" provides a commercial opportunity to expand pump choice further for applications where curve shape is important.

   Whilst the shape of the outlet flow passage controls the hydraulic performance of an impeller, it has almost no effect upon flows and pressure drops at the pump inlet. It follows therefore that, although the "best efficiency" criterion based curves and efficiencies shown in Figs. 3.3 and 3.4 are not applicable, the cavitation performance of pumps with differing area ratios can be assumed identical.

   It is useful to have in mind the likely magnitude of performance changes that can occur when a deviation from the "most efficient pump" objective is made. The interactive flows between impeller and volute throat might reasonably be expected to be close to conditions for giving optimum best efficiency for a particular specific speed when impeller outlet area and volute throat area are equal. That is when the area ratio **Y** is unity. As shown in Fig. 3.7 this is indeed the case at about $n_s = 1200$. Over the range of specific speeds applicable to centrifugal pumps the area ratio **Y** for best efficiency falls with increasing specific speed n_s.

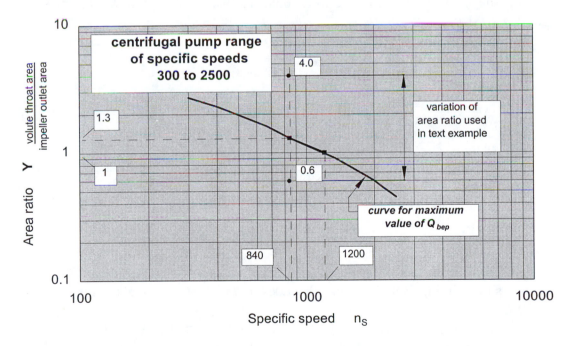

Fig. 3.7 Variation of area ratio with specific speed at Q_{bep} conditions

Anderson (Ref. 3.3) provides an example of the effect changing the area ratio has on a pump of $n_s = 840$. Figure 3.7 shows that the best efficiency at this specific speed is at about **Y** = 1.3. Changing to a pump that meets the required duty with an area ratio **Y** = 0.6 requires a larger diameter impeller and a larger casing throat area. Its performance is characterised by a 30% increase in generated head rise from the value at the best efficiency flowrate Q_{bep} to that at zero flowrate and an absorbed power curve that reaches a maximum at Q_{bep}. By contrast a pump having an area ratio **Y** = 4.0 has a significantly smaller impeller diameter and a smaller throat area (making for a cheaper pump). Its ΔH/Q curve shows a corresponding rise of between 5% and 10%, whilst the absorbed power curve continues to rise beyond Q_{bep}. No data on the magnitude of efficiency reduction are provided, but Anderson suggests that because of lower disc friction effects inside the pump the reduction in achievable best efficiency will not be as great with the **Y** = 4.0 pump as with the **Y** = 0.6 pump.

3.4 Other Influences

Secondary independent variables have significance only in special circumstances. These arise in very few centrifugal pump applications. Normally their effect can safely be ignored.

Adding kinematic viscosity ν and average surface roughness height (a centre line average height) CLA to equation 3.1 gives:

$$Q = f\ (\Delta Hg, n, D, \nu, CLA) \text{---}3.8$$

Repeating the simplified form of dimensional analysis with the new variables gives

$$\frac{Q}{n\,D^3} = f\left[\frac{\Delta Hg}{n^2\,D^2}\right]\left[\frac{\nu}{n\,D}\right]\left[\frac{CLA}{D}\right]$$

where $\dfrac{\nu}{n\,D^2}$ is a form of Reynolds number

and $\dfrac{CLA}{D}$ is a non-dimensional form of surface roughness.

3.4.1 *Kinematic viscosity*

Changes in kinematic viscosity have a very small effect on non-cavitating performance characteristics when pumping water over the temperature range 2°C to 100°C. As shown in Fig. 3.8 the change that does occur becomes apparent only below about 10°C.

However, although it is numerically very small, this change is significant. When measuring differences in generated head of the order of 3%, which will be shown to be necessary in Chapter 4, care has to be taken. In this context it is essential to discriminate between the change brought about by increased viscosity and that brought about by the presence of cavitation. A secure strategy for performance assessment would be obtained by ensuring that comparisons are only made between non-cavitating and cavitating curves at the same temperature. However, it is much simpler to eliminate altogether the risk of low temperature viscous effects compromising head-drop data by routinely specifying that the water test temperature is at least 10°C.

Highly viscous liquids and those that have unusual rheology are a problem. A competent review by Wasp, Kenny and Gandhi (Ref. 3.8) aids an understanding of this problem by providing considerable background through an unusual range of industrial applications. However, even where experience allows some confidence in forecasting non-cavitating performance characteristics, there is no associated database from which corresponding generic predictions of cavitating performance can be made. The high cost of cavitation test facilities for such special liquids exacerbates the problem.

As demonstrated by Wasp, Kenny and Gandhi, the shape of centrifugal pump performance curves can be greatly affected by unusual rheological properties. Reliance must be placed on past operating experience for a particular liquid wherever this is possible. Where no such experience exists, the estimates of NPSH to prevent cavitation causing hydraulic loss of performance or surging have, out of necessity, to follow the general rules established using cold water test data.

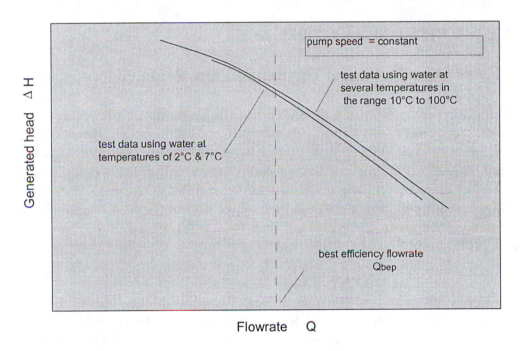

Fig. 3.8　Performance characteristic — viscosity change
　　　　 (cold/hot water comparison)

Cavitation erosion rates in a centrifugal impeller pumping a liquid that has an unusual rheology may be magnified by the combined effects of cavitation erosion and corrosion. To quantify such effects it is necessary to consider tests aimed at comparing cavitation erosion/corrosion due to the special liquid. A relatively simple test based on a Lichtarowicz Cell is described in Chapter 9.

3.4.2 *Impeller surface roughness*

This item is included for completeness and to negate the perception sometimes held that smoothness is a panacea to cavitation problems. This is not the case. Unless cavitation conditions are marginal — and the aim is to ensure they are not — the provision of surfaces that are better than "industrially smooth" is of no great advantage.

An increase in pump efficiency can be achieved when large improvements in impeller surface roughness take place. However, the increase is very small even for extremes such as commercially rough compared with highly polished or coated smooth surfaces. The variation is well within the range of uncertainty encompassing other impeller design changes affecting cavitating performance.

Smooth impeller blades are reputed to give better cavitation performance when this is measured at inception, a fact that should be expected intuitively. However, the very large amounts of cavitation — such as is present when cavitation causes a 3% generated head-drop — depend on the impeller blade angle/flowpath interaction being so turbulent that data for all pumps, rough and smooth, can hardly be distinguished apart. Using such generic data as a base to determine cavitation free conditions is not invalidated by differences in surface roughness.

PART 2.   CAVITATING PERFORMANCE

3.5 The Three Cavitating Flow Regimes

Reducing the flowrate passing through a centrifugal pump progressively from the best efficiency value gives rise to the possibility of three distinct flow regimes. These are described as stable, unstable, and transient. The most important is the stable region and this is where nearly all pumps operate. If either a pump or the adjacent plant is incorrectly designed or maloperated the other regimes may be encountered. A general characterisation of the three flow regimes, all of which are evidenced by changes in the externally measured performance of a centrifugal pump, is presented in Figs. 3.9, 3.10 and 3.11.

3.5.1 *The stable regime*

The majority of cavitating centrifugal pumps operate in this regime. In this regime the flow through the impeller is essentially unidirectional. Cavities originating at the impeller blade inlet develop progressively until they collapse further along the impeller passage. Visual observation shows such flows are — on a macro scale — essentially stable.

The inlet pressure p_1 of a pump operating in the stable regime is constant as shown in Fig. 3.10. Discussion of cavitating centrifugal pump performance in the stable regime is contained in Chapters 4, 5, 6 and 9.

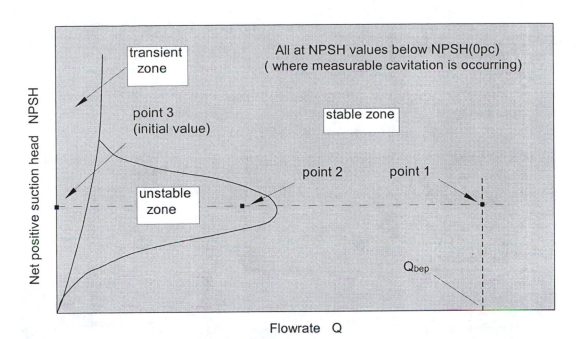

Fig. 3.9 General characterisation of stable, unstable and transient regimes
         — Zones of occurrence

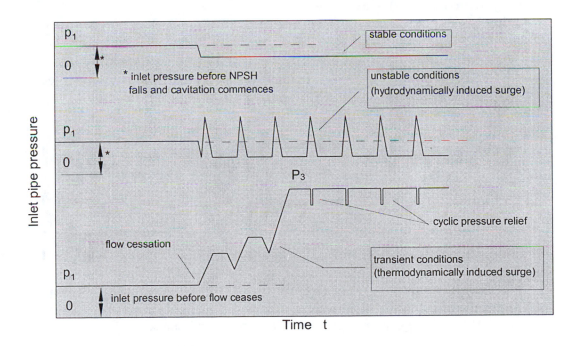

Fig. 3.10 General characterisation of stable, unstable and transient regimes
          — Inlet pressure/time relationships

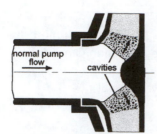

STABLE FLOW REGIME
(associated with cavitation erosion — chapter 9)

an 'any flowrate' phenomenon
flow pattern (observed); see Fig. 1.1
volume of cavities; unchanging with time

UNSTABLE FLOW REGIME
(hydrodynamically induced surge — chapter 7)

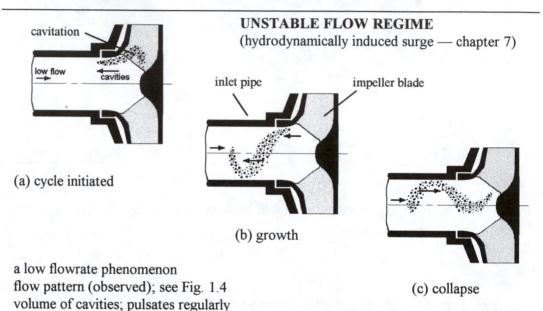

(a) cycle initiated

(b) growth

a low flowrate phenomenon
flow pattern (observed); see Fig. 1.4
volume of cavities; pulsates regularly

(c) collapse

TRANSIENT FLOW REGIME
(thermodynamically induced surge — chapter 8)

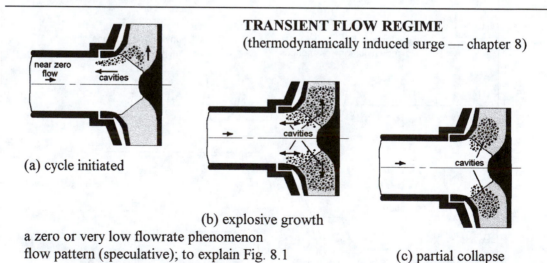

(a) cycle initiated

(b) explosive growth

a zero or very low flowrate phenomenon
flow pattern (speculative); to explain Fig. 8.1
volume of cavities; grows explosively

(c) partial collapse
(only if relieved*)

* if no relief = single pulse = vapour lock or containment boundary failure

Fig. 3.11 General characterisation of stable, unstable and transient flow regimes
— Flow patterns

3.5.2 *The unstable regime*

A few cavitating centrifugal pumps operate in the unstable regime. Usually this is brought about by a combination of unacceptably low flowrates and inadequate NPSH provision. In the unstable regime a strong recirculatory motion occurs within an impeller and the adjacent inlet pipework. This induces a spiralling motion on the cavities which are ejected along the wall of the inlet pipe before returning to enter the impeller eye. Pressure pulsations originating from this hydraulic instability can be very severe. This is exacerbated if inlet pipework and impeller inlet geometry are conducive to the propagation of recirculating cavity flows. The amplitude of pressure pulsations is determined by the cavity volumetric growth/collapse process within the impeller (NPSH dependent) and the prerotation (pump inlet and pipework geometry dependent).

The phenomenon is described as hydrodynamically induced cavitation surging and its onset, which is principally impeller design induced, is readily propagated where permitted to do so by the inlet pipe configuration.

The inlet pressure p_1 of a pump operating in the unstable regime fluctuates continuously as shown in Fig. 3.10. Discussion of cavitating centrifugal pump performance in the unstable regime is described in Chapter 7.

3.5.3 *The transient regime*

In the transient regime little or no volumetric flow interchange occurs between fluid in or near an impeller and the liquid in the surrounding pipework. Practically all the energy put into the fluid in the impeller goes to overcome hydraulic churning losses and this results in the temperature rising continuously. Where the volumetric expansion associated with this temperature rise is constrained the inlet pressure builds up. The periodic recirculations in or near impeller blade inlet associated with hydrodynamically induced surging occur at or near to zero flowrate. However, the continued input of a large amount of heat results in the processes associated with cavity growth and collapse being changed considerably. Thermodynamically induced changes dominate. The collapse brought about by vapour condensation associated with migration of heated fluid from the impeller into the relatively cool liquid in the inlet pipework region inevitably gives this regime the appearance of a pressure pulsation which increases in magnitude with time. Where this is accompanied by pressure relief through an outlet non-return valve, a plant induced surging motion is observed. Where no pressure relief is provided the surge motion is suppressed after a cycle or two. The pressure within the pump continues to rise to a level where the mechanical integrity of the pump and inlet pipework can be challenged. The observed phenomenon is described as thermodynamically induced cavitation surging. Its onset is plant design induced.

The inlet pressure p_1 of a pump operating in the transient regime builds up as shown in Fig. 3.10. Discussion of cavitating centrifugal pump performance in the transient regime is described in Chapter 8.

3.6 Typical Cavitation Characteristics

3.6.1 *Measurable head-drop curves*

Pump performance in the stable regime is of prime engineering interest. The intent is to avoid by design the cavitation surging associated with the unstable and transient regimes. The following methods apply solely to cavitation in the stable regime.

There are two methods by which centrifugal pump cavitation characteristics based on measurable head-drop can be presented. Typical curves representing them are shown in Figs. 3.12 and 3.13. The two methods have separate and particular advantages. The outcome, the determination of the NPSH corresponding to a given generated head-drop (say 3%) at a nominated flowrate, is identical.

Method 1. *Generated head vs. flowrate (NPSH constant)*

The "non-cavitating" curve AB in Fig. 3.12 is obtained by the well-established procedure of running at successively higher values of NPSH and observing that no change in shape occurred. The inference is that this demonstrates absence at all these conditions of cavitation of sufficient volume to affect volumetric performance. Choosing any one of these conditions to produce a non-cavitating $\Delta H/Q$ curve is thus demonstrated to have validity.

Cavitation tests are then carried out at a number of lower NPSH values: NPSHa, NPSHb, NPSHc, etc. During each test the value of NPSH is kept constant and the flowrate progressively increased until a deterioration in the generated head relative to the non-cavitating curve is obtained. From the generated head/flowrate curve the flowrate Q_{test} associated with any chosen head-drop criterion (x = 0%, 3%, etc.) can be determined for the net positive suction head NPSH$_{test}$. By repeating the test at a number of other NPSH$_{test}$ values the NPSH(xpc)/Q curve shown in Fig. 3.14 is constructed.

Tests over the range of flowrates of commercial interest (where the pump is to operate in service) present few problems. However, difficulties are often experienced in obtaining reliable data at low flowrates in test facility/impeller design combination where unstable flows are generated and propagated. Additionally, it is not immediately clear when using this method when the required head-drop has been attained. This means complete sets of test calculations have to be made as the data for each test point is recorded. Even with diligence it is possible that, when all the minor test corrections are made, conditions have not progressed as far as necessary and the test has to be repeated. Worse still, the test may have unwittingly progressed too far and reached a point where cavitation is extensive and the pump is at risk from vapour locking.

The features of Method 1 are typical of the observations on industrial sites where a loss of pump performance due to cavitation is a problem. Pumps usually draw from an inlet vessel where the liquid level (and hence NPSH) remains essentially constant. As the pump outlet valve is opened the pump may fail to generate the head that it had previously been shown capable of in a works test if cavitation is present.

Method 2. *Generated head vs. NPSH (Flowrate constant)*

Method 2 has particular advantages for the works test situation where it is usually more expedient to hold flowrate constant whilst progressively lowering NPSH. It is simpler and easier to apply than Method 1.

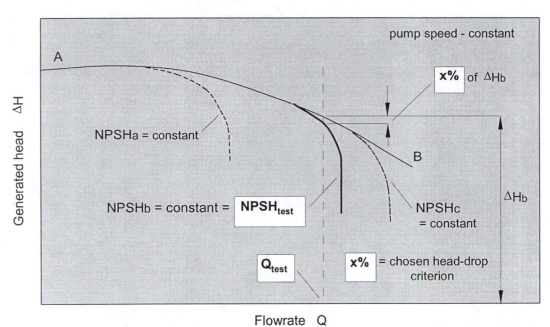

Fig. 3.12   Typical cavitation performance curves — Measurable head-drop
criterion — Constant NPSH method

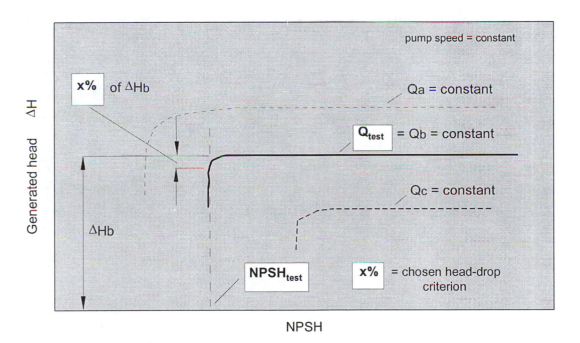

Fig. 3.13   Typical cavitation performance curves — Measurable head-drop
criterion — Constant flowrate method

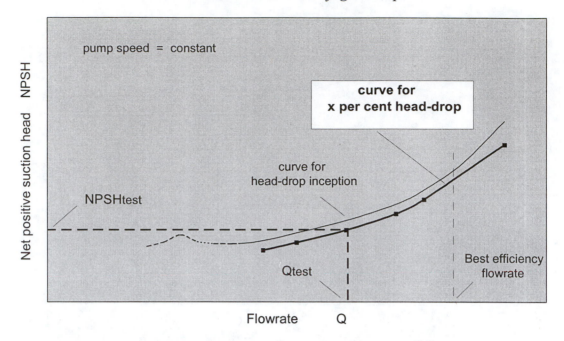

Fig. 3.14　Typical cavitation characteristic — Head-drop criterion NPSH(xpc)
(derived from either Fig. 3.12 or Fig. 3.13)

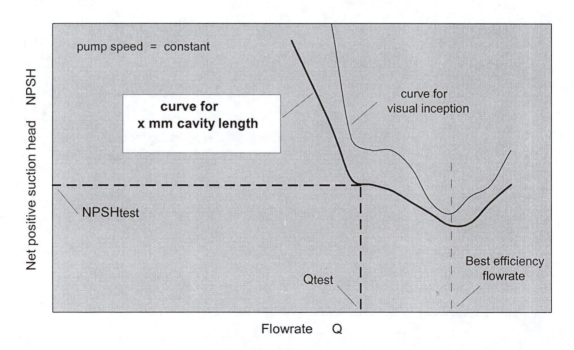

Fig. 3.15　Typical cavitation characteristic — Visual onset criterion
NPSH(xmm)

Testing starts at a constant nominal flowrate Q_{test}. It proceeds with NPSH set at a high value which is then progressively reduced. Non-cavitating conditions are in evidence if the generated head remains constant. The test continues with NPSH being reduced until head-drop conditions are reached. At a constant flowrate it is very clear when this commences. The inlet head gauge remains at the same value.

As data is accumulated during the non-cavitating part of the test where the generated head remains constant there is ample time to check what its precise value of generated head actually is and to determine the value (x% in this case) of head-drop which is required to take place. With the inlet gauge nearly stationary the test facility control system can be set to give the head-drop value of interest. This is achieved with little uncertainty. The test progresses until this head-drop value is exceeded. From the generated head/NPSH curve the net positive suction head NPSHtest value associated with any chosen head-drop criterion (x = 0%, 3%, etc.) can be determined for the flowrate Qtest. By repeating the procedure at a number of other Qtest values the curve shown in Fig. 3.14 is constructed.

Guidance in the specification of head-drop tests is given in Chapter 11.

3.6.2 *Measurable visual onset curves*

Test facilities that permit visual observation of an impeller eye are seldom available. However, where they are available a more precise understanding of cavitation performance can be obtained.

A test is carried out by setting the flowrate Qtest constant and lowering NPSH progressively. The appearance of cavitation as a measured length of cavitating flow along the impeller blade and the corresponding NPSHtest are noted. By repeating the procedure at a number of other Qtest values the NPSH(xmm) curve shown in Fig. 3.15 is constructed.

Difficulties are often experienced in obtaining reliable data at low flowrates when unstable flows are generated and propagated. Additionally, at flows much greater than best efficiency flowrate cavities develop behind the impeller blade relative to the direction of rotation. These cavities are sometimes difficult to see and be measured, particularly where the impeller blades include a three-dimensional twist.

Guidance in the specification of visual onset tests is given in Chapter 11.

3.6.3 *The relationship between head-drop and visual performance*

Engineers often have to make important decisions on the acceptability of a machine's performance from external measurements alone. This is no less true for those dealing with a cavitating centrifugal pump. The insight given by photographs in a typical pump showing the extent of cavitation greatly assists this process. Simultaneous performance measurement relates this to the magnitude of generated head reduction. This is particularly valuable in helping to diagnose the reasons for a change in performance of a particular pump.

The photographs in Fig. 3.16 were taken during head-drop tests on a low specific speed ($n_s \sim 300$) pump fitted with a plain (non-twisted) bladed impeller. The pump casing and the near-side shroud of the impeller were made of transparent material permitting observation and photography of the cavitation within the pump. Selected

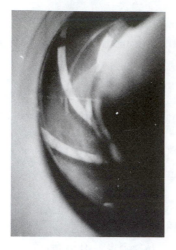

Point A

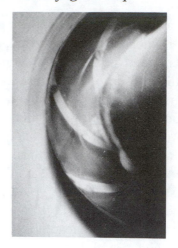

Point B

pump speed 2000 rpm +/- 2 rpm
water temperature 2°C
(cavitation close to freezing point !)

A

B

C

D

E

F

G

H

J

NPSH = 3.2 m

NPSH = 3.7 m

NPSH = 5.8 m

best efficiency flowrate
Q_{bep}

Generated head ΔH m

Flowrate Q   litres per second $m^3/s \times 10^{-3}$

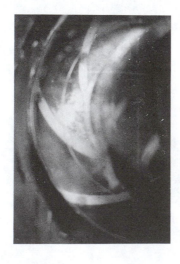

Point G

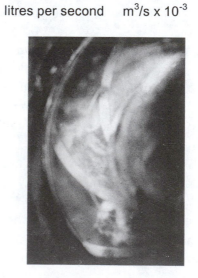

Point H

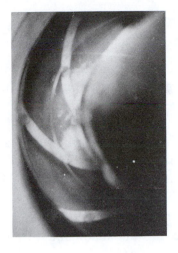

Point C

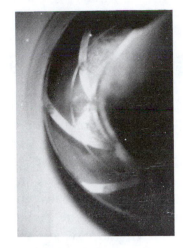

Point D

Fig. 3.16
Cavitation in a centrifugal pump
impeller at various hydraulic conditions

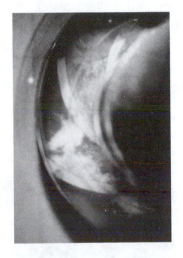

Point E

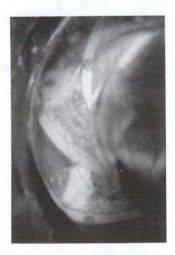

Point J

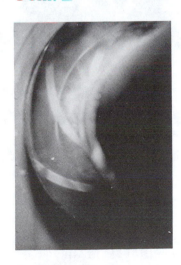

Point F

photographs are presented. All show the inlet pipe (on the right) as it leads to the impeller eye. The impeller blade inlet edge is just within the inlet pipe outer diameter. Cavitation emerges and develops along the blades starting at the blade inlet edge/inlet pipe intersection.

At NPSH = Point A no cavitation is evident. Decreasing the value to NPSH = Point B results in the generated head value reducing by about 3% near to best efficiency flowrate. A small amount of cavitation can be seen to be filling part of the flow channel between the blades. By Points C and D the generated head reduces further as yet more cavitation is seen to be produced between the blades. Finally, at Point E, the cavitation extends across the flow channel and for a considerable distance along it.

The effect of cavitation at a flowrate much greater than the best efficiency value Q_{bep} is shown as Point F. Recirculations at the back of the impeller blade appear to create a non-cavitating (transparent) high pressure region which makes the cavitating stream depart from its previous path along the back of the preceding impeller blade.

Tests at Q_{bep} with progressively lower values of NPSH produced photographs for Points G, H and J. As the generated head breaks down to less than half of its non-cavitating value, cavitation is seen to pass nearly all the way through the impeller.

Cavitation of the form described by Fig. 3.16 is almost always the cause of a generated head-drop especially when the reduction is continuous and of a constant value. An exception to this occurs with centrifugal pumps fitted with a stationary diffuser and operated at flowrates well in excess of best efficiency flowrate.

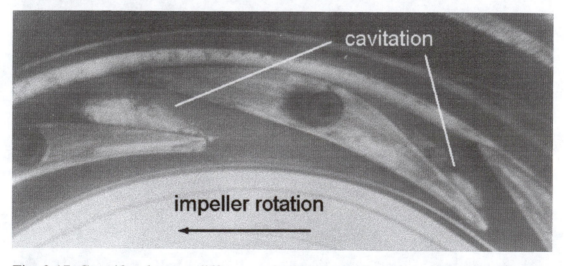

Fig. 3.17 Centrifugal pump diffuser cavitation   *Courtesy of Weir Pumps Ltd*

An example of this unusual form of cavitation is shown in Fig. 3.17. A drop in generated head from the non-cavitating curve can occur due to diffuser cavitation resulting from the gross mismatch of the direction of flow leaving the rotating impeller blades and the fixed diffuser blades. As demonstrated by Scott and Ward(Ref. 3.9) this can occur even when little or no cavitation is occurring on the impeller blades. It happens only at flowrates much greater than 120%Q_{bep}; in the tests reported it occurred above 140%Q_{bep}

3.7 Affinity Relationships and Suction Specific Speed

3.7.1 *Limits of validity*

Centrifugal pumps operating in the stable regime produce the recognisably typifying characteristic illustrated in Figs. 3.14 and 3.15. This characteristic is, for all practical purposes, unchanging for water at temperatures up to 80°C. As the majority of pumping applications are covered by this description it is convenient to take a modest water temperature as the base for comparative cavitation data. Works pump tests are usually limited to a maximum water temperature of 40°C. The standard test liquid is described as "clean cold water", a full specification for which is given in section 11.2.1.

Centrifugal pumps for water above 80°C and pumps for other liquids with thermodynamic property values that significantly affect cavity dynamics compared with "cold" water are considered separately in section 3.9.

3.7.2 *Affinity laws*

The volumetric flowrate through a cavitating centrifugal pump is known, by experience, to be dependent upon (a) the product of the net positive suction head and gravitational constant, (b) the pump speed, and (c) a characteristic dimension, the impeller diameter:

i.e. $\quad Q = f \text{ (NPSH, g, n, D)} \quad$ ------------------3.9

Dimensionally this is identical to equation 3.1, so it is evident from equation 3.3 that

$$\frac{Q}{n\,D^3} = f\left(\frac{NPSH\ g}{n^2\,D^2}\right)$$

The important finding from dimensional analysis is that centrifugal pump cavitating performance can be expressed by two independent non-dimensional groups.

These are $\qquad \dfrac{Q}{n\,D^3} \quad$ and $\quad \dfrac{NPSH\ g}{n^2\,D^2} \quad$ --------------------3.10

Equation 3.10 can be used to describe a *particular* centrifugal pump (where D = constant and g = constant):

i.e. $\quad \dfrac{Q}{n} = f\left(\dfrac{NPSH}{n^2}\right)$

From this it is evident that for any given value of $Q.n^{-1}$ there should be only one value of $NPSH.n^{-2}$. This is clearly true at cavitation inception where performance is unaffected by the presence of cavitation and the conclusions drawn from Fig. 3.2 remain valid. However, where a significant amount of cavitation is present then the relationship does not hold true. A small reduction in NPSH at higher speeds below that predicted by the dimensional analysis given above has been demonstrated to be present by Krisam (Ref. 3.10) and Rutschi (Ref. 3.11). In tests conducted by the author (Ref. 3.12) a value $NPSH.n^{-1.95}$ was observed for a near threefold change in speed where cavitating conditions corresponded to the generated head breakdown condition shown in Fig. 3.16 as Point J. This latter finding is significant for the practical application of

NPSH test results. It shows that whilst the presence of a substantial amount of cavitation distorts pump performance, it does so only slightly. Also, it shows that the use of an index for n of 2.0 produces a conservative estimate for NPSH when scaling data to a higher speed. Scaling a speed increase is usually the requirement in practice where slow speed works test results have to be translated to a higher speed for site use.

The outcome is that the affinity laws which derive from the above may be used reliably to predict performance changes based on modest amounts of cavitation — say, those associated with a 3% head-drop or about 4 mm cavity length on impeller blades — over a speed change multiple of two or less. In such cases the scaling will represent an acceptable approximation to the true value. Evidently where scaling to a higher speed is carried out the predicted requirement errs on the side of safe operation.

AFFINITY LAWS : Cavitating pumps

For a *particular* cavitating centrifugal pump
(i) **flowrate varies directly with the pump speed,**
(ii) **net positive suction head (NPSH) varies directly with the square of the pump speed.**

3.7.3 *Suction specific speed*

Equation 3.10 can be used to describe a *generic* type of centrifugal pump. Again, paralleling the approach taken for non-cavitating pumps to derive specific speed, the terms in the equation can be used to establish a dimensionless group that represents generic type.

$$\text{Suction specific speed} \quad n_{ss} = \frac{n \; Q^{1/2}}{NPSH^{3/4}} \quad \text{------------------3.11}$$

Units — Engineering usage. Practical expediency in the past prompted the use of inconsistent units for n, Q, and NPSH. The comments expressed in section 3.2.2 relating to exercising care to ensure the basis of data is known to apply equally here. In this book suction specific speed is calculated using pump speed measured in revolutions per minute (rpm), best efficiency flowrate is measured in litres per second ($m^3/s \times 10^{-3}$), and NPSH is measured in metres (m). The units for generated head and net positive suction head are the same so the conversion factors given by equations 3.5, 3.6 and 3.7 apply also to suction specific speed.

The suction specific speeds of cavitating centrifugal pumps are encompassed by extremes in value of $n_{ss} = 1000$ to $n_{ss} = 20000$. Values for typical industrial pumps that are cavitating are more likely to be in the range $n_{ss} = 2000$ to $n_{ss} = 4000$.

3.8 Performance Curves and Generic Similarity

Centrifugal pump cavitation can, by simple reasoning, be expected to be dominated by the impeller eye geometry and the inlet pressure gradient as determined by the impeller blade shape. As a first approximation the great majority of centrifugal pumps are similar in such respects so that a single suction specific speed value could, in very broad terms, be expected to have validity in describing cavitation performance for all such pumps. Experience summarised in Chapter 9 (Tables 9.5 and 9.6) will show this to be true. So, unlike specific speed characteristics, the generic curves for suction specific speed have a marked degree of similarity and cannot be segregated easily by generic constructional type. The great advantage in being able to express conditions controlling cavitation in terms of suction specific speed becomes apparent when pump selection is considered in Chapter 12. There are ways of describing pump performance without recourse to suction specific speed, but these do not facilitate such a wide understanding. The best that can be said for them is that they have advantage when looking closely at developments to gain improvements in cavitating performance for a particular impeller or for a particular closely related family of impeller designs. This can be useful for a designer in a particular pump company or a research worker following through on design theories using a single design configuration as a vehicle for analysis. They have no value to pump users who demand an understanding of the trends and uncertainties that are present in the whole range of commercially available options.

The advantages of measuring performance using specific speed and suction specific speed lies in their validity in quantifying generic trends. They have no value in forecasting the performance of the pumps to a commercial accuracy. This can only be established by test. For general engineering usage, including the production of specifications for pumping plant, the use of specific speed and suction specific speed related data provides a general understanding of the options available and the basis of a prudent methodology for evaluating them.

3.9 Other Influences — Thermodynamic Properties Affecting Cavity Dynamics

The cavitation performance characteristics of a centrifugal pump change when the thermodynamic properties of the pumped liquid differ significantly from that of clean cold water. Chapter 5 explores the reasons for this.

Practical constraints mean that, almost without exception, cavitation tests carried out at a pump manufacturer's works are conducted using clean water at a temperature close to the ambient air temperature. The high cost of test loops in special materials, the provision of safe test conditions for potentially dangerous quantities of toxic, corrosive, inflammable or cryogenic liquids, or the provision of facilities to accommodate high temperature operation usually make testing on other than cold water uneconomical.

As a consequence of the above it is necessary to establish if the same NPSH (the additional head suppressing cavitation) produces the same effect on the performance characteristic. The evidence from published literature presented in Figs. 3.18, 3.19 and 3.20 shows it does not.

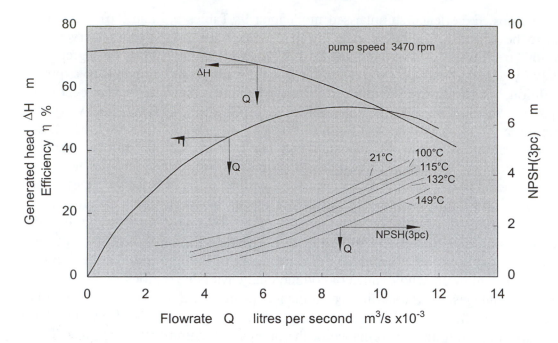

Fig. 3.18   Cavitation performance — cold/hot water
(3% head-drop)   *see Stepanoff, Ref.*
3.13

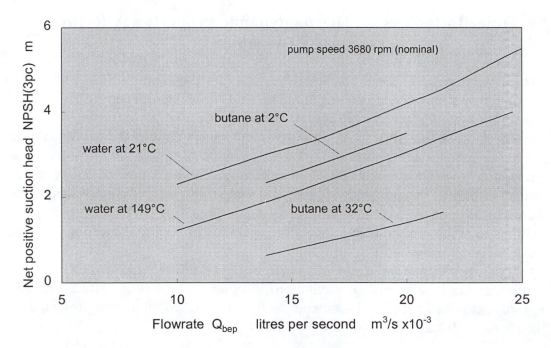

Fig. 3.19   Cavitation performance — cold water/butane
(3% head-drop)   *see Salemann, Ref.*
3.14

Stepanoff (Ref. 3.13) showed that the NPSH(3pc) value reduced by a significant proportion as water temperature increased above 100°C. Salemann (Ref. 3.14) reported on a comprehensive range of tests on other liquids, a summary of which is presented in Table 3.1.

Table 3.1 (Pump n_{ss} = 1080, Pump speed = 3585 rpm, NPSH @ datum = 3.8 m)

Fluid	Temperature °C	Reduction in NPSH m
Water	**21**	**0.0 (i. e. datum)**
	121	0.4
	149	1.1
Butane	2	0.8
	13	1.1
	32	2.7
Butane + 3% propane	2	0.9
	13	1.4
	32	3.3
Benzene	82	(-0.1)
	110	0.8
Kerosene (degasified)	20	(-0.1)
Gasoline	20	(-0.3)
Freon-12	30	0.6

Tests by Carter, Crusan and Thodal (Ref. 3.15) on a cryogenic centrifugal pump showed that the pumping capability with liquid oxygen was retained to a much lower level of NPSH than with water at ambient workshop temperature. A typical result is presented as Fig. 3.20.

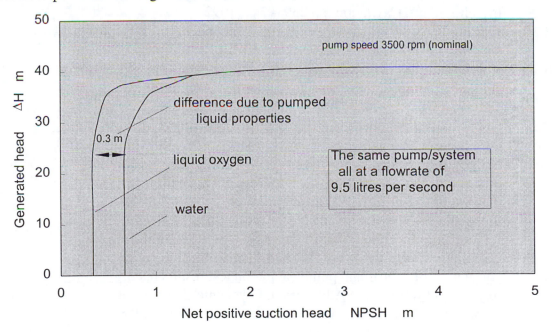

Fig. 3.20 Cavitation performance — cold water/liquid oxygen
see Carter, Crusan and Thodal, Ref. 3.15

Whilst all the test data show a significant drop as a *proportion* of the datum value, for the most part the *magnitude* of the change is very small. Certainly the particular values in the tests reported are too small to have any worthwhile influence on plant design. The important observation to make from Figs. 3.18, 3.19 and 3.20 is that the NPSH for water at 20°C is greater in value than that for hotter water and for almost all other liquids. Evidently where NPSH data for clean cold water are used without correction the predicted requirement errs on the side of safe operation.

References

3.1   Stepanoff, A. J., Centrifugal-pump performance as a function of specific speed, ASME Trans., pp. 629 – 647, 1943.

3.2   Stepanoff, A. J., Centrifugal and Axial Flow Pumps, Wiley, New York, 1957.

3.3   Anderson, H. H., Centrifugal Pumps, Trade and Technical Press, Morden, Surrey, UK, 1962

3.4   Church, A. H., Centrifugal Pumps and Blowers, Wiley, New York, 1956

3.5   Anderson, H. H., Statistical analysis of pump performance, Pumps–Pompes–Pumpen, pp. 432 – 435, No 8, 1966.

3.6   Anderson H H., The hydraulic design of centrifugal pumps and water turbines, ASME Trans., Paper 61-WA-320, 1962.

3.7   Anderson, H. H., Unpublished communication, 1966, (see Fig. 3.5).

3.8   Wasp, E. J., Kenny, J. P. & Gandhi, R. L., Solid-Liquid Flow Slurry Pipeline Transportation Trans. Tech. Publications, Germany, 1977.

3.9   Scott, C. and Ward, T., Cavitation in centrifugal pump diffusers, IMechE Conf. Cavitation, pp 239 – 247, 1992

3.10   Krisam, F., Neue erkenntnisse im kreiselpumpenbau, VDI-Zeitschrift, Vol. 95, No 11 – 12, p 320, April 1953.

3.11   Von Rutschi, K., Messung und drehzahlumrechung des NPSH – Wertes bei kreiselpumpen, Schweizer Ingenieur und Architekt, No. 59, pp 971 – 974, 1980.

3.12   Grist, E., The volumetric performance of cavitating centrifugal pumps Part 2: Predicted and measured performance, Proc. IMechE, Vol. 200, No. 59, 1986.

3.13   Stepanoff, A. J., Cavitation in centrifugal pumps with liquids other than water, ASME Trans., Jnl. Eng. Power, pp 79 – 90, 1961.

3.14   Salemann, V., Cavitation and NPSH requirements of various liquids, Trans. ASME, Paper No. 58-A-82, 1958.

3.15   Carter, T. A. Jnr., Crusan, C. R., and Thodal, F., Comparison and correlation of centrifugal pump cavitation test results handling liquid oxygen and water, Proc.Advances in Cryogenic Engineering Conference, Vol.4, pp 255 – 263, Plenum, New York, 1958.

Chapter 4

Net Positive Suction Head (NPSH) and The Pump Operational Range

4.1 NPSH as a Measure of Cavitation

4.1.1 *Role and scope*

A change in NPSH in the inlet pipe of a centrifugal pump is directly relatable to the difference in pressure driving the growth and collapse of cavities within the liquid flowing through its impeller. A datum value based on NPSH, which is demonstrable by an industrial grade test, provides a commercially acceptable basis for the quantitative assessment of the total NPSH necessary to suppress unacceptable levels of cavitation. An NPSH datum-based methodology is appropriate to the prevention of both hydraulic performance loss and cavitation erosion.

Cavitation surging that arises from impeller design is dependent upon NPSH *and* the impeller inlet pipework configuration. This complex interaction is not replicated by the layout requirements of industrial pump performance acceptance test standards, so a more general "umbrella" approach not based on NPSH alone has to be used.

Cavitation surging and vapour locking that are the result of a zero or low-flow transient related to plant design cannot be described by an NPSH value. Remedial measures via plant design changes or by improvements in component reliability/redundancy provide an appropriate method of dealing with this particular category of unacceptability.

4.1.2 *NPSH, NPSH(A), NPSH(R) and NPSH datum values explained*

NPSH   NPSH is defined as the excess of inlet total head over the head equivalent of the vapour pressure of the pumped liquid.

NPSH(A) This is defined as the NPSH *available* at the pump inlet.
(This always relates, usually implicitly, to a particular flowrate.)

NPSH(R) This is defined as the NPSH *required* at the pump inlet for operation free from unacceptable deficiencies caused by cavitation.
(This always relates, usually implicitly, to a particular pump speed and flowrate.)

To achieve acceptable operation of a centrifugal pump the value of NPSH(A) must be greater than, or at the very least equal to, the value of NPSH(R):

$$\text{i.e.} \quad \text{NPSH(A)} \geq \text{NPSH(R)} \text{ ----------------------------------- 4.1}$$

NPSH(R) is therefore the minimum permissible value of NPSH(A).

NPSH datum values. There is no way of measuring NPSH(R) directly. It has to be determined by obtaining an NPSH value at a datum condition that can be proven by a contractual test and by multiplying this value by an empirically derived figure, the cavitation protection ratio **R.**

As a generalising mathematical expression;

Cavitation protection ratio $\mathbf{R}(\text{Datum}) = \dfrac{\text{NPSH(R)}}{\text{NPSH(Datum)}}$ ----------------------
4.2

There is a choice of datum to suit the method of testing used. The value of **R** has to be appropriate to the particular datum used. The nomenclature used reflects this by a bracketed qualifying abbreviation for both **R** and NPSH. The expressions covering the two most used test methods are

for 3% head-drop test data,
$$\text{NPSH(R)} = \mathbf{R}(3\text{pc}) \times \text{NPSH}(3\text{pc}) \quad\text{-------------------------- 4.3}$$

for 4 mm visual-onset test data,
$$\text{NPSH(R)} = \mathbf{R}(4\text{mm}) \times \text{NPSH}(4\text{mm}) \quad\text{----------------------}$$
4.4

Many pump manufacturers and pump users deal only with 3% head-drop data and discard the bracketed terms. However, using the notation avoids ambiguity and is essential where data from different test methods are used. It can also be expanded to express "non-standard" test results. e.g.   10% head-drop   -------
NPSH(10pc) 20 mm visual-onset ---NPSH(20mm)

Guidance in choosing the value of **R** for the avoidance of hydraulic performance loss and the avoidance of cavitation erosion are discussed in Chapters 6 and 9 respectively.

4.1.3 *NPSH datum conditions*

Past assumptions and subsequent observations. In the years prior to 1960 many users and some manufacturers made the simple, but incorrect, assumption that until a deviation from the non-cavitating characteristic took place no cavitation was present inside a centrifugal pump. It was also assumed that a small increase above this "critical" value of NPSH would provide the necessary margin to give security against the onset of cavitation erosion, the main concern at the time. This critical value of NPSH (sometimes referred to as "breakaway" or "0% head-drop" NPSH) was the datum from which the empirical rules to calculate the NPSH necessary to secure acceptable operating conditions were then based.

It was subsequently found from visual observations that substantial amounts of cavitation are often present within a centrifugal pump impeller at much higher values than the NPSH associated with the breakaway point on the non-cavitating performance characteristic. The evidence of cavitation erosion damage to pumps operating above the critical NPSH value accumulated rapidly as a trend toward higher speed pumps meant that more pumps were operating in marginconditions. This highlighted the serious inadequacies in the way forecasting safe operating conditions was performed. It also highlighted another serious disadvantage in that the normal scatter of test points could, and did, lead to a wide range of interpretations as to the value of the critical NPSH. This made it inappropriate for use in commercial contract testing.

The choice of NPSH datum. A datum is required that is both scientifically valid and commercially acceptable. For commercial use a good test specification or test standard is one that is

    (a) appropriate in quality,

    (b) easily understood by both parties to a contract,

    (c) easily applied.

There are two NPSH values that can be shown to have commercial merit as a datum for use in defining cavitation performance:

    (i)   an NPSH measurement relating to the head-drop resulting from cavitation,

      (ii) an NPSH associated with a measurement of onset of cavitation on the impeller blade.

In Fig. 4.1 curves are presented which show various NPSH datum values over a range of flowrates. They include those for a 3% generated head-drop, designated NPSH(3pc), and those for a four millimetre cavity on the impeller blade, designated NPSH(4mm). The shape of the curves can vary quite a lot. This is principally because the chamber from which the impeller eye draws liquid varies considerably for the different pump constructional arrangements (i.e. end-suction, split-casing, etc.). Figure 4.1 shows the idealised curve shapes to which small end-suction pumps closely approximate.

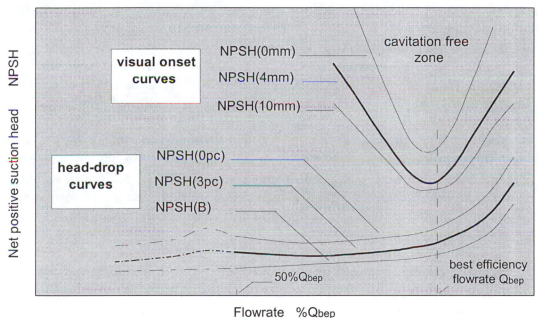

Fig. 4.1 Idealised head-drop and visual-onset NPSH variation with flowrate

NPSH(3pc). Past experience shows that establishing the value of NPSH at the flowrate where the non-cavitating and cavitating performance curves diverge (0% head-drop) is very difficult. Conversely, it also shows that an NPSH value corresponding to a 3% reduction in generated head, NPSH(3pc), is a contractually definable measurement. Industrial test methods produce NPSH(3pc) values, particularly at best efficiency flowrate, with an appropriate repeatability and accuracy.

Greater values of NPSH head-drop, up to and including the NPSH at which the generated head collapse becomes independent of flowrate, the so-called "breakdown" value NPSH(B), have scientific validity but these put the machine on test at risk from cavitation erosion attack and vapour locking. Very large values of NPSH head-drop are therefore commercially unacceptable.

The NPSH corresponding to a 3% head-drop is an engineering compromise. The cavitation present in the impeller makes operation under such conditions in service normally unacceptable. As a datum to which NPSH can be added to reduce the risk of failure NPSH(3pc) is both scientifically and commercially acceptable.

NPSH(3pc) is used very widely as a datum for forecasting NPSH requirements. However NPSH(3pc) is not used exclusively. The continuing use of head-drop values other than 3% means that care needs to be taken when interpreting results. The datum of head-drop data should always be established before it is used.

The value of NPSH(3pc) can be determined easily by test to within limits of +/- 10%. To facilitate the practical application of cavitation test data this level of uncertainty is assumed to be included within the value of **R**(3pc).

NPSH(4mm). The NPSH value that corresponds to the appearance of cavitation on an impeller blade extending 4 mm in the direction of flow, NPSH(4mm), is contractually acceptable. Smaller cavity lengths attract difficulties of interpretation. Greater cavity lengths put pumps at risk of damage from cavitation and vapour locking.

For pumps where cavitation performance is required to be known to a high degree of certainty, usually to reduce provision of NPSH to a minimum in installations where this is very expensive, an NPSH(4mm) test offers an appropriate means of demonstrating performance.

Pump operation with 4 mm of cavitation on the impeller blade is normally unacceptable. However, as a datum to which further NPSH can be added to ensure freedom from risk of failure NPSH(4mm) is both scientifically and commercially acceptable.

The practical difficulties of observing pump inlet conditions where high inlet pressure, high temperature, or opaque/chemically aggressive liquids are present mean direct measurement using transparent components is seldom possible. Options that should be considered when such circumstances arise are (i) testing at reduced speed, and (ii) testing with "cold" water. The special requirements for such tests are explored in Chapter 11.

The value of NPSH(4mm) can be determined easily by test within limits of +/- 10%. To facilitate the practical application of cavitation test data this level of uncertainty is assumed to be included within the value of **R**(4mm).

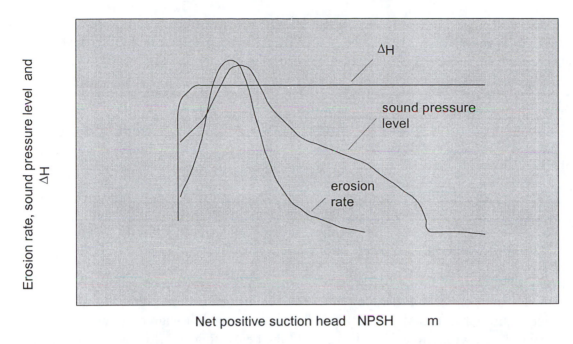

Fig. 4.2   Generated head, erosion rate and sound pressure level variation with NPSH

NPSH(spl): An unsuitable option. The violent collapse of cavitation, particularly when it appears in the form of a bubbly flow, gives rise to pressure pulsations within the pumped liquid. When such bubbly cavitation is fully developed a high rate of damage is almost certainly occurring. An audible crackling sound usually presents itself to an observer. A prima-facie case can therefore be made for an NPSH datum based on sound pressure level measurement, NPSH(spl). Using very special and expensive test facilities the relationship between hydraulic performance, cavitation erosion and hydrodynamic noise level shown in Fig. 4.2 has been demonstrated for a small pump.

The inherent difficulties in the interpretation of results and provision of test loops that are free from cavitation or free from unpredictable cavitation activity in areas other than the area in the pump being examined make NPSH(spl) unsuitable for use in commercial performance tests.

Impressive progress has been made in developing computer software for analysing "noise" signals. Limited success has been secured in boiling noise detection in nuclear power stations, similarly in sonar tracking of nuclear submarine reactor pumps in open seas, also in radio astronomy where signals can be processed to give an image of distant galaxies. All these successes have one thing in common: the background to the signal can be regarded as unchanging for the period of analysis and, as a quantifiable reference, can be used to make quantitative assessments. This is generally not the case in centrifugal pump installations.

The best that can be said for cavitation noise measurement devices is that they may have some uses in detecting the onset of cavitation and possibly some uses in detecting that cavitation erosion is occurring. It has not yet been found possible to use them as an unambiguous means of measuring cavitation erosion intensity.

NPSH(spl) is not a commercially acceptable datum for cavitation performance measurement.

4.1.4 *Industrial NPSH tests*

NPSH datum choice. The choice of NPSH datum for industrial NPSH tests is summarised in Table 4.1. Details of the procedures for conducting NPSH(3pc) and NPSH(4mm) tests are described in Chapter 11.

Table 4.1

NPSH Datum	Main features	Hydraulic performance loss	Cavitation erosion
NPSH(3pc)	Low cost	OK	OK
NPSH(4mm)	Costly Special test rig Accurate	---	OK
	OK = Suitable for commercial usage		

Note : The implied limit of measurement uncertainty for both NPSH(3pc) and NPSH(4mm) values is +/-10%.

Extent of testing. To demonstrate that a claimed performance level has been reached using either NPSH(3pc) or NPSH(4mm) necessitates that a curve be produced similar to that shown in Fig. 4.1. It is reasonable to agree on a test curve covering the flowrate range 50% to 120% of the best efficiency flowrate value Q_{bep}.

4.2 The Pump Operational Range

4.2.1 *An overall view*

The operational range is the flowrate boundary on the hydraulic performance characteristic defined by the pump curve $\Delta H/Q$ and the maximum and minimum operational flowrates Q_H and Q_L. It is necessary to be able to establish the operational range. The "worst-case" operating condition can only be identified by comparing NPSH(A) values with NPSH(R) values throughout the entire range.

The system resistance curve determines the value of the maximum flowrate for given inlet conditions. The minimum flowrate is normally determined by one of three control methods: outlet valve closure, pump speed reduction, or inlet vessel level reduction. It is fixed by design choice taking into account the adverse effects associated with low %Q_{bep} operation described in Chapters 7 and 9.

To assess whether the actual pump/system will meet the "design-basis" intent it is instructive to examine the effect of (i) overestimating the pump design duty, (ii) changing the method of flowrate control, and (iii) operating plant under conditions such as start-up or where changes in pressure at a liquid surface occur. It will become evident how a wide disparity can arise between the design-basis calculation and the actual worst-case outcome arising when unused and sometimes unnecessary margins are included. It is pertinent to note that a cavitating pump running well in excess of the design duty point flowrate is a common underlying cause of failure.

The system resistance curve equation is determined by the choices made in plant system design and as such is outside the scope of this book. However, as it is important to understand how the uncertainties that arise relate to cavitation performance, sample calculations based on a simplified plant system are included at the end of this chapter. The numerical data in Figs. 4.3 to 4.8 relate to these sample calculations.

4.2.2 *The pump design duty*

It is commonplace for the pump duty to be based on a flowrate that is fixed by process requirements. The generated head at this flowrate is then calculated. Ideally, its value is equal to that of the total system resistance to flow at the design flowrate.

To avoid the possibility of a shortfall in flowrate the system resistance is usually calculated based on a worst-case scenario of worn pipes (a high fouling factor), valve control band losses, the maximum difference in height (maximum outlet vessel and minimum inlet vessel) between liquid levels, and the maximum difference in vessel pressures. Often additional head is then included for reasons ranging from the justifiable (to account for wear in pump internal clearances) to the hopeful ("to be sure"). This certainly gives the highest generated head requirement and, if valve control of flowrate within a narrow band is always functioning (destroying surplus energy), it is possible to achieve problem-free operation if NPSH protection based on this flowrate is provided. Frequently this is not the case. Pump valve control is often crude (or nonexistent) and vessel levels for normal operation are often far from being those used in the design-basis calculations.

Pump users sometimes assume that the pump design duty will be the pump operating point. Uncertainties in the calculation processes mean that this will not be the case unless a form of flow control is activated.

4.2.3 *The pump operating point*

Non-cavitating pumps. These operate at the flowrate where the pump curve ($\Delta H/Q$) and the actual system resistance curve (h_{sys}/Q) intersect as shown in Fig. 4.3. The design intent is expressed as point A. The conservatism leading to an excess head requirement at the design flowrate Q_G means the actual system resistance is somewhat less. The actual head required is shown as point B. This means that in practice the actual intersection of the two curves is at point H. This gives the upper limit on flowrate as Q_H.

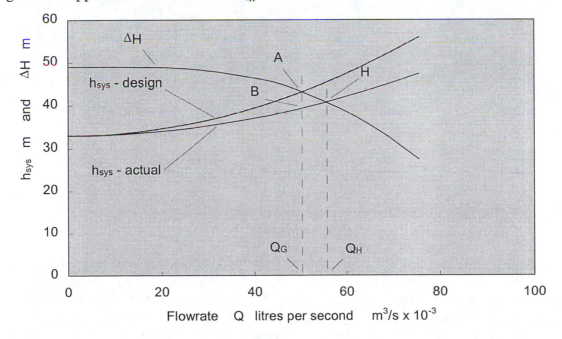

Fig. 4.3 The operating point (non-cavitating pumps)

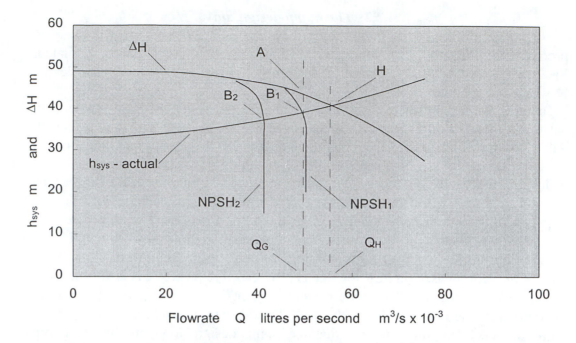

Fig. 4.4 The operating point (cavitating pumps)

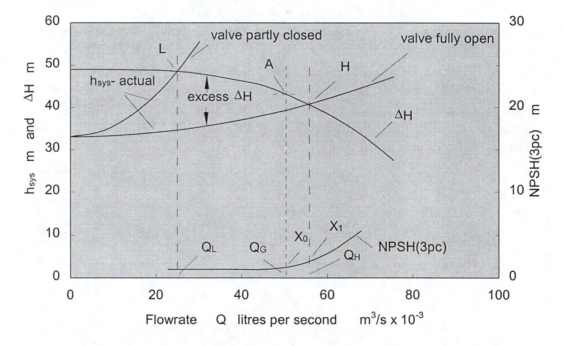

Fig. 4.5 The operating point and operational envelope
          (outlet valve control method shown)

Cavitating pumps. Pumps that cavitate sufficiently to exhibit a head-drop operate at
the intersection of the pump curve appropriate to the NPSH(A) at the pump inlet and
the system resistance curve as shown as Fig. 4.4. A pump with NPSH(A) = NPSH$_1$
will operate at point B$_1$ and will produce the required Q$_G$ albeit whilst cavitating
severely. A pump with less NPSH(A), say NPSH$_2$, will operate at point B$_2$ and will
fail to deliver the required flowrate. Raising NPSH(A) above that needed to produce a
head-drop results in the pump operating at point H (flowrate Q$_H$), the same as in
the non-cavitating mode.

4.2.4 *Methods of varying the pump flowrate*

Outlet valve control. A pump will operate at Q_H instead of the intended Q_G unless energy arising in the form of excess generated head is destroyed. Actual conditions appropriate to an outlet valve control system are illustrated in Fig. 4.5. Throttling a valve increases the system resistance. This destroys the excess generated head. Dependent upon the degree of throttling this gives an operating flowrate range of Q_L to Q_H. For the fixed speed pump shown, it is evident by inspection that the maximum NPSH(3pc) occurs at the maximum flowrate Q_H and has a value given by point X_1.

Variable speed control. The selection of a variable speed pump is worthy of consideration when the inevitable energy loss inherent in the outlet valve control method is commercially significant. This is particularly likely to be so for a plant where the system resistance curve comprises mainly pipe frictional losses and a wide range of operating flowrate is required. As can be seen from Fig. 4.6 the maximum NPSH(3pc) in the operating range, like the outlet valve method, is at the maximum flowrate Q_H and has a value given by point X_1. Should the flowrate fall so that the minimum permissible flowrate is reached a control interlock should be provided to initiate pump outlet valve closure thus obliging the pump to either run in low-flow (leak-off) protection mode or shut down.

The best efficiency flowrate reduces in direct proportion to a change in pump speed (see section 3.2.1 — Affinity laws). This means a lower value of Q_L is achievable compared with the outlet valve control method. The lower speed and lower $\%Q_{bep}$ also confers advantages with respect to protection against surging and cavitation erosion. These advantages are evaluated in Chapters 7 and 9 respectively.

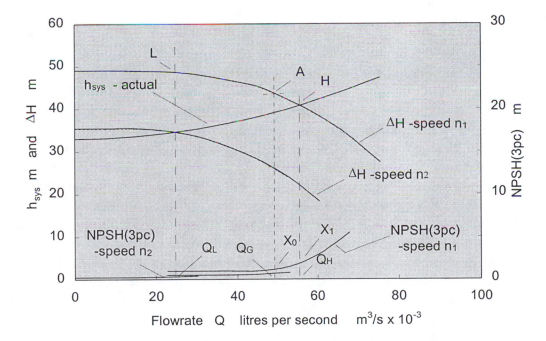

Fig. 4.6 The operating envelope (variable speed control)

Free suction control. This is a method by which cavitation is allowed to act as the regulator of flowrate. Referring to Fig. 4.7 it can be seen if the liquid is being pumped out of the inlet tank faster than it is put in, the inlet vessel level becomes lower and the NPSH(A) moves to a lower value, say, from $NPSH_1$ to $NPSH_2$. The flowrate reduces correspondingly from B_1 to B_2 on the system resistance curve. As level builds up again the reverse occurs and the pumping rate increases. Should the level build up to $NPSH_3$ then the pump operates at B_3. If maximum NPSH(A) at point B_3 is fixed, say, by a weir that prevents a further increase by having excess liquid drawn off so that the suction static head cannot increase, then the maximum possible flowrate Q_H is that appropriate to $NPSH_3$.

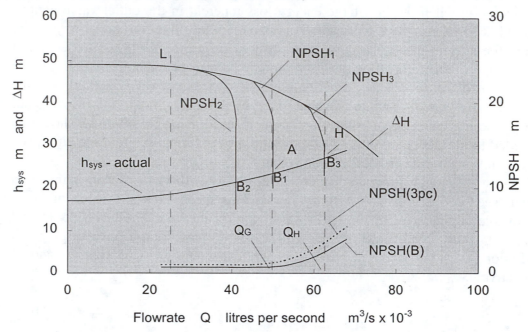

Fig. 4.7 The operational envelope (free suction control)

Free suction control has found use in applications where the inlet tank has a large liquid inventory such as the turbine condenser well from which power station condensate extraction pumps draw liquid. It also has niche applications in the nuclear and chemical industries where air entrainment through valves is unacceptable or where the leakage and unreliability of moving components has to be minimised. The pump duty is achieved with NPSH conditions defined by NPSH(B) values. Greater response to changing pump inlet conditions (e.g. the depth in a sump or inlet vessel) is given by a pump with a flat NPSH(B) curve. Such a curve is usually associated with an impeller that has a large inlet eye where the effect of large volumes of cavitation at intentionally very low local velocities is similar over a wide range of flowrates.

There are several disadvantages to free suction control. Prediction of the volumetric flowrate for a pump on free suction control necessitates a detailed knowledge of the NPSH curves throughout a wide range of flowrates and head-drops; information that is not normally available. Actual volumetric flowrate crucially depends upon NPSH(A). A small change from that used as a basis for design calculations can make a big difference to the flowrate. It is therefore prudent during plant design to incorporate measures which provide flexibility in the value of NPSH(A). The possibility of running down to very low flowrates where mechanical integrity may be challenged is also present. This is why very slow response times are preferred. More importantly, the risk of severe cavitation erosion exists where significant head-drops due to cavitation are occurring. Pumps are specifically designed for this mode of operation with, as previously mentioned, impellers having an especially large eye diameter to maximise vapour handling capabilities. Even so the risk of unacceptable cavitation damage is always present. Free suction control should not be used unless full pump cavitating performance curves are available, the shape of the system resistance curve is suitable, and acceptance tests to demonstrate freedom from an unacceptable level of cavitation erosion attack (see Chapter 12 for details) can be carried out.

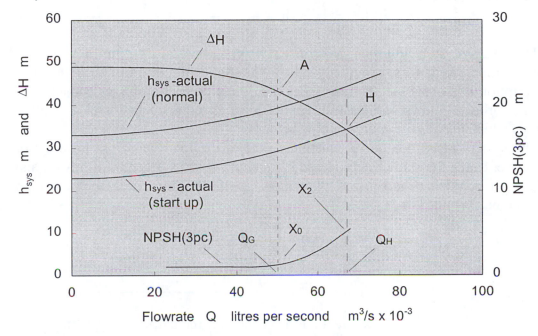

Fig. 4.8   Evaluating pump start-up conditions and/or pressure changes on liquid surfaces

4.2.5   *Start-up conditions*

A point often missed that sometimes gives rise to a long period of "at risk" pump operation is the start-up condition. Often at the commencement of a plant operating cycle the inlet vessel level is high and outlet vessel level is low, the very reason for some control systems to initiate a pump start. Where large vessels are used this may mean that the pump will take a considerable time transferring liquid before design operating levels are reached. On start-up the static head component of the system resistance is reduced as shown in Fig. 4.8 so that the pump "runs out" to a higher value for the maximum flowrate Q_H. The maximum NPSH(3pc) in the operating range increases to X_2.

A similar situation can arise when the pressure acting at the surface of liquid in enclosed vessels is "process dependent" and pressure differences adversely affect the system resistance. Such long-period swings in flowrate are analysed by regarding them as pseudo-steady-state conditions.

4.2.6 *Pressure changes at the liquid surface of inlet or outlet vessels*

A pressure change at the liquid surfaces produces an identical effect to a change in vessel level. Indeed one can be substituted for the other. A constant change in pressure throughout the operational range will produce an effect identical to that shown in Fig. 4.8 with the maximum NPSH(3pc) again being X_2 at flowrate Q_H. However, pressure changes are often "process related" and as such have the potential to occur more quickly than liquid level changes. A fully detailed example illustrates the consequences of this in Chapter 6.

4.2.7 *The system resistance curve equation*

Basic expressions. The system resistance at a given flowrate is the total head loss that arises from liquid passing through a particular system. It is expressed by an equation of the form

$$h_{sys} = h_s + h_F \qquad \text{------------------------------------4.5}$$

The total system head loss h_{sys} comprises two components, a "static" head loss h_s and a variable head loss, commonly called the "frictional" head loss, h_F. The latter description is a misnomer as h_F includes with the pipe friction losses encountered in the turbulent flows of industrial applications, all component losses and other losses which vary as the square of pipe velocity.

The system resistance curve equation describes the total head loss in terms of flowrate. Using the static heads h_{S1} and h_{S2}, and the vessel surface pressures p_{V1} and p_{V2}, and recognising the square law relationship of all other losses, then equation 4.5 can be expanded to

$$h_s = \left[\left(h_{S2} + \frac{p_{V1}}{\rho_L g} \right) - \left(h_{S1} + \frac{p_{V2}}{\rho_L g} \right) \right] \qquad \text{-------4.6}$$

$$h_F = k_2 Q^2 \qquad \text{--4.7}$$

where k_2 is a constant for the particular system.

When, as is often the case, the inlet and outlet vessels both have a free surface at barometric pressure then $p_{V1} = p_{V2}$ and the terms in which they appear cancel each other out. The system resistance curve equation is then given by

$$h_{sys} = h_s + k_2 Q^2 \qquad \text{----------------------------4.8}$$

Calculation of the static head loss h_s. The static head loss is calculated using equation 4.6. It is convenient to choose the pump centreline as the datum for measurements h_{S1} and h_{S2}. The former is then available for use in the calculation of NPSH(A) values.

Calculation of the variable head loss h_F. Two methods of calculating the variable head losses are in common use. The first, which grew up with the development of hydraulic systems, expresses the loss for each system component as an equivalent length of straight pipe, sums up the total of equivalent lengths, and then calculates the total loss. The second, developed from more general fluid mechanics and supported by a wider range of test data, is based on using loss coefficients to calculate the head loss for individual components.

To calculate values for the system resistance curve it is necessary to quantify the hydraulic losses throughout the pipework system. This includes straight pipe lengths, bends, truncations etc., valves, intermediate vessels, flow measurement devices, and all the other items that provide resistance to flow. Reference works are available which give a wide range of basic data. References 4.1, 4.2 and 4.3 are examples. These works also include calculational methods which range from a simple "rule-of-thumb" approach (useful in considering preliminary design options) to a more complicated "state-of-the-art" analysis (appropriate to final design evaluations, especially those which include large energy losses). Loss data for specific proprietary items such as strainers, etc. are available from reputable manufacturers.

Calculations are split at the inlet to the pump in anticipation of the need to calculate NPSH(A) from the same data.

4.3   Putting a Value to NPSH

The two distinct uses for NPSH are (1) in pump testing as a measure of how inlet head controls pump cavitation performance and (2) in plant design as a measure of the inlet head provided by plant design. They often appear separately but are compatible under the same definition of NPSH. Equations 4.1, 4.2, 4.3 and 4.4 describe their relationships mathematically.

4.3.1   *For pump testing — To establish datum values for NPSH(3pc) and NPSH(4mm)*

By definition, NPSH = {total inlet head} - {head equivalent of vapour }
{ pressure of pumped liquid}

From section 2.4 and ISO test specification 3555-1977,

$$NPSH = \left\{ H_1 + \frac{p_b}{\rho_L g} \right\} - \left\{ \frac{p_{vap}}{\rho_L g} \right\} \text{-----------------------------4.9}$$

The inlet total head H_1 is determined using equation 2.1. The barometric pressure is added to make the "gauge" pressure readings taken during an ISO-based pump test consistent with the vapour pressure data which are, by convention, presented in tables of liquid properties (and used in plant process calculations) as an absolute pressure.

4.3.2   *For plant design — To establish the value of NPSH(A)*

By definition, NPSH = {total inlet head} - {head equivalent of vapour }
{ pressure of pumped liquid}

This gives:

$$NPSH(A) = \left\lceil h_{S1} - h_{F1} + \frac{p_{V1}}{\rho_L g} + \frac{v_1^2}{2g} \right\rceil - \frac{p_{vap}}{\rho_L g} \text{-----------------4.10}$$

A typical calculation using equation 4.10 is included within section 4.4.

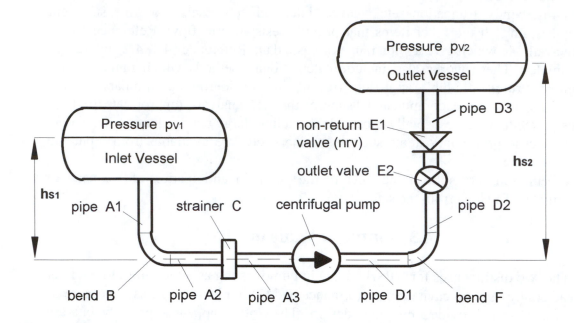

Fig. 4.9 A typical system layout

4.4 Learning by Example

The following example uses the equivalent pipe method with the simplest data. The pump system it is based upon is described by Fig. 4.9 and by the pipe size and length data included in Table 4.2. The purpose is to illustrate the overall procedure and highlight the features important in avoiding problems from cavitation. Actual pump curve and system data have been used; the NPSH(3pc) and NPSH(B) data are not test figures but are chosen to demonstrate the effect of moving from cavitating to non-cavitating conditions whilst all other pump and system conditions remain the same. A tabular format is used. This is convenient especially for the more complex systems. Calculations are used to determine the following:

      1. the pump duty (ΔH_G at a specified value Q_G)
      2. the system resistance equation ($h_{sys} = h_s + k_2 Q^2$)
      3. the NPSH(A) at flowrate Q_G

The output is then commented upon making use of earlier text figures. The sensitivity of the findings are then tested, again with the aid of the text figures.

Note: In the examples data are presented to three significant figures. This is usually more than sufficient. In large multiple component systems however a judgement has to be made if this is appropriate bearing in mind the reliability of the source data and the likely cumulative effect of errors.

4.4.1 *Design data and calculations*

Flowrate (design): $Q_G = 50$ litres per second.
Pumped liquid: Water at 95°C.
Property values: $p_{vap} = 0.845$ bar,
$\rho_L = 962$ m³/kg,
$g = 9.81$ m/s².

Table 4.2 System losses

System component		Pipe diameter	Equivalent straight pipe length	Loss per 100 m @ 50 litres per second *	System resistance
Description	Item (Fig.4.9)	mm	m	m	m
Inlet vessel velocity head loss — (200 mm diameter pipe)**					**0.13**
Inlet system losses					
Straight pipes	A1+A2+A3	200	13		
Bend	B	200	5		
Strainer (fouled)	C	200	12		
		Sub total	30	1.2	**0.36**
Outlet system losses					
Straight pipes	D1+D2+D3	150	120		
NRV and Outlet valves	E1 & E2	150	3		
Bend	F	150	5		
		Sub total	128	5.5	**7.00**

* From Ref. 4.1 the friction loss for straight pipe @ 50 litres per second is:
1.2 m per 100 m length for 200 mm diameter pipe,
5.5 m per 100 m length for 150 mm diameter pipe.

Pump inlet pipe velocity: $v_1 = \dfrac{50 \times 10^{-3}}{0.7854 \times (200 \times 10^{-3})^2} = 1.6$ m/s

** Pump inlet pipe velocity head: $\dfrac{v_1^2}{2g} = \dfrac{1.6^2}{2 \times 9.81} = 0.13$ m

Inlet system loss up to pump inlet $h_{F1} = 0.36$ m   for use in calculation of NPSH(A)
Total variable resistance: $\qquad h_F = 0.13 + 0.36 + 7.00 = 7.5$ m (approx.)

Table 4.3

Pump flowrate	Liquid level difference			Pressure difference [expressed as an equivalent head]			Static head differential
	h_{S2}	h_{S1}	$(h_{S2}-h_{S1})$	p_{V2}	p_{V1}	$[(p_{V2}-p_{V1})/\rho_L g]$	h_S
litres / second	m	m	m	bar	bar	m	m
0 (Start-up)***	14.0	6.0	8.0	2.42	1.0	15.0	23.0
25	15.0	5.0	10.0	3.17	1.0	23.0	33.0
50 (Q_G)	16.0	**4.0**	12.0	2.98	1.0	21.0	**33.0**

*** Flowrate initially Q_H.   (1 bar = 10.6 m for water at 95•C)

Liquid level above pump inlet: h_{S1} =   4.0 m   for use in calculation of NPSH(A)
Total static resistance: h_S = 33.0 m

Total system resistance: h_{sys} = $h_S + h_F$ = 33.0 + 7.5 = 40.5 m

   It is common practice to add a further 5% to cater for uncertainties, including the fact that test specification ISO 3555-1977 Class B has a permissible limit on pump total head of +/-1.5% so the possibility of a 1.5% head shortfall exists. Rounding up often follows. Applying this "to-be-sure" or "safety margin" approach gives a generated head requirement of 43.0 m.

Pump duty
As the total system resistance equals the pump generated head the pump duty is :

> generated head ΔH_G = 43.0 m   flowrate Q_G = 50 litres per second

A pump is now selected that has a $\Delta H/Q$ characteristic passing through this duty point.

System resistance curve

From equation 4.8 at 50 litres per second k_2 = $\dfrac{43.0 - 33.0}{50^2}$ = 0.0040

From equation 4.5 the "design" system resistance curve equation is
$$h_{sys} = 33.0 + 0.0040Q^2$$

The system resistance curve is plotted in Fig. 4.3 under the label "design" with the pump duty, point A, having the coordinates ΔH_G = 43.0 m, Q_G = 50 litres per second.

NPSH(A)
From equation 4.10 NPSH(A) = Total head @ pump inlet - Head equivalent of the vapour pressure of the pumped liquid

$$= \left\lceil h_{S1} - h_{F1} + \frac{v_1^2}{2g} \right\rceil + \frac{(p_{V1} - p_{vap})}{\rho_L g}$$

$$= [4.0 - 0.36 + 0.13] + \left\lceil \frac{(1.0 - 0.845)\,10^5}{962 \times 9.81} \right\rceil$$

> NPSH(A) = 5.4 m (approx.) @ 50 litres per second

Summarising, pump duty: 43.0 m @ 50 litres per second
         system resistance curve: h_{sys} = 33.0 + 0.0040Q^2
         ("design" on Fig. 4.3)
         NPSH(A) = 5.4 m @ 50 litres per second

4.4.2 Design intent and reality

Careful examination of Tables 4.2 and 4.3 soon reveals where small changes in assumed values will make a significant impact. A number of changes are postulated to evaluate typical effects.

It is instructive to rework the calculations on an "actual" basis. A cumulative effect is considered here although in practice it is most enlightening to analyse each effect in turn. Whilst an overestimate on every item may seem extreme, the evidence provided by the many oversized pumps in service shows that a general tendency to inflate values is commonplace. It will become clear that it is difficult to forecast system resistance accurately, that it is human nature to protect against a deficiency by overestimating, and, that the consequence is that the risk of unacceptable cavitation is greater than design estimates would lead one to believe. Knowledge of these facts provides a platform for assessing the adequacy of cavitation protection levels in Chapter 6 and 9.

Sensitivity Test 1 — "Actual" conditions   The "fouled" condition losses in the strainer will prove to be an overestimate when it is clean on first start-up and instead of an equivalent 12 metre may actually be, say, 8 metre. Also the assumed friction loss values may be high for new pipe with the loss per 100 m length of 200 mm diameter pipe actually being 1.1 m at 50 litres per second instead of 1.2 m and that for 250 mm pipe being 5.0 m instead of 5.5 m. Also assume that unnecessary "rounding up" of D1+D2+D3 has led to the length of 120 m when it actually is 110 m. Finally, the 5% margin for uncertainty is likely to prove unnecessary, particularly with regard to pump test data.†

Recalculation gives: $h_{F1} = 0.29$ m

$\qquad\qquad\qquad\ h_{F2} = 5.90$ m

$\qquad\qquad\qquad\ h_{sys} = 33.0 + 0.0025Q^2$ where $k_2 = \dfrac{39.3 - 33.0}{50^2} = 0.0025$

This "actual" system resistance curve is plotted in Fig. 4.3. Notice that the operating point is now point H ($\approx$ 41 m @ 56 litres per second). Notice also that the intersection point is dependent upon the shape of the $\Delta H/Q$ curve, so different specific speed pumps (or design variants) will have a different operating point. All will result in the actual pumped flowrate being higher than the intended "design" flowrate.

Summarising, pump operating point: 41 m @ 56 litres per second

$\qquad\qquad\qquad$ system resistance curve: $h_{sys} = 33.0 + 0.0025Q^2$ ("actual" on Fig. 4.3)

† It is worth considering the process by which the pump manufacturer is likely to approach pump testing. As it is very difficult (though not impossible) to get more head from an impeller with a diameter trimmed too small it is common practice to carry out an "oversize" test and then make a final trim based on the results obtained. To minimise the risk of a pump failing to meet its duty after such a final trim, human nature leads to the chosen diameter being on the high side. The end result is that a pump passing an ISO 3555-1977 Class B acceptance test will more often than not generate a head in the upper end of the acceptable range, i.e. between 100%ΔH_G and 101.5%ΔH_G for Class B tests.

Sensitivity Test 2 — "Actual conditions at start-up" Suppose the "actual" conditions in Sensitivity Test 1 apply except that at start-up the difference in vessel levels $(h_{S2} - h_{S1})$ is $(14.0 - 6.0) = 8.0$ m and the difference in vessel pressures is now the equivalent of an additional 15.0 m making a static head differential of 23.0 m compared with the "design" figure for "normal" operation of 33.0 m.

Recalculation gives: $h_{F1} = 0.29$ m $\quad\rbrace$ As Sensitivity Test 1

$\qquad\qquad\qquad h_{F2} = 5.90$ m $\quad$ $\rfloor$

$\qquad\qquad\qquad h_{sys} = 23.0 + 0.0025Q^2$

$\qquad\qquad\qquad\qquad\qquad$ since $k_2 =$ as in Sensitivity Test 1.

The system resistance curve is now as shown in Fig. 4.8. The shape of the normal and start-up curves are identical; the only difference is the static head differential h_S at start-up which is reduced from 33 m to 23 m. Notice that the actual operating flowrate, point H, moves to the much higher value of 67 litres per second.

4.4.3 Observations

By calculating the values in Table 4.2 the relative significance of various components can be assessed. This is the time to consider the cost benefit of design changes. Is the reduced friction loss associated with use of a larger diameter of pipe worthwhile? Is a suction filter with a lower head loss characteristic available? Are the vessel operating levels and pressures adequately described for all operating conditions? etc.

It is also timely to reconsider the method of varying flowrate to ensure that an acceptable running-cost/capital-cost balance will result. Continuing the previous example in Fig. 4.5 it is evident that using outlet valve control to restrict pump output to a minimum flow of 25 litres per second will result in an excess of generated head at Q_L of about 14 m, i.e. nearly one third of that produced by the pump. If long periods of low flow operation appear likely there is a prima-facie case for making additional capital expenditure that is recoverable through reduced operating costs by considering a variable speed drive. Notice that if the static head had been lower as a proportion of the generated head, the "excess ΔH" figure at Q_L would be even higher. Correspondingly if the static head had been higher, such as when pumping into a high pressure boiler, the extent of potential energy savings would be much reduced. Clearly the merits of change depend upon the particular plant design and the time spent at different flowrates within the operating range.

The hydraulic performance benefits of adopting a variable speed control are shown in Fig. 4.6. Again, the "h_{sys}-actual" curve from the earlier example is used. Reducing the pump speed from n_1 to n_2 reduces the pump curve to that shown as ΔH-speed n_2. The speed appropriate to achieving the required minimum flowrate is that which intersects the h_{sys}-actual curve at Q_L. A reiterative series of calculations scaling the curve for speed n_1 using the affinity laws ($\Delta H \propto n^2$, $Q \propto n$) will quickly establish the lowest speed. There is a wide choice of variable speed prime movers. Most can reduce speed efficiently by up to 50%. At larger reductions the efficiency and stability of some become questionable.

Calculational tip. At zero flowrate the generated head falls in direct proportion to the square of the speed reduction. As similarity extends to the shape of the $\Delta H/Q$ curve, a sufficiently accurate estimate of the speed necessary to intersect the pump curve with the system resistance curve at flowrate Q_L is soon made. Speed control can enable the pump to operate over a larger operating range with the same risk. This is because applying the affinity laws also means Q_{bep} varies directly with pump speed and that all hydraulic restrictions based on particular $\%Q_{bep}$ values now apply at lower flowrates. The value of NPSH(3pc) also reduces in accordance with affinity laws.

Returning to cavitation at the pump, it is evident that the NPSH(3pc) at or near the duty point flowrate is unlikely to be the NPSH value that determines the adequacy of NPSH provision. Run-out to flowrate Q_H with its associated NPSH(3pc) = X_1 has to be considered where the flowrate control method used to limit flowrate to Q_G is not reliable. The conditions at start-up also have to be included in an NPSH adequacy assessment by undertaking calculations at the highest flowrate where NPSH(3pc) = X_2. The outcome is evident from Fig. 4.8. The rapid increase in NPSH(3pc) at higher flowrates means that if NPSH(A) were a single value irrespective of flowrate then the risk from cavitation would be much greater than that given by an estimate which assumes operation at the duty point.

For most systems the value of NPSH(A) given by equation 4.0 falls slightly as flowrate increases. This is because, typically, the static head at inlet is much greater in magnitude than the inlet system loss h_{FI} (4.0 m compared with 0.36 m in the example) and, in turn, h_{FI} is much greater than the inlet pipe velocity head (0.36 m compared with 0.13 m). Such typical systems do not have significant variations in the pressure acting on the liquid surface in the inlet vessel. It has to be remembered that in systems where the liquid surface pressure does change, the lowest NPSH(A) value may occur at a lower flowrate than Q_H.

This brief and much simplified review of how the system design affects cavitation performance shows that, although the calculation of an NPSH(A) value is quite easy, identifying the conditions that determine the operating range, and identifying the significance of NPSH(A) within this range, necessitate detailed and careful study of all potential modes of operation.

References

4.1   Molloy, E., Pumps and Pumping, Geo. Newnes, London, 1941.
4.2   Flow of Fluids — Through Valves, Fittings and Pipe, Publication No 410, Crane Co., Chicago, 1971.
4.3   Miller, D. S., Internal Flow Systems, BHR Group Ltd, Bedford, UK, 1990.

Chapter 5

Cavity Dynamics — A Simplified Approach

5.1 Background

An exact analysis of the cavitating flows in a centrifugal pump continues to elude engineers. Until such an analysis exists it is necessary to find a simplified working model to provide an insight into how the cavity growth and collapse processes are likely to be affecting centrifugal pump performance. An approach based on the concept of notional spherical cavities is known to have some validity. Extracts from an IMechE paper (Ref. 5.1) detailing the concept are included to put into perspective the likely magnitude of benefits arising from changes in the cavitation process.

Research into the fundamental mechanisms of cavity dynamics continues to improve understanding. Relating such work to everyday pumping problems is very difficult, especially when best efficiency flowrate conditions are not present and where less than optimum hydraulic shapes are imposed on commercial designs by the need to employ manufacturing methods which are cost effective.

Acquisition of an elementary appreciation of cavity dynamics is sufficient to assist in evaluating the risks from the unacceptable effects of cavitation. In the pragmatic world of centrifugal pumps it is necessary to be content with being approximately right rather than risk being exactly wrong.

5.2   Understanding the Notional Spherical Cavity Concept

It is not essential to fully understand the theoretical analysis contained in sections 5.3 to 5.8. However, the observations and conclusions that are derived from this work are important to pump users. The conclusions in sections 5.9 and 5.10 help to define the logic to be applied when specifying cavitation test requirements.

The theoretical work presented summarises that in Ref. 5.1. It is based on an assumption that all cavities within a pump are truly spherical at their inception and remain so as they grow and collapse. Observation of cavitation in centrifugal pumps shows that truly spherical cavities are not present in practice. Whilst discreet bubbles are often observed, particularly in cold water, it is also commonplace to see "clouds" of very small vapour-filled cavities or "sheets" of cavitation. In applying the analytical work to a centrifugal pump, therefore, it is recognised that an idealised cavity shape such as a sphere may on occasion be very different from the shapes that actually occur in a particular pump. However, it is postulated that the factors included in the theoretical analysis that determine volumetric performance (including the liquid properties and the constraints imposed on cavity growth and collapse by pump geometry) are also those that, irrespective of cavity shape, produce the volumetric performance changes in a pump. Additionally the theory assumes that the aggregated surface area of spherical cavities passing any given point within a pump is directly proportional to the heat transfer surface area controlling the actual growth of cavities. That is, the radius of curvature of all the cavitating surfaces can, for comparative purposes, be represented by a single-valued radius of a number of spherical cavities.

By assuming spherical cavity growth and collapse to be representative of actual conditions in a centrifugal pump the work on bubble dynamics of Rayleigh (Ref. 5.2) and of Plesset (Ref. 5.3) can be applied, in a simplified form, to flow in the duct formed by the impeller blades and shrouds. The outcome is a method by which *comparative* assessments of performance can be made, the validity of using "cold water" NPSH data can be evaluated, and the extent of possible reductions in NPSH(R) attributable to liquid properties and their commercial significance can be determined.

The theoretical work commences with the derivation of the so-called Rayleigh and Plesset equations. This is followed by a description of how a model based upon the notional spherical cavity has been built up. Sections 5.6 and 5.7 show how this model has been used to predict changes in cavitation performance and the essential characteristics which underpin these changes. Section 5.8 compares predicted and measured performance for a small centrifugal pump.

5.3 The Rayleigh Equation

Consider a spherical cavity growing in an incompressible, non-viscous liquid due to a constant internal pressure p_x. The cavity expands to radius r in time t from initial radius r_0.

Consider now an elemental ring of thickness dR surrounding the cavity at radius R. The kinetic energy acquired by successive rings between $R = r_0$ and $R = \infty$ is equal to the energy expended in producing volumetric cavity growth at the receding cavity wall due to the driving pressure.

In this case the driving pressure P_1 equals the cavity pressure p_x.

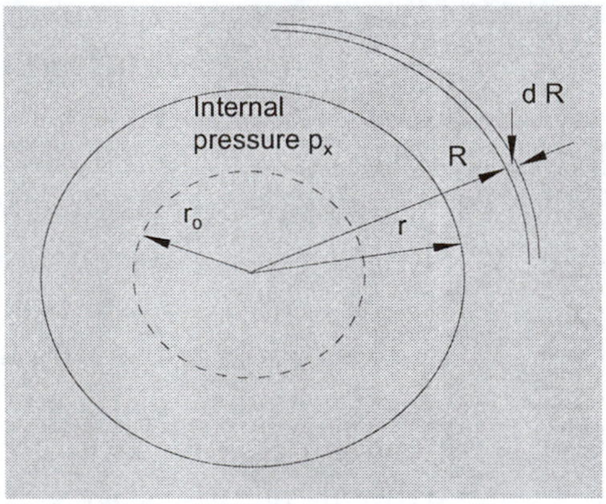

Fig. 5.1 Spherical notation— 1

Using Fig. 5.1:

$$\text{Kinetic energy acquired by successive rings} = \text{Work done expanding sphere boundary}$$

i.e.
$$\frac{1}{2} \int_{\infty}^{r} \rho_L (4\pi R^2) \left[\frac{dR}{dt}\right]^2 dR = \left[\frac{4\pi r^3}{3} - \frac{4\pi r_0^3}{3}\right] P_1$$

By continuity, the velocity at any radius R is inversely proportional to the elemental surface area and hence is inversely proportional to R^2:

i.e. $\dfrac{dR}{dt} \propto \dfrac{1}{R^2}$ and $\dfrac{dR}{dt} = \dfrac{r^2}{R^2} \cdot \dfrac{dr}{dt}$

Simplifying and integrating,

$$r\,\frac{d^2r}{dt^2} + \left\lceil \frac{3}{2} \right\rceil^2 \frac{dr}{dt} = \frac{P_1}{\rho_L} \qquad \text{-----------------------------------5.1}$$

In a spherical cavity held in equilibrium by surface tension the excess of internal pressure over surrounding liquid pressure is given by $p_{st} = 2\,\sigma_T\,r^{-1}$. The driving pressure available to produce change, P_1 in equation 5.1, is reduced.

In this case $P_1 = p_x - p_{st}$

$$r\,\frac{d^2r}{dt^2} + \frac{3}{2}\left\lceil \frac{dr}{dt} \right\rceil^2 = \frac{1}{\rho_L}\left\lceil\, p_x - p_{st} \,\right\rceil \qquad \text{-----------------5.2}$$

where $p_{st} = \dfrac{2\sigma_T}{r}$

Equation 5.2 is referred to as the Rayleigh equation. It assumes that cavity growth or collapse is controlled by the inertia of the surrounding liquid.

5.4 The Plesset Equation

The simplified analysis by Simpson and Silver (Ref. 5.4) is sufficient to show the validity of the more complex analyses carried out by Plesset (Ref. 5.3) on a spherical cavity, the rate of growth of which is restricted by conduction of heat from the liquid to the cavity wall.

Simpson and Silver showed that if conduction was restricted by an envelope of liquid which had a thickness the same as the thermal diffusion length and a "temperature driving force" occurred between the outside of the liquid envelope surrounding the cavity and the cavity wall surface, then, using Fig. 5.2,

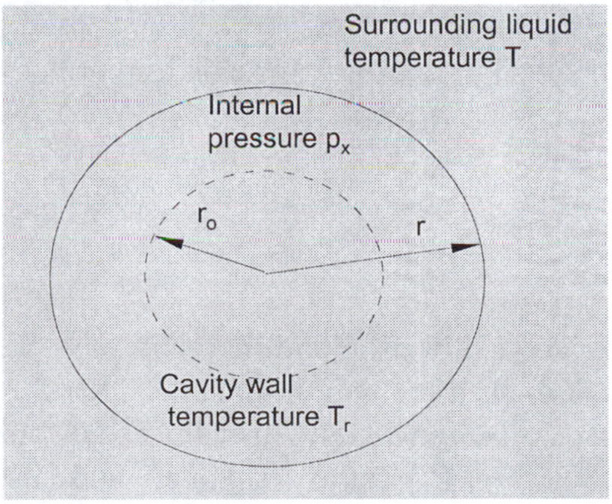

Fig. 5.2 Spherical notation — 2

$$\text{heat flow into the cavity} = \frac{(4\pi r^2)\, k_1\, (T - T_r)}{L}$$

where L = thermal diffusion length
k_1 = thermal conductivity

$$\text{heat flow to cause vaporisation} = \frac{d}{dt}\left[\frac{4\pi r^3}{3}\frac{\lambda}{V_v}\right]$$

The heat flow into the cavity must equal the heat flow to cause vaporisation:

$$T - T_r = \frac{L\lambda}{3r^2 k_1} \cdot \frac{d}{dt}\left[\frac{r^3}{V_v}\right] \quad \text{----------------5.3}$$

Consider now a cavity with a constant specific vapour volume V_v:

$$\frac{d}{dt}\left[\frac{r^3}{V_v}\right] = \frac{d}{dr}\left[\frac{r^3}{V_v}\right]\frac{dr}{dt}$$

and hence
$$\frac{d}{dr}\left[\frac{r^3}{V_v}\right] = \frac{3r^2}{V_v}\frac{dr}{dt}$$

Equation 5.3 thus becomes

$$T - T_r = \frac{L\lambda}{k_1 V_v}\frac{dr}{dt} \quad \text{----------------------------5.4}$$

By Clapeyron, $\quad \frac{\lambda}{R_o T^2}(T - T_r) = \log_e \frac{p_{vap}}{p_x} \approx \frac{p_{vap} - p_x}{p_{vap}}$

Taking V_v to be equal to the saturation value corresponding to T so that $V = R_o T_L P_{vap}^{-1}$ equations 5.3 and 5.4 can be combined as follows.

Driving pressure due to thermal $\quad p_{th} = p_{vap} - p_x = \beta t^{1/2}\frac{dr}{dt}$
conduction restriction $\quad\quad\quad\quad\quad\quad\quad\quad\quad\quad\quad\quad\quad\quad\text{--------------5.5}$

where $\beta = \frac{L\, p_{vap}^2\, \lambda^2}{k_1 R_o^2 T^3}$

Simpson and Silver in their simplified analysis took the thermal diffusion length L to be approximately $(K t)^{1/2}$ where K is the thermal diffusivity. However they acknowledged that for exact solutions this should be changed to the value obtained by Plesset of $\left[\frac{3 K t}{\pi}\right]^{1/2}$ and this is now used.

In cavities that grow when the pressure difference, inside to outside, exceeds equilibrium values the radius of any cavity r at any time t after the driving pressure P_1 is applied is given by the solution of equation 5.6.

In this case, $P_1 = p_x - p_{st} - p_{th}$

$$r \frac{d^2r}{dt^2} + \frac{3}{2}\left[\frac{dr}{dt}\right]^2 = \frac{1}{\rho_L}\left[p_x - p_{st} - p_{th}\right] \text{-----------5.6}$$

where $p_{st} = \frac{2\sigma_T}{r}$

and $p_{th} = \beta\, t^{1/2}\, \frac{dr}{dt}$ in which $\beta = \left[\frac{3K}{\pi}\right]^{1/2} \cdot \frac{p_{vap}^2 \lambda^2}{k_1 R_o^2 T^3}$

Equation 5.6 is referred to as the Plesset equation. It assumes that cavity growth and collapse can be controlled by heat transfer through the enveloping liquid.

5.5 Modelling Cavitation in a Centrifugal Pump

5.5.1 *The driving pressure producing cavity growth*

The size of cavities at visual inception is so small they do not exert a significant influence on the pressure contours around the impeller blade. At NPSH values higher than that necessary to give rise to cavitation inception the pressure contours around the pump change by the same amount as any change imposed by conditions at the pump inlet. At lower NPSH values cavity growth increasingly affects the pressure contours around the impeller blade. However, the pressure at the blade inlet is not influenced by this downstream activity and remains dependent upon the NPSH imposed at the inlet. The external pressure p_n driving the cavity growth and collapse process at the blade inlet is therefore the pressure difference obtained from the visual inception NPSH and the NPSH pertaining to the conditions under which the pump performance is being analysed when expressed as a pressure using the liquid density. To establish the significance of changes brought about by cavity growth and collapse it is useful to evaluate conditions appropriate to the greatest change. This is given by δNPSH when it is measured against the "breakdown" value NPSH(B) as shown in Fig. 5.3. This gives

pressure producing $p_n = \delta\text{NPSH}\, \rho_L\, g$ ---------------------------5.7
cavity growth

At "breakdown" δNPSH = NPSH(I) - NPSH(B)

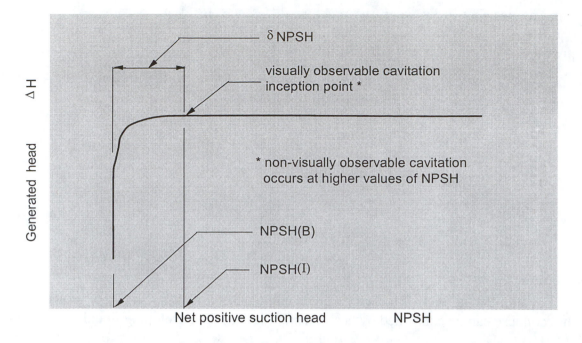

Fig. 5.3   Pictorial definition of δNPSH

5.5.2   *The pressure contour in impeller passages*

Simple kinetic energy considerations dictate that generated head is directly proportional to impeller blade velocity. Also, it is evident that the head rise across the pump ΔH is generated along a mean flow path of length l which extends from the blade inlet tip following the curvature of the blade until it reaches the impeller outer diameter. Introducing a constant of proportionality k_3 and expressing generated head in terms of generated pressure and fluid density the following relationship between generated pressure, fluid density, and path length through the impeller is postulated:

$$\Delta P = k_3 \, \rho_L \, g \, l^2 \quad \text{---5.8}$$

If the pumped fluid remains on a macro-scale a homogeneous mixture of cavities and liquid, equation 5.8 applies equally to non-cavitating and cavitating fluids since the impeller is unable to discriminate between liquids of low density and fluids which comprise higher density liquids with low density cavities in suspension. The value of k_3 is therefore constant for a particular pump. In practice this value can readily be determined from non-cavitating test data, since

$$k_3 = \frac{\Delta P}{\rho_L \, g \, l^2} = \frac{\Delta H}{l^2}$$

In a cavitating centrifugal pump the total pressure generated is the sum of the pressure generated in the cavitating zone and the pressure generated in the non-cavitating zone:

$$\Delta P = p_{gc} + p_{gl} \quad \text{---5.9}$$

where in the cavitating zone $p_{gc} = k_3 \, \rho_L \, g \, l_c^2$ $\quad \text{------------------------------5.10}$

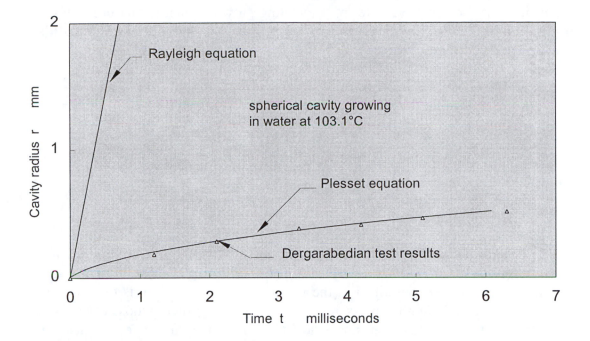

Fig. 5.4   Rayleigh and Plesset equation predictions and Dergarabedian test results

5.5.3 *Spherical cavities growing and collapsing in a stationary liquid*

Equation 5.6, developed by Plesset as a necessary extension to the work of Rayleigh, is applied to the growth of cavities. Its validity has been demonstrated experimentally for very small bubbles in a stationary liquid by Dergarabedian (Ref. 5.5). In Fig. 5.4 the test results of Dergarabedian are compared with the calculated results using both the Rayleigh and Plesset equations. It is very clear that heat transfer dominates the growth of small cavities in water at about 100°C.

With regard to the collapse of cavities it is self-evident that the retardation imposed on the rate of vapour production at the cavity wall by liquid properties does not apply when cavities collapse. An increase in pressure promotes the immediate condensation of vapour in the cavity. The time scale for such condensation is so short as to be of no practical interest in determining centrifugal pump performance.

5.5.4 *Spherical cavities in a liquid constrained to move along a duct*

Bernoulli's equation is used as a simple expedient to describe the change resulting from the interaction of notional spherical cavities and duct geometry in an impeller where fluid density at NPSH(B) conditions falls typically to about 80% of the incoming liquid density. The inclusion of vapour voidage in the continuity equation gives the velocity of the two-phase mixture with sufficient accuracy, and the energy balance inherent in the Bernoulli equation is not significantly invalidated by taking the mass to reside totally in the liquid phase since, if, as is postulated, both liquid and vapour phases move with the same velocity.

Consider a number of cavities, z per unit volume, moving in a flow Q_{bep} along a duct of constant cross-sectional area A. If the mass lost by vaporisation from the liquid into the cavities is negligible, then the only effect of the cavities is to produce a blockage to flow, which leads to an increased liquid velocity and a consequential fall in pressure local to the cavities. This fall in local pressure produces further cavity growth and, consequently, cavity size increases with great rapidity.

The additional driving pressure p_d stimulating growth which is produced by the flow being constrained by the duct can readily be derived by applying Bernoulli's equation and continuity to give

$$p_d = \frac{Q_{bep} \rho_L (2 Q_{bep} q_{vap} + q_{vap}^2)}{2 A^2 (Q_{bep} + q_{vap})} \qquad \text{since} \quad p_2 = \frac{\rho_1 \cdot p_1 + \rho_2 (v_1^2 - v_2^2)}{\rho_2} \quad \frac{}{2}$$

$$Q_1 = A.v_1 \qquad Q_2 = Q_1 + q_{vap} = A.v_2$$

$$p_d = p_1 - p_2 \qquad p_1 = 0 \text{ (the datum for change)} \qquad \rho_1 = \rho_L$$

$$\text{and} \quad q_{vap} = \text{vapour flowrate (volumetric)} = \frac{z\,4\,\pi\,r^3}{3}$$

5.5.5 *Spherical cavities in a liquid moving unconstrained in a superimposed pressure gradient which increases directly as the square of the distance travelled.*

Consider the cavities described in the previous section, but in this case subject them to an externally applied pressure p_g in a pressure gradient whose effect is to suppress cavitation. (Note that since the fluid boundaries are unconstrained the findings in the previous paragraphs do not apply.) If the density of the fluid mixture remains essentially constant equation 5.10 would apply. However, this is not the case and an adaptation must be made.

Anticipating that the method of solving the cavity dynamics equations resulting from this analysis is by an incremental step-by-step method it is convenient to calculate the pressure rise at a particular point along the flow path by summing the incremental pressure rises dp over successive incremental steps in length dl. From the particular case of cavities growing and collapsing in a centrifugal pump an "average" density calculated from the density at the beginning and end of each step has validity. (This is easily verified by comparing the results obtained using the average density with those obtained using the final density for each increment. Validity is confirmed if these values closely agree since they represent upper and lower bound possibilities.)

For a typical incremental step,

$$\rho = \frac{(\rho' + \rho'')}{2}$$

where ρ' is the fluid density at the beginning of a step
and ρ'' is the fluid density at the end of a step.

From equation 5.10,

$$dp_g = k \frac{(\rho' + \rho'')}{2} g [(l' + dl)^2 - l'^2]$$

where l' is the distance to the beginning of the step

$$p_g = \sum_{l'=0}^{l'=1} k \frac{(\rho' + \rho'')}{2} g [(l' + dl)^2 - l'^2] \qquad \text{----------------------5.11}$$

5.6 Predicting Changes in Performance Using the Model

5.6.1 *Calculational objectives*

The confidence expressed by pump designers in their ability to forecast NPSH values and NPSH curve shapes is not a satisfactory basis for pump performance acceptance. Where reliable NPSH values are required the prudent user always resorts to confirmatory contractual cavitation tests. Normally the cheapest option, a standardised "cold water" NPSH(3pc) test, is undertaken. Occasionally, where a high-speed site-based prime mover is used (e.g. a steam turbine) or the site power absorbed is too large for the works test facility, the "cold water" test is carried out at reduced speed. This identifies that the need is to be able to determine the magnitude of a *change* from cold water/slow speed "reference" conditions. The need is therefore to forecast the magnitude of such changes and to highlight where uncertainties may lie. Calculation of absolute values is unnecessary for practical applications. The following methodology provides a basis for both comparing performance under different pump operating conditions and for understanding the underlying causes of change.

5.6.2 *Calculational output*

A. Cavity characteristics. The general considerations discussed in section 5.5 are combined to produce a mathematical model of the dynamic situation inside a cavitating centrifugal pump. Taking the passages of a centrifugal pump impeller to be a duct in which a pressure gradient is superimposed, the way cavities grow and collapse is given by the solution of the following:

$$ r \frac{d^2r}{dt^2} + \frac{3}{2}\left[\frac{dr}{dt}\right]^2 = \frac{P_1}{\rho_L} \hspace{2cm} \text{------5.12} $$

where $\quad P_1 = p_n - p_{st} - p_{th} + p_d - p_g$

in which $\quad p_n = \Delta NPSH\, \rho_L\, g$

$$ p_{st} = \frac{2\sigma_T}{r} $$

p_{th} for cavity growth $= \beta\, t^{1/2} \dfrac{dr}{dt}$

p_{th} for cavity collapse* $= 0$

$$ p_d = \frac{Q_{bep}\, \rho_L\, (2\, Q_{bep}\, q_{vap} + q_{vap}^{\,2})}{2\, A^2\, (Q_{bep} + q_{vap})} $$

$$ p_d = \sum_{l'=0}^{l'=1} \frac{k\,(\rho' + \rho'')\, g\, [(l' + dl)^2 - l'^2]}{2} $$

* During collapse the thermal conductivity constraints are not applicable because the collapse mechanism depends only on the condensation process which, for cavitation in a centrifugal pump, can be regarded as instantaneous.

B.   Cavitation zone lengths.   Solving equation 5.12 by an incremental method necessitates deriving successive values of incremental step size dl. Summation of the values obtained enables the values of $l_{c'}$ (the distance along the impeller blade at which maximum cavity radius occurs) and l_c (the cavitating zone length) to be readily obtained, viz.,

$$l_{c'} = \sum_{t=0}^{t = \text{value when } r = r_{max}} \frac{(Q_{bep} + q_{vap})}{A} dt$$

$$l_c = \sum_{t=0}^{t = \text{value when } r = r_o} \frac{(Q_{bep} + q_{vap})}{A} dt$$

C.   The generated head of a cavitating pump.   This is obtained by combining equations 5.8, 5.9 and 5.10 to give

$$\Delta H = \frac{p_{gc}}{\rho_L g} + k (l^2 - l_c^2)$$

The value of p_{gc} used in the above equation is that occurring at the end of the cavitation zone length l_c.

5.6.3   *The criterion for performance comparability*

The equations summarised in section 5.6.2 enable the effect of cavitation on the volumetric performance of a pump to be calculated if all the variables are known. Obviously it is not possible to determine by observation or measurement either the number of notional spherical cavities per unit volume which represent actual volumetric performance conditions or whether the number of such cavities changes throughout the cavitating zone. However, if the criterion for volumetric performance comparability under the NPSH datum conditions is known, the volumetric performance for other pump speeds or pumped liquids under the same conditions can be calculated.

For a centrifugal pump at a given datum condition, NPSH(3pc), NPSH(4mm) or NPSH(B), it is postulated that the vapour-to-liquid volume ratio in the cavitating zone reaches the same critical value at the same distance from the impeller blade inlet for identical operating conditions. In fluids where the density of the vapour phase is much less than that of the liquid phase the critical vapour-to-liquid volume ratio value is proportional to the product of cavity site density and the maximum radius of the notional spherical cavity cubed;

$$\text{i.e.}\quad z_c\, r_{max}^3 = \text{constant}$$

5.6.4   *Numerical solutions by computer*

A computational procedure for drawing together all the above is detailed in Ref. 5.1. This does not have the status of a calculational method suitable for forecasting the performance of a particular pump. Its use is in evaluating trends. Section 5.7 provides sufficient output to enable trends important to the industrial application of centrifugal pumps to be established.

5.7 Typical Calculated Volumetric Characteristics

The results obtained from calculations based on a small end-suction centrifugal pump are presented in Figures 5.5 to 5.11.

Figures 5.5 and 5.7 show the change in cavity radius and generated pressure during the growth and collapse cycle of a notional spherical cavity in terms of path length along the impeller blade. As indicated in Fig. 5.5 this provides the reference values $l_{c'}$, r_{max}, l_c, and p_{gc} necessary for performance prediction calculations.

To provide an insight into the factors bringing about performance change, the factors influencing the cavity growth and collapse process are plotted in Figs 5.6 and 5.8. Examination of the relative magnitude of the factors, (i) surface tension, (ii) heat transfer restriction at the cavity wall, and (iii) impeller duct effect, enables the probable causes of cavity size change in a centrifugal pump to be identified.

For water at 20°C the following is evident from Figs. 5.5 and 5.6.

1. The term p_{st}, which mathematically describes the influence of surface tension, initially dominates the forces acting on the assumed truly spherical cavity. Since the pressure required to sustain a spherical cavity in equilibrium falls as the inverse of its size, once rapid growth gets under way the effect of p_{st} becomes insignificant.

2. Freed from the constraints of surface tension the cavity grows explosively under the driving δNPSH until the combined values of p_{th} and p_g, caused by heat transfer restrictions at the cavity vapour-liquid boundary wall and generated pressure in the surrounding liquid respectively, become significant. Cavity growth rate then steadies.

3. The restraining influence of p_{th} and p_g, the latter of which is rapidly increasing, soon becomes a dominant factor reducing the rate of cavity growth.

4. Growth rate falls until it reaches zero. At this time of maximum cavity size p_{th} becomes zero and the pressure p_d, the fall in pressure caused by the constraining effect of the impeller passages, reaches its maximum value.

5. The subsequent cavity collapse is very rapid. As cavity size reduces, the pressure generated with the denser fluid has a greater effect. With increasing density the velocity along the impeller passage also reduces so that measured against distance (as would be observed by eye) the collapse of the cavity appears to be quickly completed.

6. At the last moment surface tension effects begin to reappear, but by this time the generated pressure is very high and results in the cavity being rapidly crushed to less than the size it was on entering the pump inlet.

Table 1 Calculated trends in cavity growth and collapse (referenced to the end of the cavitating zone).

Ratio and % l_c value	Water temperature °C				
	2	20	43	103.1	143.3
% p_{st}/p_{gc}	Below 1% for all values between 5% l_c and 95% l_c				
% p_{th}/p_{gc}	0.2	1.7	14.7	78.6	>98.0
(at % l_c)	(33.2)	(33.6)	(28.4)	(8.7)	(< 2.0)
% p_{dmax}/p_{gc}	4.2	4.0	4.4	8.0	<< 1.0
(at % l_c)	(73.7)	(72.5)	(73.8)	(94.5)	(> 96.0)

The above values relate to the predicted performance close to best efficiency point flowrate Q_{bep}.

Comparing the results of Figs. 5.5 and 5.7 with those in Figs. 5.6 and 5.8 it is evident that at 103.1°C heat transfer effects exert a strong influence and restrict growth which, in turn, prevents impeller duct effects from becoming significant. The point at which the maximum p_{th} value occurs also becomes proportionately closer to the beginning of the cavitating zone. The data presented in Table 1 show that this is part of a trend and that in water above about 40°C heat transfer considerations dominate cavity dynamics.

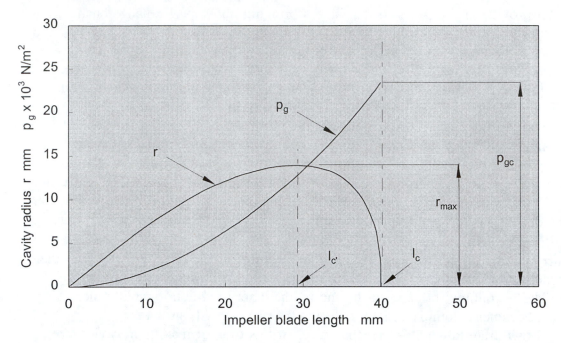

Fig. 5.5   Variation of cavity radius and generated pressure with impeller blade length (pumped liquid: water at 20•C)

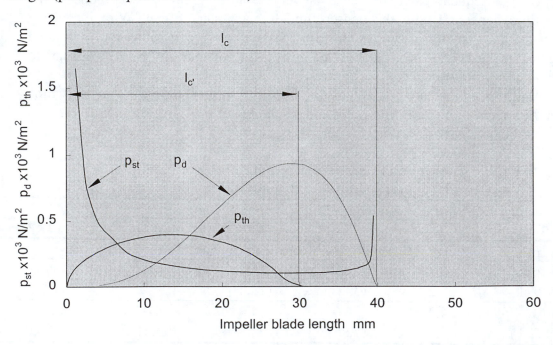

Fig. 5.6   Variation of p_{st}, p_d and p_{th} with impeller blade length (pumped liquid: water at 20•C)

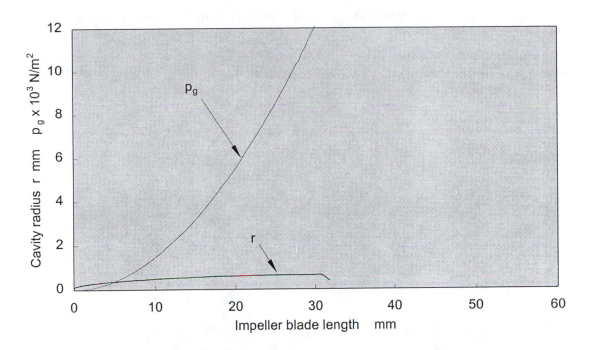

Fig. 5.7   Variation of cavity radius and generated pressure with impeller blade
length (pumped liquid: water at 103.1•C)

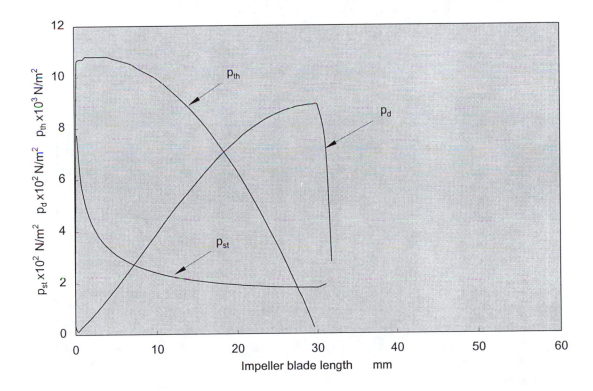

Fig. 5.8   Variation of p_{st}, p_d and p_{th} with impeller blade length
(pumped liquid: water at 103.1•C)

5.8 Predicted and Measured Performance Compared

Presentation of available data.   The predicted maximum cavity sizes at temperatures in the range 2°C to 143°C and speeds in the range 1000 rpm to 2750 rpm are shown in Fig. 5.9. The values of NPSH(B) at best efficiency flowrate predicted by the model based on calculating the change from "reference" conditions of water at 20°C and 2000 rpm are compared with test data over the temperature range 2°C to 143°C in Fig. 5.10.

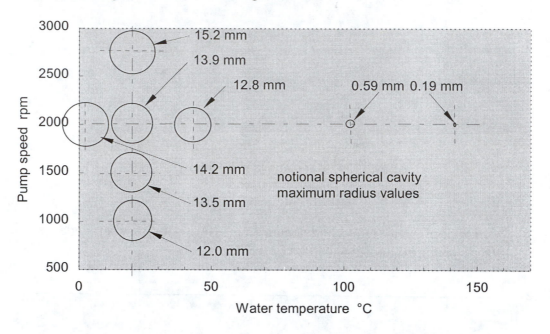

Fig. 5.9   Scalar presentation of predicted cavity size at best efficiency flowrate

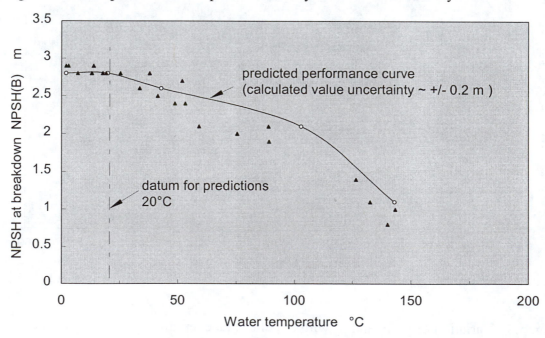

Fig. 5.10   Predicted and measured NPSH(B) at best efficiency flowrate

It is often assumed, from logic based on the success found in applying the non-cavitating affinity laws, that cavitating pump performance follows a simple square law relationship:

i.e. $NPSH \propto n^{B}$ where $B = 2$

To test the validity of this assumption predicted NPSH(B) values are used to calculate the power B: the result is presented in Fig. 5.11.

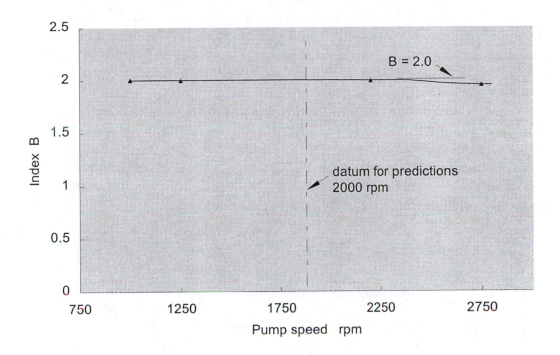

Fig. 5.11   Variation of index B with pump speed (for a particular pump)

It is evident that the predicted value of B is very close to 2 for the fully developed cavitation associated with NPSH(B); see Fig. 3.16 Point J. It falls slightly with increasing speed.

5.9   Potential Benefits — Are They Real in Practice?

The lowering in NPSH values from those measured under "datum 20°C water" conditions appears at first sight to be significant with, in the particular test pump example, a reduction of over 50% apparent for water at 150°C. As ever, this is not the whole story and other factors have to be considered.

As temperature rises so the rate of change in vapour pressure increases. Figure 5.12 shows the error in NPSH measurement caused by an underestimate of temperature of 10°C, 1.0°C and 0.1°C for water. Many centrifugal pumps have NPSH(3pc) values of 2 metres or less at the operating speed and flowrate. It is evident that for such pumps the practical implications of differences between "real" and "apparent" calculated NPSH values need to be considered for each pump installation. Even with specially designed research test rigs great difficulties arise in obtaining reliable NPSH measurements at water temperatures above 150°C. In industrial pumping systems the control and measurement of temperature is inevitably more crude, and control to better than

+/-1°C is often impractical. Under such conditions the benefits of pumping at higher temperature may be dwarfed by the uncertainty in temperature control.

Additional complications arise where hotter liquid joins the main inlet flow at or close to the pump inlet. An example of this is a pump where hydraulic balance of the pump rotor is secured by the use of a balance disk or a balance drum which necessitates returning a small flow (typically about 1%Q_{bep}) back to the pump inlet. The temperature rise of liquid passing through the pump coming from hydraulic losses means that the liquid returning from the high pressure end of large pumps can be as much as several degrees hotter than the inlet liquid it rejoins. It is evident from Fig. 5.12 that the uncertainties arising from the underestimate caused by such a temperature rise in part of the liquid flow may dwarf the difficult-to-quantify benefits arising from reductions in NPSH associated with the change to high temperature water.

Note: Good system design can significantly reduce the magnitude of the balance flow problem. Return flow should be injected well upstream, preferably before (and in the plane of) at least one bend. This facilitates mixing before the combined flow reaches the pressure drops near the impeller blades where cavitation is initiated. The flow return point should always be within any isolating valves to avoid compromising the integrity of the pump or the pump system.

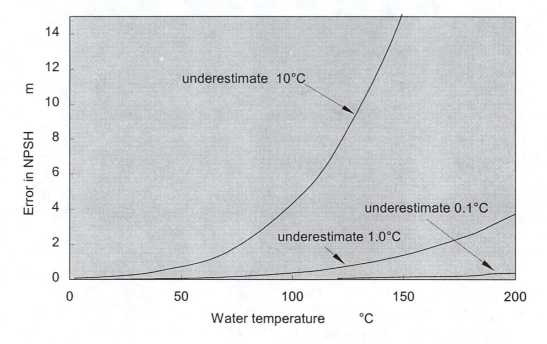

Fig. 5.12   The error in NPSH value resulting from an underestimate of
                 water temperature

5.10 Applying the Findings in Practice

The predicted volumetric performance calculated using the notional spherical cavity method shows close agreement with test results for a particular small centrifugal pump. However, the limited nature of the data means that whilst it can be used indicatively it is not a strong enough base for commercial judgements. The value of the method in indicating trends and confirming safe practices by putting reason behind them is significant. Summarising, it can be seen that the following findings are relevant to the assessment of risk of unacceptable cavitation arising when operating a centrifugal pump.

Observation **1**. Cavity growth and collapse in water changes very little over the temperature range 2°C to 40°C. Pump NPSH values, predicted and measured, confirm this finding.

Conclusion **1**. Pump tests on water below 40°C provide a safe platform for NPSH data acquisition. Temperature measurement accuracy better than about +/- 5°C is all that is needed in such circumstances.

Observation **2**. Cavities have a very small effect on the speed scaling laws even when they develop so that their aggregate volume brings about the condition that gives rise to the "breakdown" value NPSH(B). It is self-evident that cavities have no practical effect at the cavitation inception value NPSH(I).

Conclusion **2**. The two datum values of commercial interest, NPSH(3pc) and NPSH(4mm), are associated with much smaller amounts of cavities in the pump than are present at "breakdown". The assumption that NPSH(3pc) and NPSH(4mm) obey a square law relationship with speed is a reasonable one for commercial test purposes.

Observation **3**. The improvement in NPSH(3pc) in water above 100°C is probably insignificant compared with the uncertainties in temperature within the pump.

Conclusion **3**. A pragmatic approach is to base all NPSH adequacy assessments on "cold water" NPSH data and accept any change due to thermodynamic property effects as unquantifiable but beneficial.

Observation **4**. The notional spherical cavity concept is not robust enough to predict to commercially acceptable levels of confidence the changes from (a) works tests on "cold water" to (b) an application using a different pumped liquid.

Conclusion **4**. The possible variation in liquid properties (thermodynamic and rheological) and the possible variation in impeller size and operating speed of pump impellers make the commercial use of theoretical predictions of performance for particular machines unjustifiable. Reliance should only be placed on test data or previous operating experience (see section 3.9 for examples).

References

5.1 Grist, E., <u>The volumetric performance of cavitating centrifugal pumps</u>, Proc. IMechE, Vol. 200, Nos. 58 & 59, 1986.

5.2 Rayleigh, Lord, <u>On the pressure developed in a liquid during the collapse of a spherical cavity</u>, Phil. Mag., Vol. 34, pp.94-98, 1917.

5.3 Plesset, M. S., <u>The dynamics of cavitation bubbles</u>, J. Appl. Mech., Vol. 71, p.277, 1949.

5.4 Simpson, H. C., and Silver, R. S., <u>Theory of one-dimensional, two-phase, homogeneous non-equilibrium flow</u>, IMechE Conf. Two-Phase Flow, 1962.

5.5 Dergarabedian, P., <u>The rate of growth of vapor bubbles in superheated water</u>, Trans. ASME, Vol. 75, pp.537 — 545, 1953.

Chapter 6

Hydraulic Performance Loss — Duty Shortfall and Vapour Locking

6.1 Deficiencies Identified

Hydraulic performance loss arises from an incompatibility between a centrifugal pump and the plant into which it is installed. The performance deficiencies of commercial significance which appear as a consequence are listed in Table 6.1. For a given centrifugal pump the root cause of deficiency is inadequate NPSH(A).

Table 6.1   Deficiency list

	Hydraulic performance loss	
Performance deficiency	Plant operating conditions	Analysis
Pump duty shortfall	Inadequate NPSH(A)	section 6.3
Pump vapour locking	Temporarily low NPSH(A)	section 6.4

Important note. The minimum level of NPSH(A) that gives protection against hydraulic performance loss is significantly less than that needed to protect against other unacceptable forms of cavitation, particularly cavitation erosion. Such a low protection level is only acceptable for infrequently encountered operational events that are limited to periods of less than about five minutes.

6.2 Hydraulic Performance Protection — A Pump User Approach

6.2.1 *The minimum level of protection*

A prerequisite to analysing the identified deficiencies giving rise to hydraulic performance loss is to understand how performance measurements are used and margins applied to provide a secure basis for acceptable pump operation. Protection is achieved by providing an NPSH(A) somewhat greater than that defined by a datum which describes conditions which are (a) the onset of a significant reduction in generated head from that described by the non-cavitating generated head/flowrate curve and (b) a precursor to vapour locking. NPSH(3pc) matches this datum description. By being provable by an industrial grade cavitation test it also has the advantage of being a commercially suitable choice. NPSH(3pc) is therefore the datum against which all hydraulic performance loss protection levels are measured. The problem moves to determining the margin to be added to this datum to secure safe operation.

Experience (and logic applied using Fig. 4.1) shows that for any chosen flowrate the addition of a further one-third of the NPSH(3pc) value will ensure that the pumping performance is that of the non-cavitating curve. This applies to all inlet conditions whether constant or variable. It determines the lowest value that the NPSH(A) should be allowed to fall to under any circumstances.

Summarising, for all forms of hydraulic performance protection the *minimum* value is given by

$$NPSH(A) = 1.3 \text{ x } NPSH(3pc) \quad \text{----------------6.1}$$

i.e. **R**(3pc) = 1.3

Acceptability has to be evaluated for all flowrates in the operating range.

Note, at the minimum allowable value NPSH(R) = 1.3 x NPSH(3pc)

and NPSH(A) = NPSH(R) ---(equation 4.1)

For a constant pump speed a typical NPSH(3pc)/Q curve rises in the $50\%Q_{bep}$ to $120\%Q_{bep}$ range with increasing values of flowrate. At flowrates greater than the best efficiency value the rise increases rapidly. In most pumping systems the value of NPSH(A) reduces with increasing flowrate as friction losses in the inlet pipe take effect. In some systems however, particularly those where the inlet vessel is pressurised and this pressure is "process dependent", the lowest values of NPSH(A) may be at lower flowrates in the operating range as explained in Chapter 4. Care has to be taken to ensure that adequacy is assessed throughout the full operating range of the pump.

The above does not apply to the special application of pumps operating under free suction control (see section 4.2.4 and Fig. 4.7).

6.2.2 *NPSH(3pc) cavitation test requirements*

In some pump systems, including many where cold water is moved by slow speed pumps, very large values of NPSH(A) are often present at the pump inlet compared with the NPSH(3pc) values. A cavitation test is unnecessary when NPSH(A) is at least twice the NPSH(3pc) at all flowrates in the operating range.

$$\text{i.e. when NPSH(A)} \geq 2 \text{ x NPSH(3pc)} \quad \text{------------6.2}$$

A cavitation test to confirm the validity of NPSH(3pc) data and to support a hydraulic performance assessment should be undertaken if the requirements of equation 6.2 are not met. Chapter 11 details how such a test could be carried out.

The variation in NPSH(3pc) for different designs of centrifugal pump offered for the same duty can be considerable. This is particularly so when economic considerations have limited choice to existing pump designs, usually from a pump manufacturer's standard range, so that the relationship of the duty flowrate to the best efficiency flowrate for competing offers is quite different. Clearly, from Fig. 4.1, a pump with the duty point at a lower $\%Q_{bep}$ than its competitor has a greater probability of meeting the requirement of equation 6.2. This is opposite to that for cavitation erosion protection (usually the most important) discussed in Chapter 9. So whilst hydraulic performance loss protection must always be provided to secure acceptable operation, for most pumps it only has relevance in minimising the cost of NPSH provision for applications where short periods of operation at low NPSH(A) occur.

6.3   Pump Duty Shortfall

6.3.1   *The underlying cause*

A significant long-term economic loss of volumetric performance which arises from a continuing failure to deliver the required flowrate can occur solely because of cavitation in a centrifugal pump. This appears as a head-drop below the non-cavitating performance characteristic as shown in Fig. 6.1. It is the result of either (a) simply failing to provide adequate NPSH(A) and being misled by the result of a works test carried out at a higher NPSH value or (b) a failure to understand what determines the value of adequate NPSH(A).

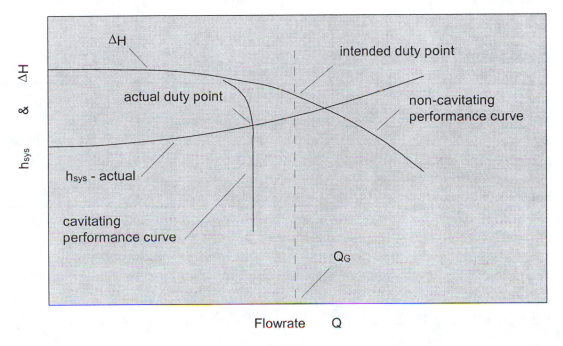

Fig. 6.1   Identification of the actual duty point

6.3.2   *An avoidance strategy*

For most pumping installations the strategy summarised in Table 6.2 meets all needs. The central objective is to calculate the value of NPSH(A) for flowrates in the operating range and, by comparison with the NPSH(3pc), determine the flowrate(s) at which hydraulic performance is at risk. Where an inadequacy is revealed remedial measures can then be taken.

Table 6.2   Strategy outline

Failure to meet the pump duty	
Action	Analysis
1 Obtain a pump $\Delta H/Q$ curve (Calculate if necessary)	(Chapter 12)
2 Quantify the "static" and "frictional" head coefficients in the system resistance curve equation	section 4.4
3 Determine the pump operating range, including the lowest (Q_L), design basis (Q_G) and worst-case (Q_H) flowrates	section 4.2
4 Calculate NPSH(A) throughout the operating range	section 4.4

New plant	Existing plant
5a Evaluate adequacy throughout the operating range (equation 6.1)	5b Evaluate adequacy throughout the operating range (equation 6.1)
6a If necessary review plant design and/or choice of pump	6b If necessary review plant operating modes (and design if practicable) and introduce most effective remedial measures
7a Add NPSH(3pc) test if necessary (equation 6.2)	

6.3.3 *Learning by example*

Consider again the example in section 4.4; in particular the "actual" system shown in Figs. 4.3 and 4.8. The NPSH(A) calculation is, using Table 6.3, as follows:

Table 6.3   System losses

System component		Pipe diameter	Equivalent straight pipe length	Loss per 100 m @ 50 litres per second	System resistance
Description	Item (Fig. 4.9)	mm	m	m	m
Inlet vessel velocity head loss — (200 mm diameter pipe)					**0.13**
Inlet system losses					
Straight pipes	A1+A2+A3	200	13		
Bend	B	200	5		
Strainer (clean)	C	200	8		
	Subtotal		26	1.1	**0.29**

From equation 4.9 NPSH(A) $= \left\lceil h_{S1} - h_{FI} + \dfrac{v_1^{\,2}}{2g} \right\rceil + \dfrac{(p_{V1} - p_{vap})}{\rho_L g}$

$$= [4.0 - 0.29 + 0.13] + \left\lceil \frac{(1.0 - 0.845)\,10^5}{962 \times 9.81} \right\rceil$$

NPSH(A) $= 5.5$ m (approx.) @ 50 litres per second

The values for NPSH(A) for the four flowrates where data is available are calculated in Table 6.4. Note that if, as would be the case in practice, more operational conditions were quantified then the calculations would be extended to include them to ensure worst-case conditions could be identified.

Table 6.4

Flowrate litres per second	Condition	h_{S1}	h_{F1}	$v_1^2/2g$	NPSH(A)
-	-	m	m	m	m
25 (Q_L)	steady	5.0	0.0073	0.03	6.6
50 (Q_G)	steady	4.0	**0.29***	**0.13**	5.5
56 (Q_H)	steady	4.0	0.36	0.16	5.5
67 (Q_H)	start-up	6.0	0.52	0.23	7.6

* "Actual" data assumed from sensitivity test 1 calculations.

Calculational tip: h_{F1} and $v_1^2/2g$ vary with the square of flowrate. Use the "actual" values at 50 litres per second calculated in section 4.4 (shown above in bold) as a base.

Finally, calculate the cavitation protection ratio **R**(3pc) using the NPSH(3pc) data presented graphically in Fig. 4.8.

Table 6.5

Flowrate litres per second	NPSH(A) m	NPSH(3pc) m	**R**(3pc)
25	6.6	1.0	6.6
50	5.5	1.4	3.9
56	5.5	2.0	2.7
67	7.6	5.0	1.5

As all values of **R**(3pc) in Table 6.5 exceed 1.3 the requirement of equation 6.1 is met. This shows that the pump will not fail to meet the required duty or encounter head-loss throughout the entire operating range.

6.3.4 *A practical method of accommodating some calculation uncertainties*
Oversized impellers. The potentially high cost consequences of adding unnecessarily large margins (the need for extra NPSH or a reliable control valve) has to be balanced against the consequences of a non-cavitating pump which is unable to meet its duty requirements because system losses have been underestimated. Sometimes the very nature of the plant into which a pump is built introduces large uncertainties. Flexibility in pump performance is required. Changing to a variable speed pump drive is seen as one possibility. Another way of introducing flexibility is to specify a pump that can accommodate an impeller with a diameter of, say, 10% greater than the duty impeller size together with a prime mover with a power rating to match this. From the affinity laws it is clear that this gives an opportunity to uprate substantially the pump capability should the need arise. The consequences of underestimating performance requirements are then manageable. Evaluating the additional capital cost of the "oversize" pump and prime mover soon shows whether this is a viable option.

An operational strategy which uses this approach is to
  (1) calculate the pump duty without adding margins for uncertainties;
  (2) install the pump with an oversized impeller and measure pump performance
       in situ to quantify the actual system resistance curve;
  (3) trim the oversized impeller to a size appropriate to the required duty.

6.4 Pump Vapour Locking

6.4.1 *The vapour locking phenomenon and the underlying cause*

Sometimes called vapour binding, this effect is observed on industrial quality instruments as a complete cessation of flow followed, on large pumps, by intermittent restoration. An example of this form of vapour locking on a boiler feedwater pump is shown in Fig. 6.2.

In the example it can be seen that as the indicated NPSH(A) fell to very low levels the recorded main pump flowrate rapidly decreased. Just prior to zero flowrate being reached vapour locking inevitably occurred. In this particular installation, loss of forward flow automatically triggered the starting of a standby pump. It can be seen from Fig. 6.2 that this pump was able to initiate and then maintain a pumping capability as NPSH(A) levels recovered. The NPSH(A) transient had significance over a period of some 300 seconds.

The depressurisation causing a vapour lock is usually the result of an injection of much colder liquid into liquid at boiling point in the inlet vessel. Figure 6.3 shows typical calculated curves for the inlet vessel pressure decay and the change in pump inlet NPSH(A) in a water filled system.

Many centrifugal pumps draw liquid from an inlet vessel in which the free surface is boiling. Typical applications are boiler feedwater pumps drawing from a deaerator and chemical pumps drawing from an inlet vessel in which evaporation is used to further the chemical process. Under such conditions a severe depressurisation of the inlet vessel gives rise to the possibility of pump vapour locking.

Pump vapour locking produces a severe disruption in pumping capability. It results from a mismatch between pump performance characteristics and plant operating conditions. One or both must be changed.

Pipework ebullition and inlet vessel nozzle cavitation can occur following inlet vessel depressurisation. They produce the same effect as pump vapour locking. They are a cause for concern since they can lead to being mistakenly diagnosed as pump vapour locking. Although outside the scope of this book, pipework ebullition and inlet vessel nozzle cavitation are the subject of a brief review at the end of this chapter. This review includes references for further reading.

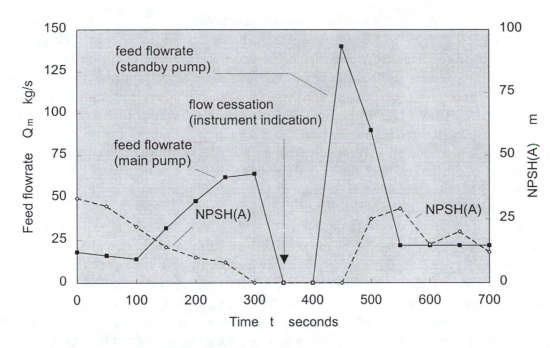

Fig. 6.2 Typical manifestation of a vapour lock
(power station pump: industrial grade instrumentation)

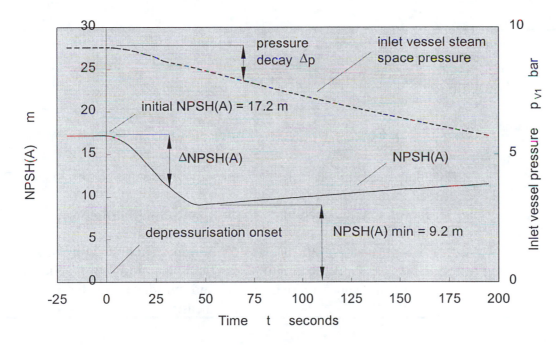

Fig. 6.3 NPSH and inlet vessel pressure transients

6.4.2 *An avoidance strategy*

The first requirement is to quantify ΔNPSH, the reduction in NPSH(A) during an inlet vessel depressurisation for a particular flowrate. Repeating this for other credible plant operating regimes enables the worst-case values to be determined for each of a number of pump inlet pipe flowrates. Subtracting these values from the steady-state NPSH(A) values (those present before the depressurisation) gives the lowest value of NPSH(A) reached for each of the flowrates considered.

The strategy provides an aid to plant design optimisation studies. Repetition of the calculations for different inlet pipe configurations and/or plant operating conditions enables both an unacceptable risk of vapour locking to be avoided and the most cost effective plant-design option to be identified.

For an existing plant, where pump NPSH(3pc)/Q performance is already fixed, the value of **R**(3pc) provided by the plant design can be calculated. If this is less than 1.3 a risk of vapour locking is indicated. It will be seen in later examples that there is often a way of reducing risk substantially by making relatively minor changes to operating procedures. Examples of this are restricting pump maximum flowrates or the permissible combinations of pumps that can operate at any one time. The merits of these options can be tested by calculation before a decision is made. The way forward in these circumstances depends upon the particular plant and pump design.

The strategy is summarised in Table 6.6. The method of implementing it is then described in some detail. Numerical examples are then used to illustrate the practical way it can be applied.

Table 6.6

Vapour lock (inlet pipe transient induced)	
1. Calculate ΔNPSH at the pump inlet	
2. Calculate NPSH(A)	
New plant	Existing plant
3a. Choose **R**(3pc) = 1.3	3b. Obtain pump NPSH(3pc)/Q data
4a. Calculate maximum acceptable NPSH(3pc) value	4b. Calculate actual **R**(3pc) value
5a. Repeat for credible operating conditions	5b. Confirm vapour lock risk
6a. Add a confirmatory NPSH(3pc) test in purchase specification if vapour locking appears a risk	6b. Review plant operating modes (and plant design if practicable) and introduce remedial measures

6.4.3 *The calculation of ΔNPSH by the step-by-step method*

Background. The pressure history of the inlet vessel vapour space following the injection of relatively cold liquid is determined by a simple enthalpy averaging technique. The method, described by Dartnell (Ref. 6.1), is easily applied and is sufficiently accurate to identify when a pump is at risk from a system induced vapour lock.

The vapour pressure in the inlet vessel falls progressively following the introduction of cooler liquid. The pump, some distance away, continues to receive the relatively hot liquid which left the inlet vessel earlier. Conditions leading to the highest risk of vapour locking at the pump, $\Delta NPSH_{max}$, usually* occur at about the time the hot liquid arrives — depressurisation in the inlet vessel is well advanced; the vapour pressure of the liquid at the pump inlet is still high. The following preliminary calculations precede a detailed description of the step-by-step procedure.

* This presumes a steady flow of colder liquid continues to be added to the inlet vessel inventory. This is not the case in some applications and the step-by-step method can take this into account.

Preliminary calculations. For the chosen plant operating mode the step-by-step method commences with two preliminary calculations.

(i) *Transit time t_p.* Until the "cold" liquid arrives at the pump the vapour pressure remains constant at a value set by the "steady-state" temperature of the liquid in the inlet vessel at the start of the transient. The time t_p at which the change occurs needs to be calculated. It represents a significant point in the transient.

The time t_p is the sum of the times to traverse the individual pipe length/diameter sections $l_1 d_1$, $l_2 d_2$,..., $l_n d_n$ that make up the inlet pipework system. In a simple single pump/pipeline system this is given by

$$t_p = \frac{\pi}{4Q} \left[l_1 d_1^2 + l_2 d_2^2 + \cdots + l_n d_n^2 \right]$$

In systems using multiple pumps and bifurcations to separate lines the appropriate flowrate/pipe diameter combinations must be used. Consistent units must, of course, be used in all calculations. Plant specifications are often based on mass flowrate, a measure of product output. A volumetric flowrate is required.

(ii) *Time increment t_x.* Doubling the value of t_p will give a time which will cover the major part of a transient including recovery beyond the maximum reduction in NPSH. A good curve can be made from 15 or more increments. Rounding up to, say, a 10-second increment or a submultiple of 10 seconds makes for ease of plotting later. Once calculations show a stable trend larger increments can be used.

The step-by-step procedure. The method continues by quantifying five variables and three values derived from them. The lines of items and property values are numbered. Values obtained from fluid property tables are lettered a, b, etc., and follow the line number which gives the value from which they are derived. This line numbering arrangement is carried forward in later examples (see Tables 6.7 and 6.8).

Line
No. Item   Derived property

Step 1 List initial values
(1) Vessel liquid mass $W_{(v)}$
(2) Vessel liquid temperature $T_{(v)}$

    (2a) Vapour pressure $p_{vap(v)}$
    (2b) Enthalpy $h_{(v)}$

(3) Flow into vessel — mass flowrate $Q_{(m)in}$
(4) Flow into vessel — temperature T_{in}

    (4a) Enthalpy h_{in}

(5) Flow out of vessel mass flowrate $Q_{(m)out}$

Step 2 Calculate inlet vessel average enthalpy after time increment t_x
The cycle of calculations for each step then proceeds as follows.
(6) Total heat at t = 0 (see Note 1) $q_{(v)} = [W_{(v)} \times h_{(v)}]$
(7) Mass change $\Delta W = t_x (Q_{(m)in} - Q_{(m)out})$
(8) Mass after time increment t_x $W_{(t)} = W_{(v)} + \Delta W_{(v)}$
(9) Heat content change $\Delta q = t_x [(Q_{(m)in} \times h_{in}) - (Q_{(m)out} \times h_{(v)})]$
(10) Heat content after time increment t_x $q_{(t)} = q_{(v)} + \Delta q$
(11) Inlet vessel average enthalpy $h_{(t)} = \dfrac{q_{(t)}}{W_{(t)}}$

Step 3 Determine vessel conditions after time increment t_x
Then by interpolation further property values are derived:

    (11a) Temperature $T_{(t)}$
    (11b) Vapour pressure $p_{vap(t)}$
    (11c) Specific volume $S_{(t)}$

Step 4 Calculate pump inlet conditions after time increment t_x
(12) Vapour pressure (at pump inlet) $p_{vap(p)}$
(13) Inlet vessel pressure change $\Delta p =$ Pump inlet pressure change
          $= p_{vap(p)} - p_{vap(t)}$
(14) Pump inlet NPSH reduction $\Delta NPSH = S_{(t)} \times \Delta p$

Step 5 Assign new values for next increment
  This is best explained by reference to Example 1.

Notes
  1. In vessels having substantial heat retaining materials of construction the total heat should include the heat equivalent of such material:
    i.e. additional heat equivalent = mass of material x specific heat of material.
  2. The vapour pressure at pump inlet $p_{vap(p)}$ has the line (2a) value until time t_p; subsequently it takes the value at time $t - t_p$. In the step-by-step method vapour pressure values are simply transferred forward by the appropriate time increment.

Being a step-by-step procedure based on increments of time the method enables changes in controlling parameters to be introduced as appropriate. These include variations in injection flowrate and the taking into account of any transient preheating received by the injected liquid. A further refinement which can be added is the effect of heat and liquid mass recirculated to the inlet vessel via a pump low-flow protection system.

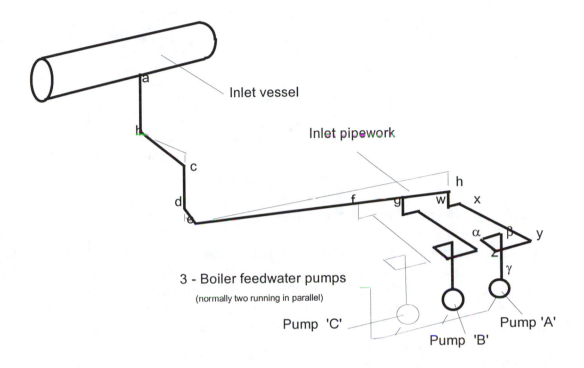

Fig. 6.4   Typical power station boiler-feedwater-pump inlet pipework system
Courtesy of Western Power Corporation

6.4.4   *Learning by example*

The following five examples demonstrate how to apply the step-by-step method. They are based on an actual power station feedwater pump system (Fig. 6.4) simplified somewhat to enable the methodology to be presented clearly. The output required for assessing the risk of vapour locking during actual normal operating regimes, shown in Fig. 6.5 and summarised in Table 6.9, is more complex and is not detailed. However, the same methodology has been used.

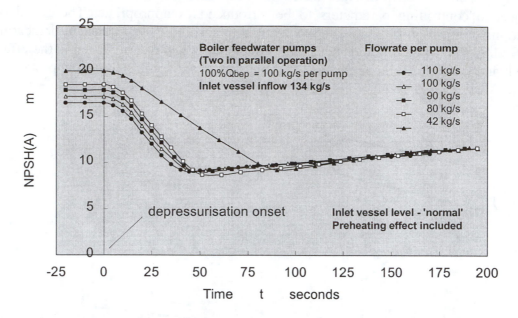

Fig. 6.5   The effect of changing pump flowrate

For lengthy industrial calculations where many modes of plant operation are to be analysed the use of a tabular method proves advantageous. An abridged form of results for two of the particular numerical examples are given in Tables 6.7 and 6.8.

Example 1 is a very basic overview to enable the procedure to be fully understood. Example 2 illustrates the effect of including changes to the heat content of the incoming liquid causing the transient. Example 3 shows the sensitivity of NPSH at the pump inlet to inlet vessel liquid level.

Plant particulars.   The step-by-step analysis is carried out for pumps in a 3 x 50% plant rating configuration. Each has a best efficiency mass flowrate of 100 kg/s.

The NPSH available at any point in the inlet pipework up to the pump is directly affected by a change in inlet vessel depth. Choosing the bottom of this vessel as the datum and adding liquid depth as a separate item enables the effect of vessel depth to be investigated or, in some circumstances, allows the minimum depth consistent with retaining vapour lock protection to be assessed. In this example "normal" inlet vessel depth is 2.77 m.

Choosing an operating mode for analysis.   As shown in Fig. 4.8, the most demanding NPSH(3pc) for single pump operation can occur at the highest flowrate. Conversely, a lower velocity of liquid in the inlet pipe keeps hot liquid longer at the pump inlet as the pressure in the inlet vessel falls and this produces a greater transient reduction in NPSH(A). It is not self-evident which pump operating mode is the most onerous case. A full and detailed evaluation must be made of all modes. However, a single pump running at a high flowrate is often a good starting point.

Preliminary step — (1) Calculate transit time t_p

For the plant shown in Fig. 6.4 and each pump running at $Q_{bep} = 100$kg/s;
with pumps A and B running in parallel $t_p \approx 80$s*,
with pump A running alone $t_p \approx 60$s*.

* Inversely pro rata for other flowrates.
The method of calculation is given in Table 6.10.

.

Preliminary step — (2) Choose a time increment t_x

Time increment $t_x = 5$s is used as a basis for all the cases analysed; steps are opened
up to 10s where progressive and steady changes are evident.

Example 1

Plant description
Boiler feedwater pump system Fig. 6.4
Pumps in operation — A and B
Pump flowrate — 42 kg/s each
Residual preheating effects — ignored

Step 1. List initial values

(1)	Vessel liquid mass	$W_{(v)}$	168,000 kg
(2)	Vessel liquid temperature	$T_{(v)}$	155.8°C
(2a)	— vapour pressure	$p_{vap(v)}$	5.548 bar
(2b)	— enthalpy	$h_{(v)}$	657.2 kJ/kg
(3)	Inflowrate (mass)	$Q_{(m)in}$	169 kg/s
(4)	Inflow temperature	T_{in}	30.8°C
(4a)	— enthalpy	h_{in}	129.8 kJ/kg
(5)	Outflowrate (mass)	$Q_{(m)out}$	84 kg/s

Step 2. Calculate inlet vessel average enthalpy after time increment t_x

(6) Heat in vessel
(at $t = 0$)

$$q_{(v)} = (1) \times (2b)$$
$$q_{(v)} = 168,600 \times 657.2$$
$$= 110.8 \times 10^6 \text{ kJ/kg}$$

(7) Mass change

$$\Delta W_{(v)} = t [(3) - (5)]$$
$$\Delta W_{(v)} = 5 (169 - 84)$$
$$= 425 \text{ kg (addition)}$$

(8) Mass (after time t)

$$W_{(t)} = (1) + (7)$$
$$W_{(t)} = 168,600 + 425$$
$$= 169,025 \text{ kg}$$

(9) Heat content change $\Delta q = t \{[(3) \times (4a)] - [(5) \times (2b)]\}$
$$\Delta q = 5[(169 \times 129.8) - (84 \times 657.2)]$$
$$= 0.166 \times 10^6 \text{ kJ} \quad \text{(reduction)}$$

(10) Heat content $q_{(t)} = (6) + (9)$
$$q_{(t)} = (110.8 - 0.166)10^6$$
$$= 110.6 \times 10^6 \text{ kJ}$$

(11) Inlet vessel
average enthalpy $h_{(t)} = (10) / (8)$
$$h_{(t)} = 110.6 \times 10^6 / 169025$$
$$= 654.3 \text{ kJ/kg}$$

Step 3. *Determine vessel conditions after time increment t_x (Fluid property table interpolations)*

(11a) Temperature $T_{(t)} = 155.1°C$
(11b) Vapour pressure $p_{vap(t)} = 5.450$ bar
(11c) Specific volume $S_{(t)} = 1.097$ dm³/kg (steam table values are relative to 4°C)

Step 4. *Calculate pump inlet conditions <u>after</u> time increment t_x*

(12) Vapour pressure $p_{vap(p)} = 5.548$ bar
     (at pump inlet)
(13) Inlet vessel pressure $\Delta p = (12) - (11b)$
     change $\Delta p = 5.548 - 5.450$
$$= 0.098 \text{ bar (reduction)}$$
(14) Pump inlet NPSH $\Delta NPSH = 10.2 \times (11c) \times (13)$ [1 bar = 10.2 m water @ 4°C]
$$\Delta NPSH = 10.2 \times 1.097 \times 0.098$$
$$= 1.10 \text{ m (reduction)}$$

Step 5 = Step 1 *(Next column). Assign new values for next increment*

	Variable		Numerical Example	
(1)	New $W_{(v)}$		169,025 kg	(was $W_{(t)}$)
(2)	New $T_{(v)}$		155.1 °C	(was $T_{(t)}$)
(2a)	New $p_{vap(v)}$		5.450 bar	(was $p_{vap(t)}$)
(2b)	New $h_{(v)}$		654.3 kJ/kg	(was $h_{(t)}$)
(3)	New $Q_{in} =$	Data for	169.0 kg/s	
(4)	New $T_{in} =$	(3), (4), (4a) and (5)	30.8 °C	
(4a)	New $h_{in} =$	unchanged throughout	129.8 kJ/kg	
(5)	New $Q_{out} =$	this example	84 kg/s	

As the step-by-step calculations progress it is convenient to collect the data produced as shown in Table 6.7.

Table 6.7   Example 1

Line no.	Item	Time into transient t seconds							
		0	5	10		**75**	80		180
(1)	$W_{(v)}$	168,600	169,025	169,450		174,975	175,400		183,900
(2)	$T_{(v)}$	155.8	155.2	154.5		146.4	145.9		135.9
(2a)	$p_{vap(v)}$	**5.548**	**5.450**	5.360		4.320	4.263		3.216
(2b)	$h_{(v)}$	657.2	654.3	651.5	Repeat	616.7	614.6	Repeat	571.7
(3)	Q_{in}	169	169	169		169	169		169
(4)	T_{in}	30.8	30.8	30.8		30.8	30.8		30.8
(4a)	h_{in}	129.8	129.8	129.8	at	129.8	129.8	at	129.8
(5)	Q_{out}	84	84	84		84	84		84
	t_x	+5	+5	+5		**+5**	+5		+10
(6)	$q_{(v) \times 10^6}$	110.8	110.6	110.4		107.9	107.7		105.1
(7)	$\Delta W_{(v)}$	425	425	425	five	425	425	five	425
(8)	$W_{(t)}$	169,025	169,450	169,875		175,400	175,825		184,750
(9)	$\Delta q_{\times 10^6}$	-0.166	-0.165	-0.164		-0.149	-0.148		-0.261
(10)	$q_{(t) \times 10^6}$	110.6	110.4	110.2	second	107.7	107.7	second	104.9
(11)	$h_{(t)}$	654.3	651.5	648.7		614.6	612.5		567.6
(11a)	$T_{(t)}$	155.1	154.5	153.8		145.9	145.4		135.9
(11b)	$p_{vap(t)}$	5.450	5.360	5.271		4.263	4.206		3.129
(11c)	$S_{(t)}$	1.097	1.096	1.095	interval	1.086	1.086	interval	1.075
(12)	$p_{vap(p)}$	**5.548**	**5.548**	**5.548**		**5.548**	**5.450**		3.922
(13)	Δp	0.098	0.188	0.277		1.285	1.244		0.793
(14)	$\Delta NPSH$	1.10	2.10	3.09		**14.2**	13.8		8.7

$\Delta NPSH_{max} = 14.2$ m after 80 seconds (i.e. given by the column $75 + t_x$ calculation).

Cooler liquid first arrives at the pump at this time and the vapour pressure at the pump inlet $p_{vap(p)}$ starts to fall. Consequently its value at 85 is that of $p_{vap(v)}$ at t = 5; subsequent values fall correspondingly as shown by the underlined figures in Table 6.7.

The simplicity of applying the step-by-step method can be appreciated by continuing to carry through the calculations shown above from the "10"- second column through the "75" up to the "180". The full set of results enable curves Δp/time and $\Delta NPSH$/time to be produced. These are presented in Fig. 6.6 as the curve "preheating effect ignored".

Example 2

Engineering systems are much more complex than the one portrayed in Example 1. For example, the boiler feedwater system used to calculate the inlet vessel and pump NPSH transients assumed a cold water inflow at temperature T_{in} = 30.8°C throughout the transient. In practice the inflow was delivered from the turbine generator condenser well (at 30.8°C) through two low-pressure heaters that were active until the transient started. The residual heat produces for a short time a "preheating effect" compared with the previously assumed immediate injection of cold liquid.

Using an estimate of the temperature variation to the inflow (hot liquid already in the pipework swept into the inlet vessel followed by progressively cooler liquid) a more realistic value for the inlet pressure change and pump inlet NPSH change is obtained. Comparison in Fig. 6.6 of the results obtained from Example 1 with those of Example 2 shows the effect of residual heating is to significantly delay the pressure decay. The change in inflow conditions that brings this about is highlighted (line 4a) in the abridged calculations presented in Table 6.8.

Notice how by using suitably small increments the tabular method allows the calculations to be "fine tuned" to accommodate such changes. Any available information on temperature variation rates for particular plant subcomponents can be incorporated readily. To appreciate the sensitivity by which this can be introduced in the examples all the associated graphical presentations (Figs. 6.5 to 6.8 inclusive) show the calculation points as markers on the curves.

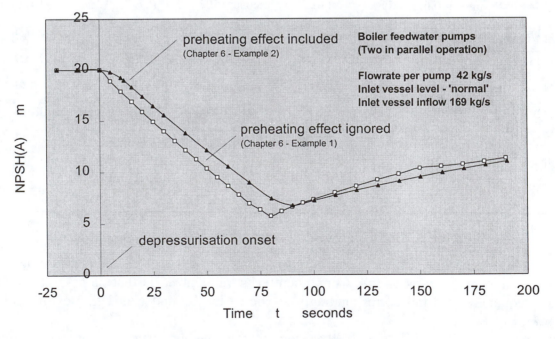

Fig. 6.6 Effect of changing heat content of inlet vessel inflow

Plant description
Boiler feedwater pump system Fig. 6.4
Pumps in operation — A and B
Pump flowrate — 42 kg/s each
Residual preheating effects — included

Table 6.8   Example 2

Line no.	Item	0	5	10	15		80		180
				Time into transient t seconds					
(1)	$W_{(v)}$	168,600	169,025	169,450	169,875		175,425		18391
(2)	$T_{(v)}$	155.8	155.7	155.3	154.7		147.2		136.9
(2a)	$p_{vap(v)}$	5.548	5.529	5.480	5.397		4.418		3.312
(2b)	$h_{(v)}$	657.2	656.7	655.2	652.7	Repeat	620.2	Repeat	576.0
(3)	Q_{in}	169	169	169	169		169		169
(4)	T_{in}	131.7	-	-	30.8		30.8		30.8
(4a)	h_{in}	**553.5***	**368.0***	**182.5***	**129.8**	at	129.8	at	129.8
(5)	Q_{out}	84	84	84	84		84		84
	t_x	+5	+5	+5	+5		**+10****		**+10****
(6)	$q_{(v) x 10}{}^6$	110.80	110.99				108.06		105.95
(7)	$\Delta W_{(v)}$	425	425	Calculate		five	850	ten	
(8)	$W_{(t)}$	169,025	169,450				176,275		183,951
(9)	$\Delta q_{x 10}{}^6$	+ 0.192	+ 0.035		as in		-0.302		- 0.264
(10)	$q_{(t) x 10}{}^6$	111.0	111.0			second	108.50	second	105.69
(11)	$h_{(t)}$	656.6	655.2				615.5		571.9
(11a)	$T_{(t)}$	155.7	155.3				146.1		136.9
(11b)	$p_{vap(t)}$	5.529	5.480	Example 1			4.289		3.221
(11c)	$S_{(t)}$	1.097	1.097			interval	1.087	interval	1.076
(12)	$p_{vap(p)}$	5.548	5.548				5.480		
(13)	Δp	0.019	0.068				1.191		0.82
(14)	$\Delta NPSH$	0.21	0.76				**13.2**		9.0

* Residual heat transferred to inlet vessel estimated from site measurements.

$\Delta NPSH_{max}$ = 13.2 m after 90 seconds.
** Larger increments when $\Delta NPSH_{max}$ passed to reduce the number of calculations.

6.4.5 *Dealing with uncertainties and options*

The effect of changing the value of an assigned variable, individually or collectively, can only be established by conducting a sensitivity analysis. That is, by making a change, repeating the calculation, and thereby quantifying the effect of the change. When the worst case deemed acceptable has been established for the chosen system design, the minimum protection level being provided for the pump to prevent vapour locking can be determined.

Example 3

As an example Fig. 6.7 shows the effect of operating with a low level in the inlet vessel. The mixing of "cold" inflow with a smaller inlet vessel inventory produces a more severe reduction in NPSH at the pump. This is exacerbated by the initially lower steady-state NPSH caused by low water level.

Plant description

Boiler feedwater pump system Fig. 6.4
Pumps in operation — A and B
Pump flowrate — 42 kg/s each

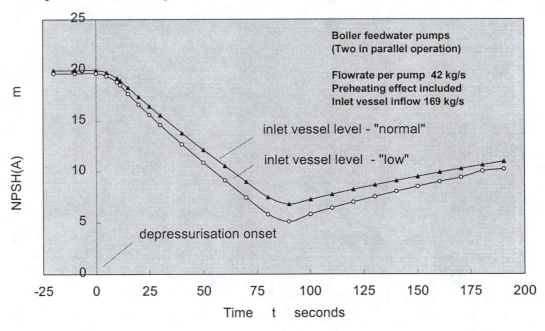

Fig. 6.7 Effect of low inlet vessel level

Example 4

In determining the minimum value of NPSH(A) it is necessary to evaluate ΔNPSH for all operating conditions. In a system with a single fixed-speed pump the characteristic shape of the NPSH(3pc)/Q curve indicates the value of NPSH(3pc) associated with the maximum flowrate that is likely to occur which will determine the lowest value of NPSH(A) to provide security against vapour locking for all operating conditions. Where pumps run in parallel and/or a variable speed prime mover is used, or where, as in Fig. 6.4, multiple inlet piping runs converge on a common manifold prior to more than one pump, then calculations to determine ΔNPSH need to be carried out over a range of flowrates, pump speeds and number of pumps in service as appropriate. The effect of choosing one rather than two pumps to meet a given flowrate is demonstrated in Fig. 6.8.

Plant description

Boiler feedwater pump system Fig. 6.4
Pumps in operation - (A and B) *or* A
Pump flowrate - 42 kg/s each *or* 84 kg/s for a single pump

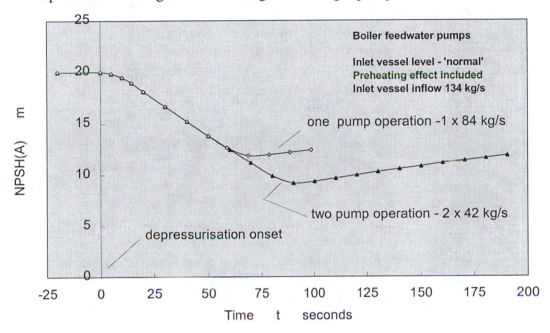

Fig. 6.8 Effect of pump duty/number of pumps in service

Example 5

Comparing Figs. 6.5 and 6.8 with Figs. 6.6 and 6.7 allows the effect of changing the inlet vessel inflow from 134 kg/s to 169 kg/s whilst two pumps are operating to be evaluated. This was an operational option in the particular plant from which the examples are drawn.

Plant description

Boiler feedwater pump system Fig. 6.4
Pumps in operation — A and B
Pump flowrate — 42 kg/s each

6.5   Vapour Locking Protection — A Pump User Approach

The differences in plant conditions for the various pump operating modes shown in Table 6.9 are typical of power generation, chemical, and other types of process plant. The wide ranging effect of secondary changes make early "across the board" judgements irresponsible. In the power plant example, feedwater temperature is much reduced when the generating plant is on standby; inflow temperatures also have new values as various subsystems (spillway bypass, leak-off, etc.) are included/excluded from operation. Operating different pumps (permutations of parallel operation with pumps A, B, and C) produce yet more possibilities. This is the "real world" where solutions to problems have to be found.

A comprehensive plant operation description is the cornerstone to identifying (1) for a new plant the essential NPSH provision to prevent vapour locking, (2) for an existing plant the adequacy of NPSH provided and, if necessary, the areas where restrictions on the operating envelope may reduce the risk of vapour locking to an acceptable level.

In assessing adequacy of protection against cavitation for the particular plant described in section 6.4 it is important to note from Fig. 6.5 that inlet pipe pressure transients usually reach a maximum value in a matter of tens of seconds and the total transient is, in any practical sense, ended in a couple of minutes. The risk of unacceptable cavitation erosion on such a timetable can be ignored.

A value of $R(3pc)$ is required which will cater for uncertainties in both (a) defining conditions from which is obtained the calculated value of $\Delta NPSH_{min}$, — this includes the inflow/outflow temperatures and flowrates plus the effects of inlet tank mixing, and (b) the NPSH(3pc) works test value. The former takes guidance from the sensitivity analysis just described, the latter from the test standard for the NPSH(3pc) test.

The cavitation protection multiplier value of $R(3pc) = 1.3$ provides adequate protection. Values of $R(3pc)$ of less than 1.3 should not be contemplated. Lower values expose a pumping plant to the possibility of severe surging and other accompanying unacceptable flow disturbances.

Table 6.9 is used to compare the minimum available NPSH during the various transients. The $\Delta NPSH$ values are taken from Fig. 6.5 and are used to determine the value of $R(3pc)$. Since all the values of R up to the 2 x 100 condition are greater than 1.3 the system is not at risk from inlet pipe pressure induced vapour locking under such conditions. Running above this condition should be avoided.

The calculated pressure decay is likely to be pessimistically high since mixing in the inlet vessel is never complete and boiling in pockets of high temperature liquid prevails thereby giving marginally higher vessel pressures than enthalpy averaging indicates. However, this advantage may well be lost when set against the other uncertainties encountered in quantifying the factors making up the heat balance assessment.

Further reading on inlet vessel pressure transients is given in Refs. 6.2, 6.3 and 6.4. This relates mainly to power station boiler feedwater systems and addresses in some detail through the views of specialist authors how the secondary effects particular to such systems might be dealt with.

Table 6.9 Power station operating conditions reviewed

			Number of pumps x flowrate (%Qbep)				
Item			2 x 42	2 x 80	2 x 90	2 x 100	2 x 110
(1)	$W_{(v)}$	kg	168,600	168,600	168,600	168,600	168,600
(2)	T_{out}	C	155.8	176.3	176.3	176.3	176.3
(3)	Q_{in}	kg/s	134	134	134	134	134
(4)*	T_{in}	C	131.7	149.3	149.3	149.3	149.3
(4)**	T_{in}	C	38.9	38.9	38.9	38.9	38.9
(5)	Q_{out}	kg/s	42	80	90	100	110
NPSH		m					
(steady state)			20.0	18.5	17.9	17.2	16.5
$\Delta NPSH_{max}$		m	10.8	9.8	8.8	8.0	7.4
$NPSH_{min}$		m	9.2	8.7	9.1	9.2	9.1

Use $NPSH_{min}$ to assess vapour lock protection needs

Assessment of vapour locking protection to meet nominated flowrate conditions

(a) New plant

(What is the maximum NPSH(3pc) a new pump could have and not be "at risk" ?)

		2 x 42	2 x 80	2 x 90	2 x 100	2 x 110
R(3pc)		1.3	1.3	1.3	1.3	1.3
Max. allowable						
NPSH(3pc) m•		7.1	6.7	7.0	7.1	7.0

(b) Existing plant

(How "at-risk" to vapour lock – R(3pc) = 1.3 – is the plant in Fig. 6.5 ?)

	2 x 42	2 x 80	2 x 90	2 x 100	2 x 110
NPSH(3pc) m••	2.0	3.5	4.7	6.4	7.4
R(3pc)	4.6	2.5	1.9	1.4	1.2
Verdict	[----------------Safe from vapour lock------------------]				[At risk]

• Pump offer value — to be proven by test.

•• Actual NPSH(3pc) figures taken from cold water works test data. No account of the benefit in performance associated with liquid property changes described in section 3.9 has been included.

Notes:

(1) Actual power plant values are used. In line (3) the value of Q_{in} is lower than that in Examples 1 and 2. This change significantly affects the outcome.

(2) All pumps operate at maximum (constant) speed
           with residual heating included
           with inlet vessel level normal (2.77 m above datum)

*   Inlet vessel inflow initial temperature
** Condensate well temperature

6.6 Ebullition in the Inlet Pipework (Vapour Bubble Formation)

Inlet vessel transients, particularly those caused by the sudden injection of relatively cold liquid, give rise to the possibility of ebullition in the inlet pipework to a pump. As the pressure in the inlet vessel rapidly collapses, the hot liquid on its way down to the pump may, in a poorly designed system, commence boiling if the local static pressure is insufficient to suppress it. Long horizontal pipe runs exacerbate the risk of local boiling. The consequence of such ebullition is a transient reduction or a total loss of inlet flow at the pump.

The hydrostatic head acting on a particle of liquid travelling from the inlet vessel tank to the inlet to the pump is easily calculated from knowledge of the pipework geometry and pipe friction loss data. The pressure decay in the inlet vessel is calculated by using the step-by-step method described earlier.

Learning by example.   The risk of vapour lock from ebullition in the inlet pipework is calculated using a knowledge of the plant layout and the rate of pressure decay in the inlet vessel. The method of doing this is best illustrated by continuing the earlier numerical example.

Plant description

     Boiler feedwater system Fig. 6.4
     Pumps in operation — A and B
     Pump flowrate — 100 kg/s each
     Residual preheating effects — included

The pipework layout from inlet vessel "a" to pump inlet "γ" is shown in Fig. 6.4. The longest pipe length serves Pump A. This gives rise to the greatest risk of ebullition since the long horizontal lengths give no static pressure increase during which time the inlet vessel pressure continues to decay. For industrial calculations, where many lengths of pipe with differing diameters and where more than one pump may be operating, the tabular method shown in Table 6.10 proves advantageous.

The static head rise experienced by a particle leaving the inlet vessel nozzle at the commencement of the transient is obtained for various values of time t. This appears as a "stepped" curve "A" in Fig. 6.9. The ΔNPSH decay data (from Tables 6.5 and 6.6) are plotted as curve "B" on Fig. 6.9. The decay data are displaced on the vertical (ordinate) scale by an amount corresponding to the head of water in the inlet vessel at the start of the transient.

The distance H_T separating Curve A and Curve B at any time t represents the head preventing ebullition. Ebullition is predicted to take place if the curves meet. That is when $H_T = 0$. In the example shown it clearly does not. By choosing the inlet vessel nozzle (at the bottom of the vessel) as the measurement datum the effect of lowering the inlet vessel depth is immediately apparent since such a change merely displaces curve B upward.

Calculations are conducted an order of magnitude more accurately than required for the final answer to avoid cumulative errors from rounding up.

Table 6.10   Ebullition calculations for 2 x 100 kg/s

Pipe section		Vertical drop m	Cumulative drop m (1)	Time taken s	Cumulative time s	Friction loss m	Cumulative friction loss (2)	Static head H_T m (1) - (2)
Main pipe	a-b	3.39	3.39	3.36	3.36	0.034	0.034	3.36
	b-c	0.60	3.99	6.36	9.72	0.045	0.079	3.91
	c-d	6.50	10.49	6.44	16.16	0.040	0.119	10.37
	d-e	0.75	11.24	1.05	17.21	0.026	0.145	11.10
	e-f	0.48	11.72	14.56	31.77	0.081	0.226	11.50
	f-g	0.17	11.89	5.10	36.87	0.046	0.272	11.62
Branch pipe	g-h	0.17	12.06	5.92	42.79	0.041	0.313	11.75
	h-w	1.28	13.34	1.47	44.26	0.005	0.317	13.02
	w-x	0	13.34	2.24	46.50	0.029	0.346	12.99
	x-y	0.34	13.68	15.74	62.24	0.072	0.418	13.26
	y-z	0.12	13.80	5.52	67.76	0.039	0.457	13.34
	z-α	0.11	13.91	4.94	72.70	0.0375	0.495	13.42
	α-β	0.07	13.98	3.22	75.92	0.032	0.527	13.45
	β-γ	3.8	17.81	4.40	t_p = 80.32	0.014	0.540	17.27

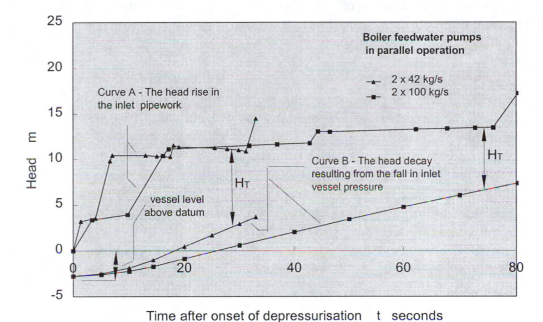

Fig. 6.9   Inlet pipe ebullition risk evaluated

Table 6.9 shows that at a total mass flowrate of 2 x 100 kg/s cool liquid reaches the pump in about 80 seconds. Further inspection shows that the head available to suppress ebullition H_T reaches a low value at the end of each of the long horizontal pipe runs. This is to be expected since a particle travelling along such a pipe experiences very little change in static head whilst the superimposed head decay from the inlet vessel goes remorselessly on.

Two plant operating conditions are plotted in Fig. 6.9: one for the flowrate of 2 x 100 kg/s and another for 2 x 42 kg/s. Both are covered by Table 6.9. It should be noted that to be certain that plant cannot be compromised by ebullition all modes of plant operation must be evaluated.

6.7   Inlet Vessel Nozzle Cavitation

When liquid from an inlet vessel passes into the pump inlet pipework the flow paths can be quite complex. The possibility of low-pressure zones which give rise to cavitation exists. This is accentuated where (a) no provision is made (by way of a pipework entry nozzle, say) to accelerate liquid progressively toward the inlet pipework, (b) the liquid level in the inlet vessel is very low thereby giving rise to vortexing, and (c) liquid is discharged into the inlet vessel in close proximity to the outflow point.

Inlet vessel nozzle cavitation is not a problem for most systems, particularly where rudimentary measures have been taken to mitigate the effect of poor entry flow distribution. Difficulties arise from flowpaths that result in the liquid entering the nozzle with an uneven circumferential distribution, particularly when such flows include a local high velocity path across the inlet vessel "floor". Where flow cessation is observed in an existing plant and the calculations described in sections 6.4 and 6.5 show the pump to be well clear of vapour locking, the suitability of the inlet pipe design and location is worthy of investigation. Further information on these and related issues can be obtained from Refs. 6.1, 6.2, 6.3 and 6.4.

References

6.1   Dartnell, L. M., The thermal-hydraulic design of main feedwater pump suction systems for large thermal power plant, Proc. IMechE, Vol. 199, No. A4, pp. 215 – 227, 1985.

6.2   Gage, A., Criterion for the suction conditions of feed pumps Fourth Int. Congress on Industrial Heating, Paper 36, 1952.

6.3   Liao, G. S. and Leung, P., Analysis of feedwater pump suction pressure decay under instant turbine load rejection, ASME J. Engineering for Power, Paper No. 71-WA/Pwr-2, pp. 83 – 90, 1972.

6.4   Liao, G. S., Analysis of power plant deaerator under transient turbine loads, ASME J. Engineering for Power, Paper No. 72-WA/Pwr-4, pp. 1 – 9, 1973.

Chapter 7

Cavitation Surging —
Hydrodynamically Induced

7.1 Deficiencies Identified

Cavitation surging in a centrifugal pump is a thermodynamic process with boundaries controlled by hydrodynamic flow patterns. Cavities grow only where low-pressure regions exist in which the pressure is below vapour pressure. Local pressure gradients and local "hot spots" determine where these regions are and how long cavities can survive before collapsing back into the main body of the liquid.

Hydrodynamically induced cavitation surging can occur at low values of $\%Q_{bep}$ in almost all centrifugal pumps. The onset and intensity of cavitation surging is primarily determined by the shape of the impeller blades at inlet. The extent to which cavities travel away from the impeller is determined by inlet pipe geometry. In most impeller/pipework combinations the low pressure regions are very localised and the backflow recirculation giving rise to cavitation surging hardly leaves the inlet blade region. In some pumps however, particularly those for which the inlet pipework has axial symmetry with the impeller eye, the associated backflow extends well into the inlet pipe. In pumps where a backflow recirculation device is fitted, the intensity and extent of surging is much reduced over a wide flowrate range down to very low $\%Q_{bep}$ values.

Impeller design induced cavitation surging arises from the propensity of centrifugal pump impellers to generate a backflow. This occurs at flowrates much less than the best efficiency point value. Where the inlet geometry is conducive to its propagation and cavitation is occurring a pulsating volume of vapour extends well into the inlet pipe preceding the impeller. Visual observation reveals cavities which appear to dart very quickly out from the impeller. A rotational component, following the direction of rotation of the impeller, is also apparent before the large volumes of cavitation collapse back into the impeller. The surging can be very violent and result in mechanical damage. Cavitation surging induced by impeller design therefore poses an operationally unacceptable risk. The performance deficiencies of commercial significance which appear as a consequence are listed in Table 7.1.

In this chapter the root causes of impeller induced cavitation surging are identified and a strategy which will secure acceptable pump operation is described.

Table 7.1   Deficiency list

Cavitation surging — hydrodynamically induced	
Performance deficiency	Plant operating conditions
Cyclic pressure pulsations ⎫	⎧ Low NPSH
Mechanical component failure following ⎬	⎨ Low flowrate (below ~ $50\%Q_{bep}$)
consequential repetitive severe vibrations ⎭	⎩ Conducive inlet pipe geometry

7.2 The Surge Mechanism

7.2.1 *The underlying cause*

When a non-cavitating centrifugal pump operates at significantly below its best efficiency flowrate, say below 50% Q_{bep}, the flow entering the pump inlet encounters impeller blades at a very unfavourable angle. Strong vortices are set up by the high rotational speed of the blades encountering liquid which is constrained by pump outlet resistance to move slowly in the radial direction. In all but the narrowest of impellers (blades less than ~ 10 mm wide) a mismatch arises between the angle of liquid flow entering the impeller eye and the blade inlet edge. This varies significantly across the width of the blade. The difference in pressure across the blade produces a strong recirculation. Impeller geometry and low %Q_{bep} combine to eject liquid from the impeller as a backflow into the inlet piping at the outer diameter of the impeller eye. The momentum of this backflow mass imparts prerotation to the liquid approaching entry to the impeller. This prerotation is steady and continuous. With no cavitation present any perturbations that occur have no effect on pump performance and are usually small enough to go unnoticed in the general background turbulence of low %Q_{bep} conditions. In fact with careful measurement the onset of a non-cavitating backflow can be detected by a kink in the ΔH /Q curve.

In a cavitating centrifugal pump operating at low %Q_{bep} the conditions on the impeller blade are radically different. Cavities grow in a region near the inlet blade tip before being "folded" back into the backflow as shown in Fig. 7.1. The low-pressure gradients within this backflow recirculation enable them to grow rapidly for considerable distances as shown in Fig. 7.2. Their occurrence is probably propagated in

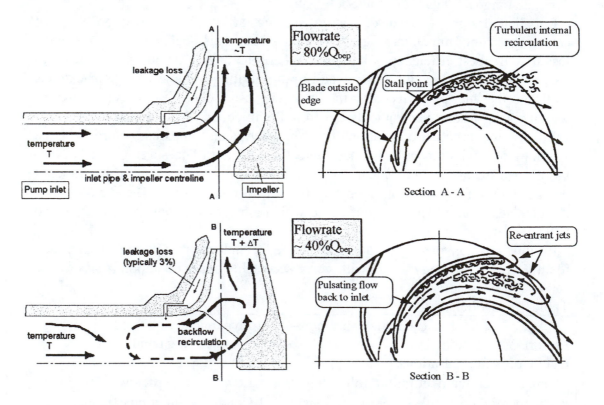

Fig. 7.1 Impeller backflow recirculation

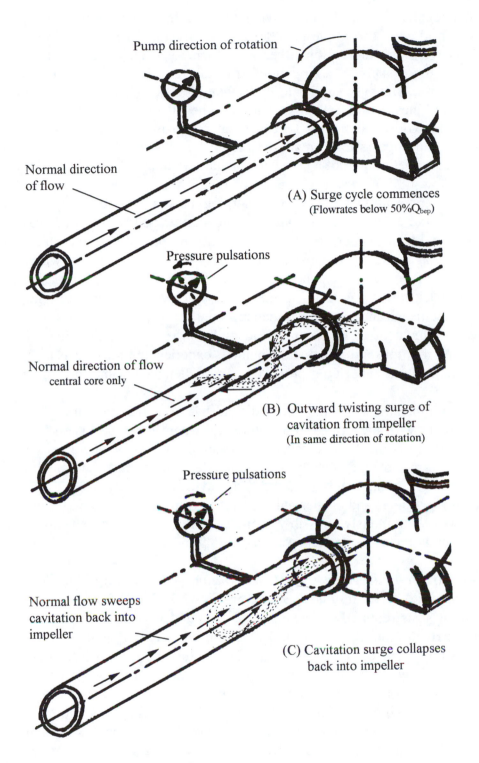

Pump direction of rotation

Normal direction
of flow

(A) Surge cycle commences
(Flowrates below 50%Q_{bep})

Pressure pulsations

Normal direction of flow
central core only

(B) Outward twisting surge of
cavitation from impeller
(In same direction of rotation)

Pressure pulsations

Normal flow sweeps
cavitation back into
impeller

(C) Cavitation surge collapses
back into impeller

Fig. 7.2
Hydrodynamically induced cavitation surging.
One cycle = (A) ∏ (B) ∏ (C) ∏ (A) typically
takes less than 0.5 second

the turbulent region between liquid moving outward from the pump and liquid moving as a solid core toward the impeller eye. A "fracture" surface is produced along which cavitation can develop either as filamentary "ropes" which may be a single sheet or, more likely, as a large number of discrete sites where cavities can grow and collapse. It is very similar to the cavitating ropes observed in the draft tube of water turbines. However, unlike water turbines, the growth and collapse of the recycling backflow cavities in a centrifugal pump is much more dynamic. Cavitation surging appears to be much less rapid than the blade cavitation seen (Fig. 1.1) on an impeller at $100\%Q_{bep}$ where cavitation collapse takes place over a distance of a few millimetres. The reason for this is the relatively low pressure gradient that is present where cavitation surging occurs. On returning to the impeller blades the central core of vapour/liquid mixture disrupts further cavity production. The added vapour volume in the backflow partially blocks the inlet pipe. This blockage leads to a transient increase in local axial velocity which in turn brings more favourable incidence conditions. These conditions bring about a reduction in backflow. The flow at the blade inlet edge reverts to one of unfavourable incidence and the cycle is then repeated.

The pressure fluctuations in the inlet and outlet pipe and the simultaneously observed plant vibrations indicate that the pump is experiencing the effects of large changes in momentum. The force produced can challenge the integrity of axial thrust bearings.

In cavitating impellers where the pumped fluid has a vapour pressure which is increasing rapidly with increasing temperature and where local "hot spots" occur the cavitation surge is much more severe. An example of this is where neck ring recirculation enters the inlet pipe from the impeller outlet region within boiler feedwater pumps handling water at 150°C. The cavities evidently take advantage of low pressure gradients and the destabilising effect of pressure pulsation "spikes" to grow more voluminously, possibly from a larger number of sites, before collapsing at about the same frequency as "colder" liquid.

7.2.2 *Research and development test observations — Centrifugal impellers*

Inlet pipe surging.   The results presented in Figs.7.3, 7.4, 7.5 and 7.13 were obtained during tests reported by Sloteman, Cooper and Dussourd (Ref. 7.1) and by Kasztejna, Heald and Cooper (Ref. 7.2) on an end-suction centrifugal pump with a long straight (transparent) inlet pipe. The pictures are taken from a video recording. They show vividly the extent to which cavitation surge can occur.

At $40\%Q_{bep}$ the cavities are carried some 10 pipe diameters back down the inlet pipe during a surge with a frequency of about 2 Hz. A complete surge cycle is shown for this flowrate. The highly dynamic nature of the movement is difficult to convey from the static frames of a video. The following description is given to help bring the pictures to life.

Throughout the picture sequence a substantial flow is moving along the inlet pipe, left to right, into the eye of the impeller within the pump casing. In real time the surging motion appears to be a whiplash-shaped cloud of cavities shooting out of the impeller against the normal flow direction before once more collapsing backward into the pump. To an observer standing close to the inlet pipe the oscillating movement is akin to watching a tennis net-cord rally from the net-cord judge seat. It appears rapid and staccato. It also has a twisting motion superimposed.

With the benefit of "stills" from 25-frame-per-second video photography the motion can be examined more closely. As is usual with centrifugal pump low $\%Q_{bep}$ conditions the main flow pattern is influenced by changing secondary flows. Large organised structural movements are continually breaking down and reforming. Figure 7.3 is a sequence of twelve consecutive photographs taken during a period when a particularly clear view of the twisting (helical) motion was evident. The "outward" journey from the pump is read by viewing clockwise frames 1 to 6. The "return" journey continues with frames 7 to 12. Frame 1 shows a clear pump inlet. By frame 5, only 0.16 seconds later, cavities are streaming out of the impeller in a curling flow close to the pipe wall. The stream continues to progress out from the pump until by frame 7 after just 0.24 seconds it stops and turns to collapse back. By frame 12 the pump inlet is clear again. The "out and back" movement covering all of ten pipe diameters (some 2.5 metres each way) is completed in about half a second. Remember the tennis rally comparison to get an appreciation of the timing! The rapidity of this movement, with a seemingly short period for vapour production on the impeller blades, is the root cause of wildly fluctuating mechanical forces on pump components.

To be able to discern clearly the twist on a 25-frames-per-second video means that the surging volume is not at this time discharging from all the impeller blades simultaneously, although it may do so on other occasions. Also, the velocity of the surge "front" is evidently moving much faster than the local liquid velocity indicating rapid growth of the cavitating region rather than cavities being carried along on the backflow. It seems probable that cavities in the recirculation behind the impeller blade find an opportunity to propagate by expanding along a "fracture" path at the interface between the relatively low pressure backflow moving along the inlet pipe wall where liquid is attracting some prerotation from the flows and the central core of liquid entering the impeller.

The photographs in Fig. 7.3 are black/white reversed to show the stream of cavities as a black twister. Some of the darkness on the underside of the inlet pipe is reflected local light.

Frame 1   *t = 0.00 seconds*

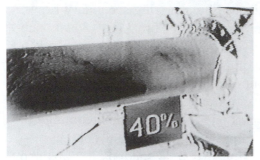

Pump inlet clear

Frame 12   *t = 0.44 seconds*

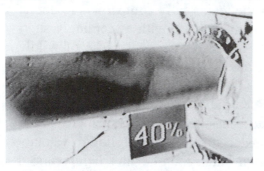

Cavitation surge cycle ends

Frame 11   *t = 0.40 seconds*

Pump inlet almost clear again

Frame 10   *t = 0.36 seconds*

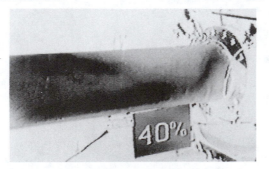

Movement → continues

Frame 2   *t = 0.04 seconds*

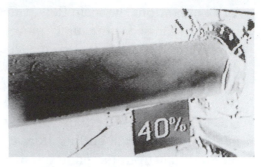

Surge emerges ← from impeller
(bottom right of pipe)

Fig. 7.3

Cavitation surging
in an end-suction
centrifugal pump

Frame 9   *t = 0.32 seconds*

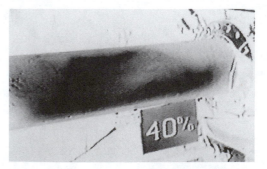

Cavities move → sweep up nearside wall

Frame 3   *t = 0.08 seconds*

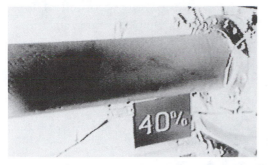

Cavities ejected ← along pipe bottom

Frame 4   *t = 0.12 seconds*

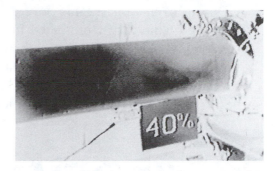

Cavity cloud ← gathering strength

Frame 5   *t = 0.16 seconds*

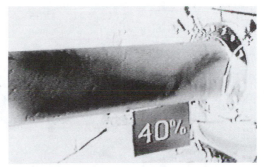

Outward motion ← rotates

*Pictures courtesy of
Ingersoll — Dresser Pump Company*

Frame 6   *t = 0.20 seconds*

Strong outward ← 'whiplash' motion

Frame 8   *t = 0.28 seconds*

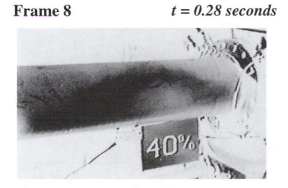

Cavity stream starts to move → to inlet

Frame 7   *t = 0.24 seconds*

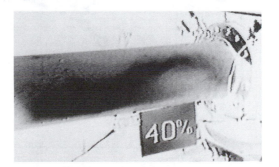

'Coil' stops growing axially

During cavitation surging severe vibrations on the test rig were recorded. The measured axial thrust presented in Fig. 7.4 shows changes that are cyclical and correspond for the most part to the frequency of the observed pressure pulsations. From the large "spikes" in axial thrust it is evident that the rate of change and severity of axial loading could easily challenge the capability of proprietary thrust bearings sized on near-steady-state load assessments.

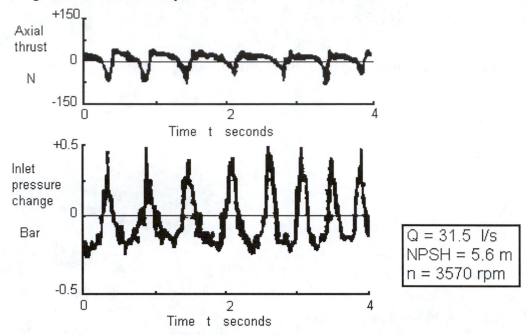

Fig. 7.4   Axial thrust and inlet pressure change measurements produced under low NPSH conditions at 20%Q_{bep}

Courtesy of Ingersoll — Dresser Pump Company

The magnitude of pressure pulsations and generated head recorded at several different values of NPSH during cavitation surging are shown in Figs. 7.5(a) and (b). For these cold water tests it is worth recalling that a 1 bar pressure swing is approximately equal to a 10 m change in NPSH. It can be seen that for this particular

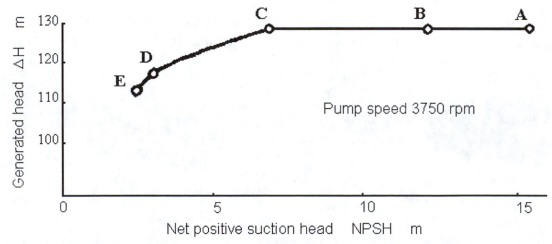

Fig. 7.5(a) Generated head produced during cavitation surging at 20%Q_{bep}

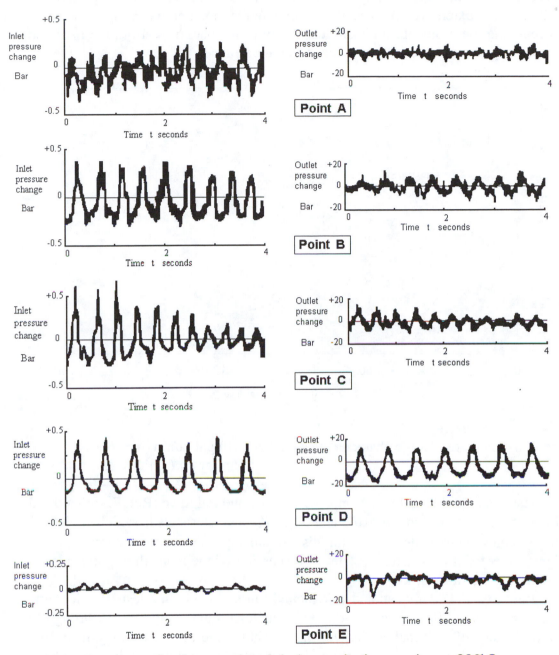

Fig. 7.5(b) Pressure pulsations produced during cavitation surging at 20%Q$_{bep}$
Courtesy of Ingersoll — Dresser Pump Company

pump inlet pipe surging occurs at an NPSH(A) value well above that for NPSH(0pc), that is, in a region where no head breakdown is present. Kasztejna found that, as shown by the Point A trace, in excess of three times the NPSH(3pc) value was required to begin to suppress the major pressure pulsation. At and below NPSH(0pc) the pressure in the outlet pipe pulses in unison with the inlet pipe indicating that the surge extends through the impeller to the outlet side of the pump. At very low NPSH values extensive volumes of cavitation in the impeller eye dampen the surge by reducing the momentum transfer capability of the backflow. Both the inlet and the outlet pressure traces take on a smoother appearance.

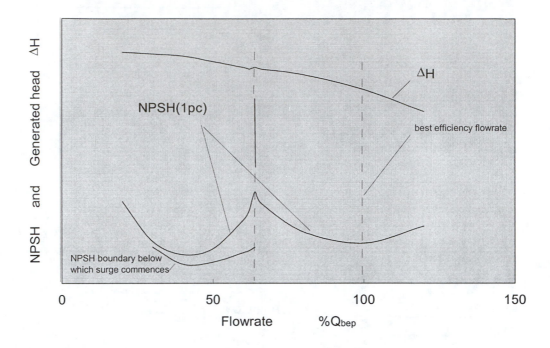

Fig. 7.6   Cavitation surge characteristics for a centrifugal pump ($n_s = 660$)
see Massey and Palgrave, Ref. 7.3

In tests on a centrifugal pump impeller Massey and Palgrave (Ref. 7.3) identified (Fig. 7.6) the regions on the ΔH/Q and NPSH/Q characteristics in which cavitation surge occurred. Also noted was the discontinuity present in the two curves; the significance of this is taken up later in discussion of the axial flow pump results reported at the same time.

Fraser (Ref. 7.4) demonstrated convincingly in non-cavitating tests that in some circumstances the flow recirculation in a centrifugal pump could travel through to the outlet side of an impeller. Fig. 7.7 shows that the pressure pulsations, measured at the inlet and outlet of a centrifugal pump impeller, exhibited almost uniform values over the 60%Q_{bep} to 120%Q_{bep} range. The rise in pressure pulsations at the inlet demonstrated the potential of pumps to commence cavitation surging below 60%Q_{bep}. Of interest is the recorded "sudden and dramatic" increase in the magnitude of pressure pulsations on the outlet side which occurred at 45%Q_{bep}. Clearly in this particular machine if cavitation surging was initiated in the inlet pipe there was a strong possibility of it being transmitted through to the outlet side if the pressure gradient was insufficient to suppress the cavities as they were carried along in the recirculating flows.

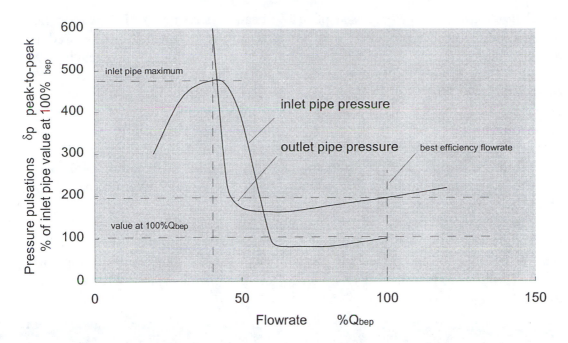

Fig. 7.7   Centrifugal pump impeller inlet and outlet pressure pulsations
see Fraser, Ref. 7.4

7.2.3 Research and development test observations — Inducers and axial flow impellers

Both inducers and axial flow impellers are, by any strict definition, outside the scope of this book. However, their propensity to develop cavitation surge at the "low pressure gradient" extreme of pump design provides useful information. These "open" designs also make for easier visual observation of the surge mechanism and, within the wider picture, aid an understanding of centrifugal impeller surge.

Cavitation surge is more intense and more voluminous in high specific speed pumps. This is in part explained by the low pressure gradients which are a feature of inducers and axial flow pump impellers.

Inducers. An inducer can be regarded as a low generated head axial flow impeller. Its function is to immediately precede a centrifugal impeller and to provide it with sufficient inlet NPSH. It discharges directly into the eye of the centrifugal impeller to which it is physically coupled. With a low head rise this results in blades being very lightly loaded thereby tolerating large cavity volumes without impairing the head generating capability. The open nature of the axial flow passages that form the inducer makes it able to handle larger volumes of cavitation than the centrifugal impeller could on its own. By generating sufficient head to enforce collapse of any cavities, the inducer allows the centrifugal pump impeller to generate a substantial head free from the risks posed by the presence of cavitation. The inducer/centrifugal impeller combination therefore provides a hydraulically sound means of extending the range of centrifugal impellers to very low NPSH applications.

Axial flow pumps. These consist of a low head generating axial flow impeller (often described as a propeller) and an axial diffuser. The diffuser is used to recover energy from what would otherwise be a spinning liquid mass leaving the impeller. Occasionally multistage axial flow pumps are produced.

Inlet/outlet pipe surging. In 1968 Gasiunas (Ref. 7.6) showed a severe pulsating cavitating flow that centred on an axial flow impeller. In a description that has come to characterise high specific speed machine experience he wrote:

"The violent pressure fluctuations commence approximately 12 inches (300 mm) into the delivery pipe and penetrate a similar distance into the suction pipe. This process is accompanied by a very intense noise. The piping system moves by at least 3/8 inch (9 mm) with every pressure shock."

Figure 7.8 shows the extent of cavitation. The description accompanying the particular frame describes how the surges "shrink" from 12 inches (300 mm) either side of the impeller and diffuser to the confines of the impeller only during each cycle. The low frequency (2.5 Hz) intense "chuffing" sound is now a recognised indicator of cavitation surging.

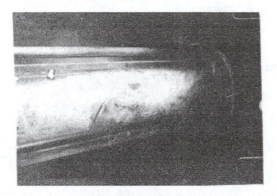

(a) Impeller swamped with   (b) View with less cavitation to show
    cavitation surging (see text)   impeller/diffuser arrangement

Fig. 7.8   Cavitation surge in an axial-flow pump   *see Gasiunas, Ref. 7.6*

In tests on a high specific speed axial flow pump Massey and Palgrave (Ref. 7.3) confirmed the presence of a discontinuity in the NPSH/Q curve for cavitation inception described earlier by Minami, Kawaguchi and Homma (Ref. 7.7). This is clearly demonstrated in Fig. 7.9 as a measurable feature for NPSH(1pc) values. Additional vapour makes the discontinuity less evident in the NPSH(3pc) curve. As the breakdown value NPSH(B) is reached the surging becomes disorganised and effectively ceases. As could be reasonably deduced, the NPSH(breakdown) curve shows no discontinuity throughout the flowrate range.

From the same tests Massey and Palgrave show (Fig. 7.10) that as flowrate is reduced the surging motion changes from a short stroke of 0.25 pipe diameter at a sharp sounding 3.5 Hz to much greater than 3 diameters at frequencies less than 1.3 Hz. It is clear that at the lowest flowrates (much lower than $100\%Q_{bep}$) the reduced momentum of the incoming liquid coupled to the low frictional resistance gradient enables the surging cavitation to travel spectacular distances if the incoming flowpath (in this case a long straight pipe) is symmetrical.

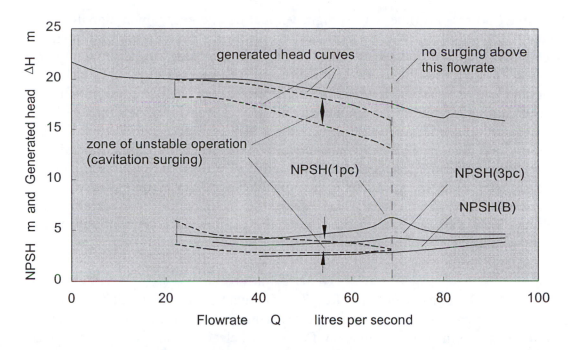

Fig. 7.9   Cavitation surge characteristics for an axial-flow pump ($n_s = 4700$)
see Massey and Palgrave, Ref. 7.3

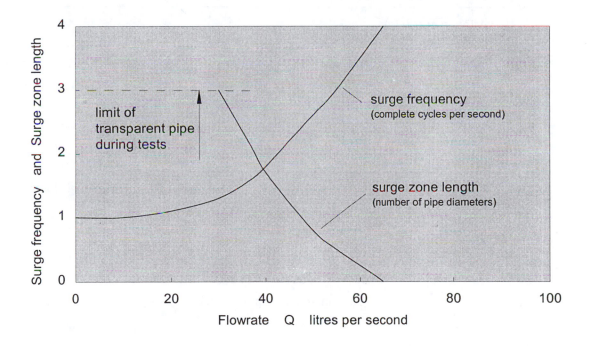

Fig. 7.10   Surge frequency and surge zone length (Corresponding to Fig.7.9)
see Massey and Palgrave, Ref. 7.3

As inducers became used on large power utility pumps from about 1970 onward the consequences of cavitation surging became apparent. A typical example of this was a 1200 kW (1600 hp) x 1470 rpm deaerator lift pump which exhibited a 3 Hz intense "chuffing" noise from the long straight vertical downcomer which signalled the presence of hydraulically induced cyclic loading. At the time the cavitation surge mechanism was only just becoming understood and users were having great difficulty in discriminating between inducers that were giving good service and those that were not. In 1975 the major U.K. power utility decided to exclude large inducers as an option in specifications for new power plant designs until an acceptable rationale was available.

Pump manufacturers subsequently developed improved designs of inducer utilising special visual test facilities. The inclusion of an inducer as an option in large pumping plant specifications is now regarded as an appropriate choice where low NPSH(A) conditions prevail *and* the pump manufacturer can demonstrate from test data that the boundaries of cavitation surging are clear of the pump operating range.

7.2.4 *Generic analysis*

Massey and Palgrave deduced in Ref. 7.3 that, to a first approximation, the lower the suction specific speed the lower the cavitation surge onset flowrate. Fraser in Ref. 7.4 came to the same conclusion. Fraser gave clear guidance on how to select pumps to avoid cavitation surging together with a description of the phenomenon which reinforced the findings of earlier workers summarised above. He describes how he examined "hundreds of pumps" before developing the recommended limits of operation particular to centrifugal pumps shown in Fig. 7.11.

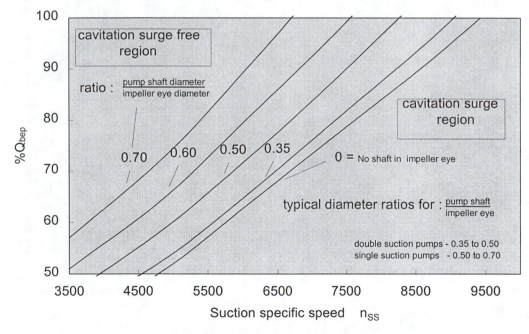

Fig. 7.11   Cavitation surge limits for centrifugal pump impellers

see Fraser, Ref. 7.4

As discussed in Chapter 4, dimensional similarities in impeller eye geometry give validity to a generalising assumption that a single suction specific speed value can be used to characterise inlet performance of all centrifugal pumps. It is not surprising therefore that Fraser found rules for the initial selection of pumps could be drawn up on this basis, albeit with a concession to the effect of introducing a shaft through the impeller eye — a feature of many designs. The work of Fraser is a significant contribution to linking an understanding of the cavitation surge mechanism to a practical, if rudimentary, means of avoiding it when considering plant design options. However, this early work, conducted mainly on end-suction pumps, does oversimplify matters and Palgrave (Ref. 7.5) points to the need to discriminate between performance measured at zero incidence angle and that measured at best efficiency point. A true measure of cavitation surge is one related to incidence angle.

Users have no way of knowing if zero incidence and best efficiency are the same for a particular pump. Viewed from "the outside" without the benefit of data from research test observations the variations in internal design appear as scatter in the graph of $\%Q_{bep}/n_{ss}$ when all pumps are included. Fortunately this does not matter. The regions in which cavitation performance occurs is well away from where pumps should operate, and an avoidance strategy for new pumps need not be concerned with this area of centrifugal pump internal design. It does however help in the understanding of existing plant problems and adds impeller redesign to the list of remedial measures to be considered when cavitation surging occurs.

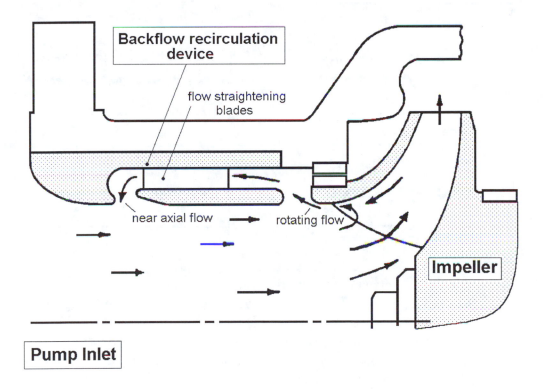

Fig. 7.12 A backflow device — Principal features   (see patents, Refs. 7.9 and 7.10)

Courtesy of Ingersoll— Dresser Pump Company

7.2.5 *Backflow recirculators*

Remarkable progress was made by Cooper, Dussourd and Sloteman (Ref. 7.8) in developing a backflow recirculator suitable for commercial applications. After observing the motion of the cavitation surge in the tests such as those shown in Fig. 7.3, the patented anti-swirl devices (Refs. 7.9 and 7.10) were developed and are now used extensively. This type of device is, of course, most effective where inlet pipe axial symmetry precedes the pump. The remarkable improvement produced by the used extensively. This type of device is, of course, most effective where inlet pipe axial symmetry precedes the pump. The remarkable improvement produced by the device (Fig. 7.12) in reducing axial thrust is shown in Fig. 7.13. In all cases the fluctuations are much less even though at each flowrate the NPSH value was, if anything, lower and so to the disadvantage of the backflow recirculation device. The associated inlet pressure pulsations for all these tests were the same or less than those at best efficiency flowrate where, of course, no cavitation surging was occurring.

Well-designed backflow recirculators or devices that prevent the development of inlet pipe recirculation are a simple and effective method of extending the safe operating flowrate range of a centrifugal pump.

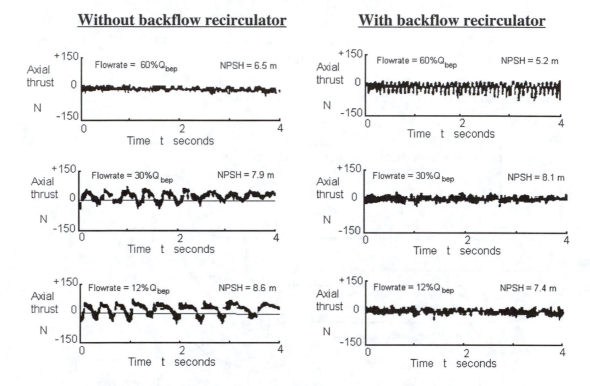

Fig. 7.13 Cavitation surge intensity (axial thrust value) with and without
              backflow device   *Courtesy of Ingersoll — Dresser Pump Company*

7.2.6 *The influence on cavitation surging of pump design configuration*

The choice of a pump that inherently avoids axial symmetry of inlet flows to an impeller will reduce cavitation surge potential. Examples of this are vertical-in-line pumps, split-casing double-entry pumps, and barrel-case boiler feedwater pumps. Preceding an end-suction pump with a bend has much the same effect although such a practice may well distort inlet flows to such an extent that pump performance or mechanical integrity is impaired.

Avoiding inlet symmetry will not eliminate cavitation surging — only reduce *some* of its effects. Clearly where the mismatching of inlet flow angles to impeller blade angles is the same the onset of cavitation surging will be repeated. Only the extent to which it is propagated by way of backflow down the inlet pipe is reduced. All centrifugal pumps drawing from unrestricted inlet pipework probably exhibit some degree of cavitation surge where low $\%Q_{bep}$ occurs and where NPSH is reduced toward the NPSH(3pc) value. In pumps with an unsymmetrical inlet this may be no more than minor perturbations. Chapter 8 explores the possibility that these minor perturbations are a latent source for a more damaging form of cavitation surging.

7.2.7 *Summary of findings*

The test data described have been obtained using a variety of pump types and sizes. Few inconsistencies are present in the observations. Reviewing all the information, it can be seen that the common features are as follows.

1. Cavitation surging can be very violent and can lead to mechanical damage of the pump and pipework system.
2. Cavitation surging is a low $\%Q_{bep}$ phenomenon.
3. Cavitation surging exists in its most destructive form below NPSH(1pc) for flowrates down to $20\%Q_{bep}$.
4. Cavitation surging can be propagated significant distances (typically up to 10 pipe diameters) where inlet pipes with axial symmetry directly precede the eye of a centrifugal pump impeller.
5. The cavitation surge zone length tends to increase as flow is reduced.
6. The cavitation surge frequency (typically 1 Hz to 10 Hz) tends to reduce as flowrate is reduced.
7. The value of $\%Q_{bep}$ marking the onset of backflow can, if care is exercised, be detected by a change in the NPSH characteristic based on very small head-drop values (~NPSH(1pc)). This is the maximum flowrate at which cavitation surging can occur.
8. Cavitation surging at NPSH(3pc) and below cannot be detected reliably on the NPSH/Q characteristic.
9. Cavitation surging for all centrifugal pumps can be broadly characterised by a single suction specific speed value.
10. Below a suction specific speed value of 3500 all centrifugal pumps operate free from cavitation surge down to $50\%Q_{bep}$.
11. At very high values of suction specific speed (greater than 5000) surging can present itself at very high $\%Q_{bep}$ values.
12. The intensity of cavitation surging can be reduced substantially by a backflow recirculator.

It is evident from the above that there will be very little risk of cavitation surging for centrifugal pumps where NPSH(A) is greater than the minimum 1.3 x NPSH(3pc) provided to avoid vapour lock risk (see Chapter 6), and the operating range is within the $50\%Q_{bep}$ to $120\%Q_{bep}$ band.

7.3 Cavitation Surging (Hydrodynamically Induced) — An Avoidance Strategy

An avoidance strategy for both new and existing plants can be drawn up from the knowledge gained from section 7.2. The objective for new pumps is to establish the parameters that determine the boundaries for safe operation. The objective for existing pumps is to identify remedial measures to avoid cavitation surging where it is positively demonstrated to be the cause of unacceptable plant operation.

A convincing method of predicting by calculation the intensity of surging for all pumps is not available. No general method exists that ties pump-impeller/pump-inlet design to the magnitude and frequency of pressure pulsations in the inlet pipe. Most pump manufacturers who have addressed the problem have adopted a semi-empirical approach which has concentrated, quite understandably, on developing a methodology for reducing the effects of surging to a tolerable level at very low values of $\%Q_{bep}$. A notable success in this respect has been the development of the backflow recirculator.

User specifications for new pumps cannot discriminate between design methods — except where the overriding case of identical operating service can be claimed. Such replication of pump and pump inlet pipework geometry is likely to exist only in rare instances.

The difference in surge intensity caused by variations in impeller design and, more importantly, the strong influence of local pipework configuration make it difficult to characterise performance. The large number of permutations means that it is not possible to eliminate the risk of cavitation surging by being prescriptive in pump specifications. Fortunately this is in any case unnecessary. Most pumps will be free from unacceptable cavitation surging if NPSH(A) is provided to prevent vapour locking using the rule developed in Chapter 6 and the operating range is limited to the $50\%Q_{bep}$ to $120\%Q_{bep}$ band.

To deal with pumps that fall outside the above NPSH(A)/$\%Q_{bep}$ description it is necessary to solicit the information that will enable it to be determined under what circumstances a pump is at risk and what restriction on operation is likely to make hydraulic performance acceptable. This establishes a basis for completing an assessment on the suitability of a pump to operate free from cavitation surge.

Where cavitation surge is a problem in existing pumps the understanding gained earlier in this chapter of the underlying mechanism giving rise to cavitation surging enables the remedial measures that are likely to be efficacious to be deduced.

A step-by-step avoidance strategy is listed in Table 7.2. A check list to aid in dealing with problems in existing plant is given in Table 7.3.

Notes   1. Given that the potential for damage is so great when cavitation surging takes hold, contrary to the advice given by Fraser (Ref. 7.4), it is prudent not to make a concession to intermittent operation.

2. Assessment of the conditions necessary to avoid cavitation surging should be made against the minimum values of NPSH(A) for several flowrates throughout the operating range. For pumps where an NPSH transient occurs the associated minimum NPSH(A) value should be used since even a very short duration transient can trigger the onset of an unacceptable cavitation surge.

Table 7.2 Data collection and assessment — New plant

Cavitation surging — Hydrodynamically induced — New plant

Information required to commence an assessment

Plant design/plant operation data (Supplied by plant designer or user)

(i) Q, ••, and choice limitations on pump configuration (e.g. end-suction, etc.)

(ii) Pump operating flowrate range   (see Chapter 6)

(iii) NPSH(A) throughout the pump operating flowrate range
   including transients   (see Chapter 6)

Pump design data* (Supplied by pump manufacturer)

(iv) Pump impeller eye/shaft diameter ratio for designs where the shaft passes through
   the impeller eye, pump speed and Q_{bep} value

Assessment

Stage 1

1 All pumps

Calculate suction specific speed n_{ss} (equation 3.6)

Stage 2

2A Pumps where $n_{ss} \leq 3500$	**2B Pumps where $n_{ss} > 3500$**
If $n_{ss} \leq 3500$ and operating range is within the 50%Q_{bep} to 120%Q_{bep} band cavitation surge risk is negligible. **Decide if this is acceptable**	If $n_{ss} > 3500$ Recalculate n_{ss} using NPSH(A) = 1.3 x NPSH(A) at Q_{bep} Use Fig. 7.11 to determine %Q_{bep} for the lowest flowrate where cavitation surge risk is negligible. **Decide if this is acceptable**
If $n_{ss} \leq 3500$ but a lower value than 50%Q_{bep} is required, the available options are given below	If the operating range restriction is unacceptable the available options are given below

Stage 3

3 All pumps not cleared by 2(A) or 2(B)

(i)   Explore the possibility of adding a backflow recirculator.**

(ii)   Evaluate methods of reducing individual Q_{bep} including choosing a pump with
       Q_{bep} at a lower value of Q and putting an additional pump in parallel.

(iii)   Reject because of the risk of unacceptable cavitation surging.***

---

*   For early plant design studies where data are not available, (i) assume the diameter ratio is 0.5 (a pessimistic possibility for most applications), (ii) estimate a pump speed and Q_{bep} using Chapter 3 (see how this is done in Chapter 12).

** The pump manufacturer to supply evidence that the backflow recirculator is effective over the required operating flowrate range at the appropriate NPSH(A) values. The data must be reliable and relevant to the particular application. If in doubt reject this option.

*** Review plant design choice particularly the need to operate individual pumps at low flowrates. (See the example in Chapter 12 for a way of approaching this.)

Table 7.3   Dealing with problems in an existing plant

Cavitation surging — Hydrodynamically induced — Existing plant

Checklist

1   Evaluate the possibility of raising the lower limit of %Q_{bep} in the operating range by changing plant operating procedures or the pump control system; e.g. run intermittently at higher flowrates rather than long periods at low flowrates where surging can be triggered.

2   Check if the pump is oversized (Chapter 6) and if the lower limit of Q_{bep} can be raised by reducing the impeller diameter.

3   Check if NPSH(A) can be raised; e.g. increase the control level in the inlet vessel.

Most improvements in pump performance come from aggregating the effect of the small changes which prove to be a practical possibility. For cavitation surging the dominant feature which gives most opportunity for an improvement is the minimum operating flowrate.

References

7.1   Sloteman, D. P., Cooper, P. and Dussourd, J. L., <u>Control of backflow at the inlets of centrifugal pumps and inducers</u>, First Int. Pump Symp., Texas A&M Univ., pp. 9 – 22, May 1984.

7.2   Kasztejna, P. J., Heald, C. C. and Cooper, P., <u>Experimental study of the influence of backflow control on pump hydraulic-mechanical interactions</u> Second Int. Pump Symp., Texas A&M Univ., pp. 33 – 40, 1985.

7 3   Massey, I. C. and Palgrave, R., <u>The suction instability problem in rotodynamic pumps</u>, NEL Conf. Pumps & Turbines, Vol.1, Paper 4 – 1, Sept. 1976.
(The name of co-author R. Palgrave was omitted from the published paper)

7.4   Fraser, W. H., <u>Recirculation in centrifugal pumps</u>,
 ASME WAM, pp. 65 – 86, 1981. (Also in World Pumps, Vol. 188, 1982.)

7.5   Palgrave, R., <u>Avoid low flow pump problems</u>,
BPMA Tech. Conf., pp. 124 – 139, April 1994.

7.6   Gasiunas, A. A., <u>Visual observation of supercavitating flow in rotating machinery</u>, BPMA Tech. Conf., pp. 221– 238, Dec. 1968.

7.7   Minami, S., Kawaguchi, K. and Homma, T., <u>Experimental study on cavitation in centrifugal pump impellers</u>, JSME Bulletin, Vol. 3, No. 9, pp. 19 – 35, 1960.

7.8   Cooper, P., Dussourd, J. L. and Sloteman, D. P., <u>Stabilisation of the off-design behaviour of centrifugal pumps and inducers</u>, IMechE Conf., Fluid Machinery for the Oil, Petrochemical and Related Industries, pp. 13 – 20, March 1984.

7.9   US Patent 4,375,937, <u>Roto-Dynamic Pump with Backflow Recirculator</u>, 1983.

7.10   US Patent 4,375,938, <u>Roto-Dynamic Pump with a Diffusion Backflow Recirculator</u>, 1983.

Chapter 8

Cavitation Surging —
Thermodynamically Induced

8.1 Deficiencies Identified

Cavitation surging in a centrifugal pump is a thermodynamic process controlled by hydrodynamic flow patterns. In circumstances where thermodynamic considerations dominate events the phenomenon is described as thermodynamically induced cavitation surging.

Thermodynamically induced cavitation surging results from deficiencies in plant design which permit pump operation at zero or near-zero flowrate conditions. At zero flowrate the cavities are an ever-growing, time-dependent phenomenon. At near-zero flowrate, heating magnifies the hydrodynamic surging taking place within the pump.

Normally, centrifugal pumps generate the head to match the system resistance over a range of flowrates which, for economic operation, are usually near to the best efficiency flowrate. During such an operation power in the form of hydraulic losses is dispersed mainly as heat to the liquid passing through the machine. This is evidenced by a temperature rise of the pumped liquid as it passes from inlet to outlet branch. The temperature rise is independent of time until wear increases internal clearances and affects hydraulic performance. Even then the change is usually very small and slow.

In a cavitating pump operating at or near to zero flowrate, the heat generated is not carried away and the temperature of the fluid rapidly builds up. During this transient the progressive build-up of temperature can magnify the effect of the hydrodynamically induced surging described in Chapter 7. An explosive growth in the total volume of the cavities within the impeller may occur. Where the rate of heat input is large and the design of the plant restricts the expansion of the pumped fluid, very high inlet pipe pressures can result. The performance deficiencies of commercial significance which arise as a consequence are listed in Table 8.1.

Table 8.1   Deficiency list

Cavitation surging — Thermodynamically induced	
Performance deficiency	Plant operating conditions
(A) At zero flowrate	
Progressively growing pressure pulsations	Fault conditions:
Mechanical component failure	zero flowrate
(B) At near zero flowrate	
Cyclic pressure pulsations	Fault conditions:
Mechanical component failure	very low $\%Q_{bep}$

The potential consequences of this form of cavitation surging are serious. Breach of the pressure containment boundary may be a possibility; mechanical damage to an axial thrust bearing or to any "less than robust" design of non-return valve is highly likely. It is clearly prudent to examine the outcome of such a surge to ensure that plant safety is not compromised. The consequence of pressure containment boundary failure must be etched in the mind of those making the assessment.

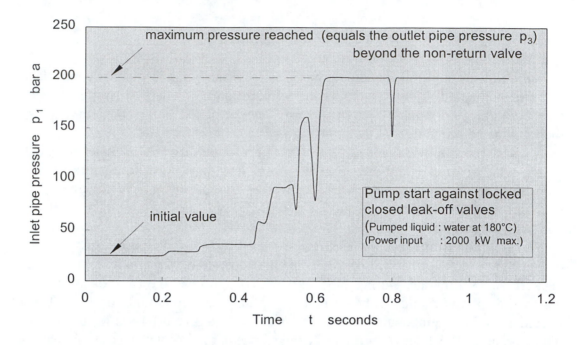

Fig. 8.1   Thermodynamically induced cavitation surging (test record)

For the particular example of thermodynamically induced surging shown in Fig. 8.1 a maximum inlet pipe pressure of less than 50 bar was expected; 200 bar was recorded.

The risk of catastrophic failure has to be considered for all pumps. However, the magnitude and rate of change of pressure is only likely to reach unacceptable levels in "high-energy" pumps. A high-energy pump is defined as one with a power input into the pumped fluid exceeding 100 kW per litre of impeller blade swept volume.

8.2   The Pressure Pulsation Phenomenon

8.2.1   An outline description

Chapter 7 showed that simple hydrodynamically induced surging is present in all centrifugal pumps to a greater or lesser extent at low $\%Q_{bep}$ flowrates. Where the local pump/pipework geometry is not conducive to the fully developed form of inlet pipe surge or where a backflow recirculation device is fitted the surge is confined to a circulation in the immediate vicinity of the impeller blade inlet tips. The presence of added heat at very low or zero flowrate magnifies hydrodynamic surging and this results in it taking on a potentially more dangerous form. A model of the pressure pulsations associated with zero flowrate is now described.

Consider a pump in which the through flowrate has just become zero. In this condition the power supplied to overcome hydraulic losses heats the liquid contained within a volume which approximates to little more than the swept volume of the impeller blades. In high-energy pumps the magnitude of power input is such that the temperature of this liquid rapidly increases. Consequential expansion of the liquid occurs which, when no forward flowrate is possible, results in the inlet pipework pressure increasing against either a closed inlet non-return valve or the liquid inertia in the inlet pipework.

In practice all high-energy input pumps operating at very low or zero flowrate are cavitating since it is economically impractical to provide the NPSH necessary to suppress it. In multistage pumps cavitation is usually limited to the first stage impeller. The heat added to the fluid in a cavitating pump impeller leads to further volumetric growth at the cavity vapour/liquid interface. The violent mixing of liquid and vapour that occurs in an impeller at zero flowrate ensures that heat transfer between liquid and vapour is not restricted by thermal conductivity effects. However, this simplifying fact is offset by (i) the restriction on vapour volumetric growth imposed by the increase in local vapour pressure associated with increases in local static pressure and (ii) the reduction in power input as the vapour volume begins to reduce the pressure generation capabilities of the pump.

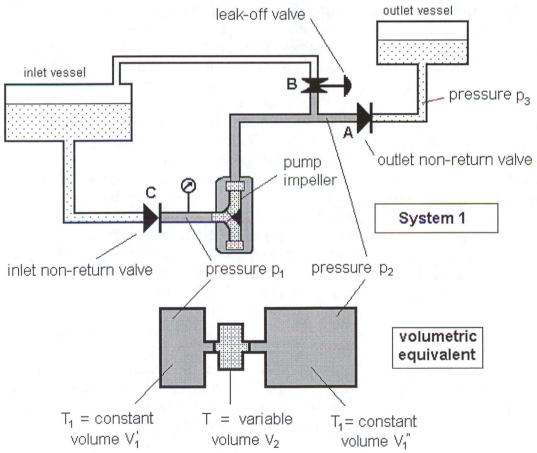

Fig. 8.2 Basic pump and pipework arrangement — System 1
         (restricted inlet and outlet pipework)

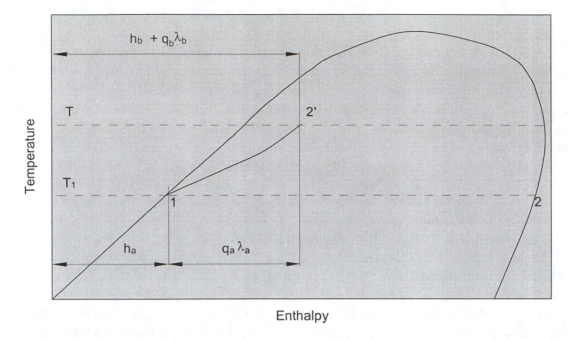

Fig. 8.3　Fluid temperature/enthalpy in a restricted inlet pipework system

In the typical high-energy pump system shown in Fig. 8.2 the outlet non-return valve A and the leak-off recirculation protection valve B are closed thus bringing about zero forward flow conditions. In a well-designed system this is a fault condition caused by a failure of the low-flow protection to operate. With no throughflow the inlet non-return valve is also closed. Continuing heat input by the pump results in expansion of the "trapped" volume of fluid between valves A, B and C. This in turn pressurises the liquid in the pump and pipework, which then presses closed more firmly the inlet non-return valve. As the liquid within the pump impeller becomes isolated the liquid pressure local to the cavity/liquid interface is that of the vapour pressure. This is Point 1 on Fig. 8.3. As further heat is added an increased volume of vapour is produced. For many high-energy pump systems and for most vapours (and certainly for steam), long before dry saturated conditions are reached at Point 2, the additional volume produced has exerted a self-pressurising effect on the system volume enclosed by valves A, B and C. This results in Point 2' being reached at the new temperature T where the added volume dv causes a change in local vapour pressure. The change in vapour pressure associated with the change in temperature from T_1 to T equals the change in static pressure dp. This appears, of course, as an increment of pressure in both the inlet pipework and the outlet pipework. In high-energy pumps this process continues until the static pressure change dp associated with increasing values of T raises the pump pressure to the outlet non-return valve lift pressure p_3.

At zero flowrate the vapour volume rapidly grows and with favourable system geometry could reach the point where the generated pressure begins to fall before pressure p_3 is reached. However, although pump input power may rapidly diminish as an impeller becomes vapour locked, even in this case heat continues to be added to the fluid in the impeller. At zero flowrate the change in heat loss arising from the change in power to the centrifugal pump impeller is directly proportional to the change in fluid density.

For single stage pumps the pulsation description applies directly. For multistage pumps a summation of the changes in each impeller cell, taking account of the progressive vapour locking of successive stages, is a necessary refinement. Using the model a single inlet pipe pressure transient is forecastable.

Once relief has been secured by lifting the outlet non-return valve the pump impeller(s), operating against a zero head rise system, displace some of the vapour/liquid mixture forward. The inlet pressure falls. The generated pressure and the pump power input both rise. A pulsating motion is the result. It seems reasonable to assume this cyclic motion takes its erratic period from the partially formed hydrodynamic surging flow patterns within the impeller blades.

8.2.2 *A mathematical model*

The following method may be used to obtain an approximate value of the inlet pipe pressure rise as the constraining effect of the system causes the vapour pressure to rise to that associated with temperature T. It is sufficiently accurate to demonstrate by numerical example the likely consequences of operating a pump under such conditions. However, it is only an approximation. This and uncertainties in values such as the volume of heated liquid mean it should not be the basis of design calculations.

The following method of estimating the order of magnitude of pressure changes in inlet pipework and identifying a likely reason for swing check valve failures originated in 1988 (Ref. 8.1) in response to feedrange vibration problems on power station feedwater pumps.

At Fig. 8.3 Point 1 vapour is forming within the liquid as heat is added. The heat added by the impeller in time Δt to liquid at initial temperature T_1 would, in an unconstrained system, produce a small additional volume of wet vapour for which the dryness fraction can be calculated.

Heat added per unit mass
of fluid in incremental time $\Delta t \quad = \dfrac{\phi \, . \, \Delta t}{m} = q_a \, \lambda_a$

where ϕ = rate of heat input to fluid from pump power losses.
Subscript a denotes values appropriate to an unconstrained system.
Subscript b denotes values appropriate to a constrained system.

Making the crude assumption that the mass of fluid in the heated zone remains constant (a good approximation in the first steps and not too erroneous at higher values of T where the value of V_V approaches that of V_L), gives

mass of fluid in heated zone $m = \dfrac{V_2}{V_{La}}$

Therefore, $q_a = \dfrac{V_{La} \, . \, \phi \, . \, \Delta t}{V_2 \, . \, \lambda_a}$

-----------------------------8.1

From Fig. 8.3, assuming that heat put into the remaining liquid can be neglected, then:

$$h_a + q_a \cdot \lambda_a = h_b + q_b \cdot \lambda_b \qquad \text{giving} \quad q_b = \frac{h_a + q_a \cdot \lambda_a - h_b}{\lambda_b}$$

Change in vapour volume $\quad dv = \dfrac{V_2}{V_{Va}} \cdot q_b \cdot V_{Vb}$

Assuming again that only the liquid in the unheated zone V_1 (much larger than V_2) must absorb this increase in volume, then

$$dp = \frac{k_4 \cdot dv}{V_1} \qquad \text{where } V_1 = V'_1 + V''_1$$
$$\text{and } k_4 = \text{bulk modulus of the liquid in volume } V_1$$

Combining the equations gives

$$dp = k_4 \cdot \frac{V_2 V_{Vb}}{V_1 V_{Va}} \left[\frac{h_a + q_a \lambda_a - h_b}{\lambda_b} \right] \qquad \text{------------------------------------ 8.2}$$

Since liquid continues to evaporate to produce cavities at the new temperature T corresponding to the vapour pressure p_{vapb}, then

$$dp = p_{vapb} - p_{vapa} \qquad \text{------------------------------------ 8.3}$$

For thermodynamic equilibrium the values of dp in equations 8.2 and 8.3 must be identical. The nature of fluid properties makes it necessary to solve these equations graphically for each incremental step in time Δt to give values T, p_{vap} appropriate to Fig. 8.3 Point 2′. The instantaneous volumetric growth rate is obtained from the value of dv and the calculational time increment. The value of interest is that when dp is such that the outlet non-return valve is just lifting.

8.2.3 *Outlet non-return valve response*

At the moment that the pressure transient reaches the value which lifts the outlet non-return valve (i.e. when p_2 reaches p_3) the potential rate of increase in cavity volume within the pump and inlet pipework reaches a significant finite value. As the non-return valve starts to lift the pressure build-up is relieved. The rate of relief is dependent upon the inertia and compressibility of the liquid in the outlet pipework. Where the contained volume beyond the outlet non-return valve is large, vapour volumetric growth in the pump and adjacent pipework continues against a pressure of approximately p_3. Inevitably, however, whatever the discharge pipework system design, a pressure overshoot will occur. This pressure overshoot is directly proportional to the forces opening the outlet non-return valve. Where resistance to the volumetric growth within the pump is insignificant extremely rapid opening of the non-return valve can ensue.

The instantaneous value of volumetric growth attained when non-return valve lift commences is very dependent upon the value of pressure change dp it has reached. This in turn is dependent upon how the pump is operated. To understand why differences occur consider the pump in Fig. 8.4 reducing flowrate from that of the duty Point A.

In normal service the leak-off system would be expected to open in good time to allow a pressure relieving flow to commence in the leak-off line. As speed is reduced and flow through the outlet non-return valve ceases the pump changes to operating on the leak-off system resistance line B-F. The "off-load" Point G at speed d can be used as a holding position prior to switching the pump off. For economic (power absorbed) and pump wear (cavitation erosion) reasons this time should be minimised.

Following a leak-off system failure to open the pump would operate at zero flowrate (speed d and Point D). It would generate the pressure rise F-D. To lift the outlet non-return valve a pressure pulse dp in excess of magnitude D-C is required.

If conditions are such that the pump impeller becomes vapour locked and generated pressure falls to a value F-E, the pressure pulse then required to lift the valve is E-C.

The maximum possible pressure rise in the inlet pipe is the difference between the relieving pressure beyond the outlet non-return valve and the pressure at the pump outlet.

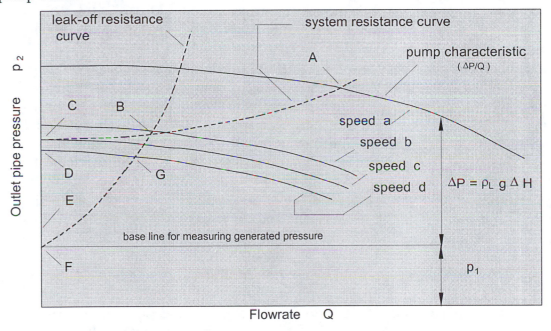

Fig. 8.4 Variable speed pump operating data with system and leak-off resistance curves (typical for boiler feedwater pumps)

8.2.4 *Practical significance explored through numerical examples*

The method of dealing with thermodynamically induced surging is not by attempting to calculate the maximum pressure and frequency of the surge as a basis for pump/plant design calculations. The calculational uncertainties brought about by having to estimate the volume of fluid being heated in the impeller and the wide range of possible operating conditions make this route untenable. An alternative way forward is discussed later. However, it is useful to quantify numerically, albeit with reservations on uncertainties, where thermal effects are likely to impinge on pump and plant design in the hope that this will give guidance on how safe operating practice might be established.

The analytical model of the thermodynamically induced cavitation surging phenomenon is based on the simple fact that fluids expand when heated. However, it only assumes importance when it leads to either (i) the associated pressure transient challenging the structural integrity of pressure containing plant items or (ii) the change in outlet volumetric flowrate caused by relieving transients being greater than the operating capability of the system components. To explore this further numerical examples are presented. These are based upon a typical high energy system with an inlet non-return valve such as that described by Fig. 8.2 — System 1. They take the cavitating impeller to be pumping water. For simplicity these examples also assume no system leakage loss.

For most fluids the vapour/liquid specific volume ratio and the liquid bulk modulus reduce significantly with increasing temperature. Therefore, the value of the initial liquid temperature T_1 has an influence on the rate of inlet pressure rise and on the volumetric growth rate attained when discharge non-return valve lift occurs. Calculated values for inlet pipe pressure rise and instantaneous volumetric growth rate for a pump operating in a system initially containing water at 20°C, 150°C and 180°C have been derived for two values of power input, 2000 kW and 6000 kW. In the examples the swept volume of impeller blades and associated "washed" volume V_2 is taken to be 0.01 m³ and the volume of pump inlet and outlet pipework V_1 to be 10 m³.

Calculated pressure change and volumetric growth rate transients are presented in Figs. 8.5 and 8.6 respectively. The pressure change curves shown in Fig. 8.5 reflect the rise in vapour pressure associated with the temperature increase in the impeller fluid. The low rate of change of vapour pressure with temperature is effective in delaying rapid pressure growth at lower initial temperatures. Inevitably, whatever the initial conditions, as more heat is added the more the pressure rise continues toward the outlet non-return valve lift pressure. Examining the curves for 2000 kW input power more closely, it is evident that with 20°C initial temperature, if leak-off recirculation commences within less than three seconds, the pressure transient will be limited to less than 10 bar. This time requirement is easily achievable if the leak-off valve is operable. As the initial temperature rises, however, the time for leak-off to react is significantly reduced and for a 150°C initial temperature the same transient pressure change occurs in about one second. With 6000 kW input power at the same 150°C initial temperature this time is reduced still further to less than 0.4 seconds. It is evident that as conditions become more arduous acceptable plant operation depends more and more on the leak-off signal being set to anticipate and prevent low-flowrate conditions arising.

If the pump was operating at zero flowrate such that the pressure difference C-D in Fig. 8.4 equals 10 bar, failure of the leak-off system to operate would result in a non-return valve opening "blip" occurring on the time scales indicated. As Point D falls, either due to reduction in pump speed or vapour locking of the impeller, larger values of pressure rise take place in the inlet and outlet pipework.

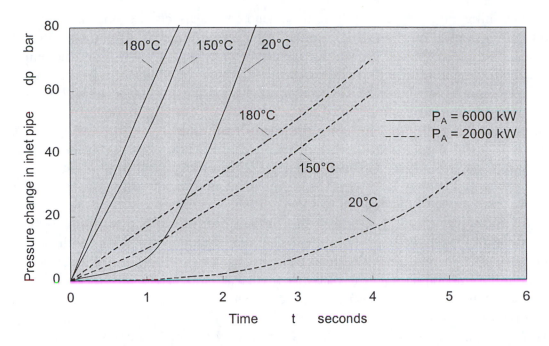

Fig. 8.5   Inlet pipe pressure rise with time at zero flowrate (calculated)

The trend in Fig. 8.5 shows that if the power input were to be reduced to 1000 kW then for water temperatures below 150°C the pressure change in the inlet pipe can be kept to modest levels if the leak-off system functions within 3 or 4 seconds. This is approaching the practical limit for some commercially available leak-off valves. The 1000 kW in 0.01 m³ corresponds with the definition of the description "high-energy".

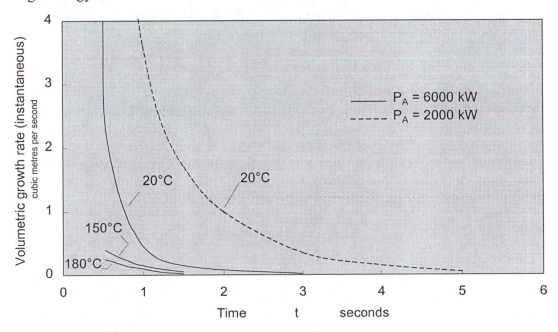

Fig. 8.6   Instantaneous volumetric growth rate with time at zero flowrate (calculated)

Examining Fig. 8.6 reveals how the phenomenon affects the operation of the outlet non-return valve. At a low initial water temperature of 20°C the predicted instantaneous volumetric growth rate is very large, typically many times the pump design flowrate, so that if conditions arise in which the valve lifts, a very rapid opening can occur. Under 150°C initial conditions the instantaneous volumetric growth rate has a much reduced value. However, even in these less onerous conditions the ability of less robust designs of non-return valve (e.g. swing check valves) to withstand such pulses must be placed in doubt. Clearly, a valve which responds adequately to the consequences of low-flow protection failure is a requirement. The very large errors of a simplistic analysis do not put in doubt the likely crucial role of the non-return valve.

Reviewing the above it is evident that in the system described, with its 1000 :1 pipework/impeller volume ratio (V_1/V_2), zero flowrate operation with water initially at 20°C and 2000 kW input power poses a risk of rapid non-return valve opening if the pump is held just below discharge pressure p_3. However, if a pressure only a few bars lower is reached the pump can withstand several seconds of operation without this risk arising. In an initially hotter system (150°C or 180°C) the pressure rise due to 2000 kW input power is much more rapid. In these circumstances the lifting of the non-return valve is, in most systems, more likely and although the instantaneous volumetric growth rate is very much less than that at 20°C it is still comparable with the system design flowrate. Failure of the leak-off protection to respond promptly on 150°C/180°C systems could result in rapid non-return valve opening. If a higher initial power is applied (e.g. 6000 kW as shown in Figs. 8.5 and 8.6) the rate of pressure rise increases approximately in direct proportion.

In the examples the latent heat of vaporisation is large so that the vapour never becomes dry. If the analysis were to be extended to calculate values for liquids other than water this aspect would have to be considered further. The avoidance strategy in section 8.6 makes this unnecessary.

8.3   The Influence of Plant Design

For the purposes of considering the effect of plant design on thermodynamically induced cavitation surging the available variations can be reduced to four basic system arrangements. These are described by Figs. 1.6 and 8.2 (System 1) and by Fig. 8.7 (Systems 2, 3 and 4). In the following paragraphs the word "restricted" is used to describe the presence of any device (e.g. a valve) which prevents the free expansion of fluid in a pump.

Restricted inlet and outlet pipework arrangements — System 1. The arrangement described by Fig. 8.2 — System 1 has been dealt with in detail in section 8.2. It is commonly used where pumps are installed with standby pumps in parallel. An inlet non-return valve is necessary to prevent recirculation back through the standby pump. In this system the integrity of the pressure containment is at risk.

Where the restriction on the outlet side of the pump is caused by a downstream blockage, say, by closure of a valve as shown in Fig. 1.6, then no relief path through the outlet pipe is available. In this system the integrity of the pressure containment is at an even greater risk.

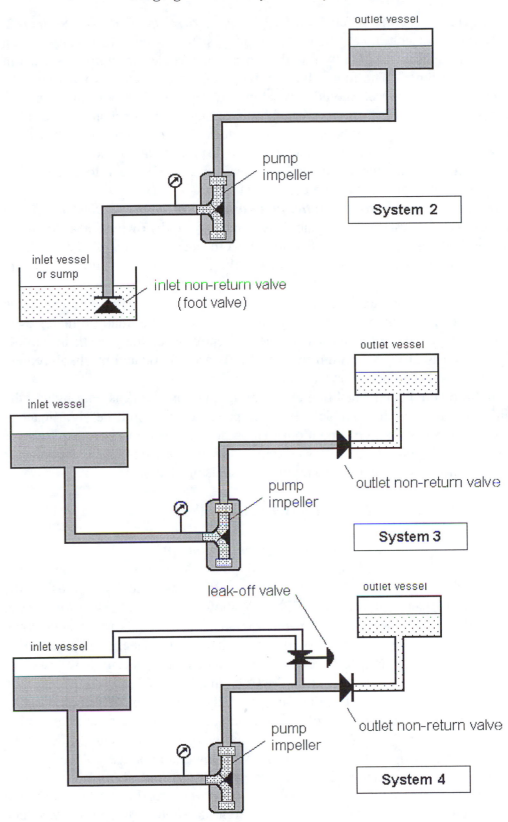

Fig. 8.7 Basic pump and pipework arrangements — Systems 2, 3 and 4
Outlet control valves omitted (assumed open)

Restricted inlet — unrestricted outlet pipework arrangements — System 2.
Possibly the most common system in pumping is that using a foot valve (often
together with a strainer) to draw liquid from an inlet vessel or sump as shown in
Fig. 8.7 — System 2. The foot valve is a type of inlet non-return valve specially
designed for this purpose. Should flow stop (say, by a faulty foot valve or by a
blocked strainer) then the progressive addition of heat to the unchanging liquid in
the pump impeller leads to it becoming vapour locked. The liquid is free, however,
to expand progressively up the outlet pipe by displacement of the liquid in the
pipe. This means that the pump will progress to a full vapour lock condition
without any significant inlet pipe pressure build-up.

*Unrestricted inlet — restricted outlet pipework arrangements — Systems 3 and
4.* In this type of system there is no impediment to fluid flowing back down the
inlet pipe. Systems 3 and 4 in Fig. 8.7 both meet this description and when leak-off
valve(s) in the latter remain closed, which is the case of interest, the systems are
identical.

For low power centrifugal pumps where heat input at zero flowrate raises the
temperature of the pumped fluid slowly an impeller may take many minutes even
hours to become vapour locked. Under these circumstances the growth in volume
of vapour as it is produced within the impeller is accommodated by displacement
of the inlet pipe liquid back down the inlet pipe.

For high-powered centrifugal pumps where the inlet pipework is lengthy and the
contained liquid has considerable inertia the pressure rise could be significant in
terms of inlet piping design limitations. The acceleration head necessary to drive
the liquid back into the inlet tank is determined by the vapour volume growth rate
and the driving pressure within the pump which this produces.

8.4 Vapour Locking Tests

8.4.1 *Tests on high-energy pumps*

The development of the "Advanced Class" pump (Refs. 8.2 and 8.3) led to the
inclusion of a "dry run" capability which was demonstrated by test using the
special pipework and control valves X and Y shown in Fig. 8.8. The objective of
including the requirement was to enable vapour locked pumps to be shut down
without risk of rubbing, seizure, or other mechanical damage. It is relevant to
cavitation surge evaluations to review the observations made during these tests.

The description recorded by Leith, McColl and Ryall (Ref. 8.4) states, "On
completion of the cavitation test the pump continued to run under fully cavitating
conditions for a further period of 10 minutes, during which time the suction
isolating valve was progressively closed until completely shut. The brake
horsepower absorbed by the pump fell to a very low value, some 10% of normal
value, and the pump remained in service for a further period of 10 minutes. Both
suction and discharge pressure gauges indicated zero readings; the pump ran
smoothly, emitting a mild not unpleasant siren-like note. The pump was then
stopped, still in the "dry run" condition." The pump was reported undamaged and
subsequent hydraulic tests showed performance to be unaffected.

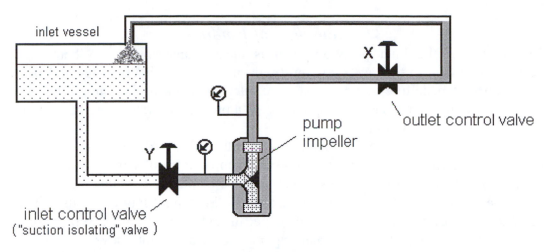

Fig. 8.8   Vapour locking test arrangement

The vapour lock test arrangement shown in Fig. 8.8 is in essence identical to that shown in System 2. The test experience reported confirms that even for high-energy pumps with an unrestricted outlet no problems arise if the liquid inertia in the outlet pipe is small. The same pump design in a Fig. 8.2 — System 1 arrangement gave rise to the inlet pipe pressure trace shown in Fig. 8.1. Evidently, and as is to be expected from the analysis of thermodynamically induced cavitation surging, an appreciation of the importance of addressing the design of a certain plant arrangement and ensuring adequate leak-off system operational reliability is critical to maintaining adequate levels of plant safety.

8.4.2   *Tests on small low-powered pumps*

During development of a 0.5 kW vertical-in-line circulator the pump was inadvertently left running against a closed outlet valve for over 48 hours. The heat build-up in the trapped liquid was sufficient to vapour lock the pump. In a fortuitously low ambient temperature and with no restriction on the inlet pipe the pump survived. The mechanical seal was "cooked" but operable. Evidently vigilance is needed even on small pump systems if costly repair bills are to be avoided.

8.4.3   *Re-establishing safe operating conditions*

As an appendage a cautionary note needs to be added. It should be remembered that if a pump becomes vapour locked the temperature of the fluid inside it will have been raised considerably. Even a low-power water circulator will contain steam at about 100°C, so venting a running pump to atmosphere through air-cocks can be hazardous. To secure a safe condition switch off the power and keep clear whilst the pump cools down especially if it was not designed for operation at the temperature reached. Large diameter, thin-walled casings can distort or crack under the induced thermal stress. A thorough dimensional and concentricity check on all running clearances is to be recommended before running the pump again.

8.5 *Summary of Findings*

A rupture of the inlet pipework wall is clearly unacceptable. Where high pressures that can do this are a possibility, even where this is consequential on a protection system failure, the pump and the inlet pipework design should be of a rating that could withstand such a challenge. If a protection system can fail then at some point in time it will fail.

In a system safe from pressure containment failure, thermodynamic surging will lead to vapour locking. For most designs of pump this results in seizure in any running clearances that depend upon the presence of pumped liquid. Neck rings, interstage bushes, and mechanical shaft seals are all potential failure points. Shaft breakage may arise as consequential damage. In high-energy pumps seizure may happen within seconds. On much smaller pumps seizure may take several hours.

The provision of a low-flow protection system which has appropriate reliability and redundancy will minimise the number of times seizure can occur. Indeed such systems should protect most pumps for their entire operating life. However, the possibility of their failure always has to be addressed. All pumps should have low-flow protection. This must be appropriate to the pump and the system in which the pump is installed. Chapter 10 explores how this may be achieved.

The very rapid closure of a valve on the outlet side of a pump in a system incorporating an inlet non-return valve (as in Fig. 1.5) can defeat almost all practical protection measures if the leak-off valves are unable to respond quickly. It is clearly essential in these circumstances that the low-flow protection is actuated in a timely manner. Consideration should be given to inhibiting full closure of downstream valves until an adequate leak-off flowrate has been established. Where this is not possible the provision of a bursting-disc or vent-tank that is sufficient to accommodate the expanding fluid and removing the drive power source should be evaluated, although in most cases because of the rapidity of pressure build-up these will have only a small influence on the outcome.

A step-by-step avoidance strategy is listed in Table 8.2.

8.6 Cavitation Surging (Thermodynamically Induced) —
An Avoidance Strategy

The strategy is presented in tabular form. Table 8.2A lists all the data needed. Table 8.2B details an assessment procedure.

Table 8.2A   Data collection

Information required to commence an assessment

Plant design data   (supplied by plant designer or user)

(i)   Layout drawings showing pipework, valves, and all devices which could restrict flow

(ii)   Pressure rating and hydrostatic test rating of all components within the boundary defined by "restrictions".

Pump design data

(iii)   Pump casing pressure rating and hydrostatic proving-test rating

Table 8.2B   Assessment procedure

Assessment

Stage 1

1   All pumps
      Identify type of system by inlet/outlet configuration   (see section 8.3)

Stage 2

2A   For System 1 (i.e. a restricted inlet and outlet system)
      (a) Ensure the pump and inlet pipework as far as any upstream restriction to flow (usually a non-return valve) is of an adequate pressure rating. It should be at least equal to the outlet pipe rating.
      (b) Proceed to Stage 3.

2B   For Systems 2, 3 and 4 (i.e. inlet or outlet but not both are unrestricted)
      No special requirements. Proceed to Stage 3.

Stage 3

3   For all systems
      (a) Ensure isolating valves cannot be closed on a running pump nor a pump started with isolation valves closed.*
      (b) Ensure adequate low-flow protection is provided.   (see Chapter 10)

*   By electrical control interlocks, mechanical locking or operator instructions, or any combination of these.

References

8.1 Grist, E., <u>Pressure pulsations in cavitating high-energy centrifugal pumps and adjacent pipework at very low or zero flowrate</u>, IMechE Conf., Part Load Pumping, pp. 143 – 152, 1988.

8.2 Otway, F. O. J., <u>The design principles for boiler feed pumps for CEGB 660MW units</u>, IMechE Conf., Advanced-Class Boiler Feed Pumps, pp. 11 – 17, 1970.

8.3 Grist, E., <u>Advanced class boiler feed pumps – Operational experience</u>, British Pump Manufacturers' Association, Seventh Tech. Conf., pp. 1 – 21, 1981.

8.4 Leith, T. O., McColl, J. R. and Ryall, M. L., <u>Advanced-class boiler feed pumps for 660MW generators</u>, IMechE Conf., Advanced-Class Boiler Feed Pumps, pp. 36 – 47, 1970.

Chapter 9

Cavitation Erosion

9.1 Deficiencies Identified

Cavitation erosion is the removal of material from a surface by forces produced by the collapse of cavities adjacent to it. The occurrence of cavitation is determined by local pressure gradients and, by definition, is always where the local pressure has been low enough to allow vapour to form. Cavitation erosion damage can be severe enough to impair the mechanical integrity of a centrifugal pump. The various common forms of commercially unacceptable damage encountered are shown as a composite picture in Fig. 9.1. They usually occur individually; they nearly all occur on the impeller.

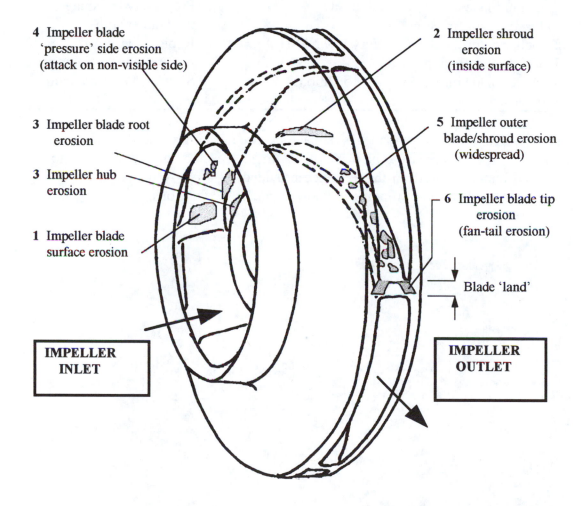

Fig. 9.1   The location of common forms of impeller cavitation erosion damage
(numbers relate to Table 9.1 listing)

A knowledge of where cavitation attack might occur is a prerequisite to drawing valid conclusions (a) during visual cavitation tests in which the location of collapsing cavities is to be observed, and (b) from the examination of pumps after running (on test or in service) to deduce the presence or otherwise of an unacceptable level of cavitation erosion attack. Table 9.1 summarises the forms of cavitation erosion and their damage characteristics.

Table 9.1 Deficiency list

Performance deficiency	Cavitation erosion Plant operating condition	Damage characteristics
1. Impeller blade surface erosion	All conditions	Progressive* damage over the blade width
2. Impeller shroud erosion	All conditions	Rapid localised damage
3. Impeller hub/blade root erosion	All conditions	Rapid localised damage
4. Impeller blade "pressure" side erosion	High %Q_{bep}	Progressive* damage over the blade width
5. Impeller outer blade/shroud erosion	Very low NPSH(A)	Rapid widespread circumferential damage
6. Impeller blade outlet tip erosion	All conditions	Slow self-limiting damage

* Progressive in this context means damage from a rate of erosion that enables an industrially useful forecast to be made of the time to reach a defined level of more advanced damage based on the evidence provided by a partly damaged impeller after a measured period of running in known operating conditions.

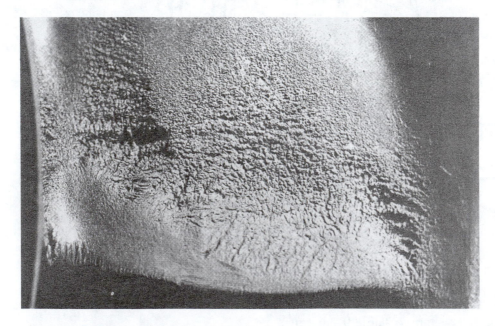

Fig. 9.2 Impeller blade surface erosion

9.1.1 *Double shrouded impellers*
1. *Impeller blade surface erosion*

In centrifugal pumps cavitation erosion is most often observed on the "suction" side of impeller blades. In a material that possesses some ductility the surface is roughened by cavitation. Steel and bronze impellers are examples of this. This roughness is progressively increased with time until, as shown in Fig. 9.2, it begins to take on the appearance of very rough sandpaper. In more brittle materials such as iron, large grains of material are removed and holes appear so that eventually the area under attack either looks like a lace curtain or loses a large portion of impeller blade as in Fig. 9.3.

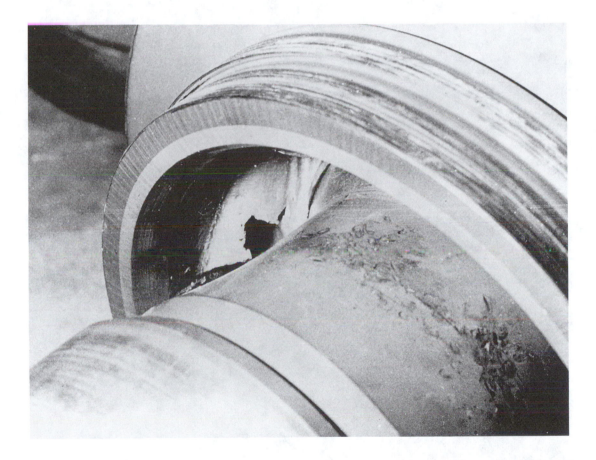

Fig. 9.3   Impeller blade edge failure

Close examination of the damage patterns on pump impellers that operate continuously at best efficiency flowrate conditions with the same NPSH(A) and speed shows cavitation erosion to be remarkably consistent in the location of attack on each blade. Often the shape and texture of major areas of pitting are repeated identically on successive blades. An example of an advanced stage of this "progressive" type of damage is shown in Fig. 9.4. The crescent-shaped damage pattern is clearly repeated on successive blades. Even with such a marked amount of damage it can be seen that metal has only been removed from regions which are, in a relative sense, lightly loaded so that even now the impeller is not structurally at risk.

Fig. 9.4 Impeller blade surface erosion — An advanced damage pattern
Courtesy of Weir Pumps Ltd

Fig. 9.5   Impeller outlet edge damage following inlet edge failure

Pumps that operate away from Q_{bep} are sometimes damaged in a similar manner. Occasionally the damage is only on every other blade. This alternate blade cavitation damage is caused by flowrate recirculations associated with partly loaded blades. This unusual damage pattern is not often seen because pump designers normally choose an odd number of blades (5, 7, 9, etc.) to avoid the even-multiple harmonics initiated by the pressure pulses set up at off-design flowrates influencing shaft dynamics by volute-tongue/impeller-blade or diffuser-blade/impeller-blade passing frequencies. Choosing an odd number of blades does not preclude alternate blade cavitation erosion taking place. Examples of, say, a nine-bladed impeller being damaged by an intense local attack repeated identically on blades 1, 3, and 8 with moderate attack on blades 5 and 6 do occur. The flow regime which leads to local flow fluctuations which produce the alternating blade damage pattern evidently breaks down and reforms within a once-per-revolution cycle or a multiple of this. Small differences in impeller blade geometry appear to determine where this breakdown occurs within the cyclic flow pattern. A knowledge of the sensitivity of the location of cavitation erosion damage to minor changes in inlet blade geometry and the damage pattern repeatability blade-to-blade is important when specifying the extent of cavitation tests.

The surface roughening produced by cavitation erosion in itself has no significant effect on hydraulic performance. However, as damage progresses the structural integrity of the inlet blade edge becomes at risk. Centrifugal forces acting on the front shroud of an impeller coupled with the pressure difference across the blade combine to weaken the blade. Eventually a piece of blade becomes detached. Impeller passages are often quite large so even a large piece of material can, with luck, pass through the pump. However, the possibility of the piece becoming jammed between the rotating element and the stationary casing, particularly in pumps with close fitting diffusers, is very real. When jamming does occur the outlet edge of the impeller blade can be broken off between the shrouds. An example of this is shown in Fig. 9.5 — the near view shows the broken outer length of impeller blade; the inner shows the severe cavitation attack and leading edge of the impeller blade broken off.

The progressive nature of impeller blade surface erosion means that it is a form of cavitation that can be tolerated. Indeed, for very large impellers, such as those used in hydro-electric pumped storage schemes where operating flowrate is carefully controlled by adjustable guide vanes, a policy of planned periodic repair is often part of the plant maintenance programme. This is feasible where pressure gradients within the impeller are low and damage is as a consequence spread over a wide area. In high-speed impellers, where pressure gradients are severe and damage is more localised, a repair policy is nearly always impractical. An appropriate maintenance policy is one in which cavitation erosion damage is monitored until it approaches an unacceptable level. The damaged impeller is then replaced. The criterion for unacceptability, first introduced in 1974 (Ref. 9.1), is when a piece of material with an area greater than 1 cm^2 becomes detached.

2. *Impeller shroud erosion*

This type of damage is most often, although not exclusively, seen on narrow impellers. As is shown in Fig. 9.6 a deep groove is produced. This is the result of a narrow stream of cavitation coming into contact with the shroud as it reaches a region in which it collapses. It is nearly always observed on the inside front impeller shroud although occasionally it has been seen on the back shroud.

3. *Impeller hub erosion and impeller blade root erosion*

These forms of damage are associated with high-speed impellers. Visual cavitation testing of impellers on which it occurs shows the presence of a narrow form of intense vortices which come into contact with the impeller surface.

Fig. 9.6　Impeller shroud erosion

Hub erosion damage is often difficult to observe and photograph. Its location on the "inner" side of the hub is protected from view by the outer shroud and neck ring. Figure 9.7 gives an impression of the damage. Impeller blade root erosion is very

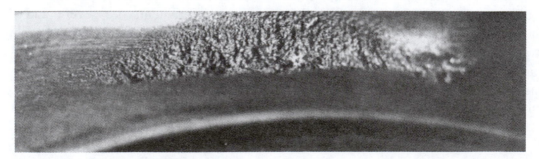

Fig. 9.7　Impeller hub erosion

Fig. 9.8   Impeller blade root erosion   *Courtesy of Weir Pumps*
Ltd

similar. The fillet at the blade root initiates a stream of vortices at flowrates lower than Q_{bep}. Under low NPSH conditions these cavitate. In well-designed impellers the cavities leave the blade and collapse in the main body of the liquid away from the impeller walls. This is difficult to achieve when the impeller hub curvature dominates the local geometry. A typical example of cavitation erosion damage at the blade root is shown in Fig. 9.8.

On rare occasions cavitation damage in the form of a groove has been seen on the rear of impellers where the rear impeller shroud and the pump casing wall come into close proximity and liquid throughflow is practically zero. An intense circumferential vortex flow which cavitates as a result of the heat generated by disc friction reducing the local NPSH is thought to be the cause, although no conclusive proof has been obtained. Both impeller hub and impeller blade root cavitation often produce a deep groove very quickly, and although it does not immediately pose a threat to the structural integrity of the impeller urgent action must be taken by either a change to the impeller design or a change in the pump operating conditions to prevent the groove progressing in depth to the point where unacceptable damage occurs. Usually this form of attack is rapid and its effect on impeller material is well described by the saying, "It cuts like a hot knife through butter".

4. *Impeller blade "pressure" side erosion*

A centrifugal pump operating at a flowrate higher than best efficiency flowrate induces a strong recirculatory flow on the "pressure" side of the impeller blade, that is, on the side away from direct view when looking into the impeller eye. When NPSH(A) is low cavitation occurs. Often the first evidence of the pressure side erosion caused by this cavitation is when it breaks through the impeller blade. This form of damage often occurs further down the impeller blade than the blade surface erosion on the "suction" side described earlier. Figure 9.9 shows impeller blade pressure side erosion damage on a large boiler feedwater pump; Fig. 9.10 shows it on a small water pump.

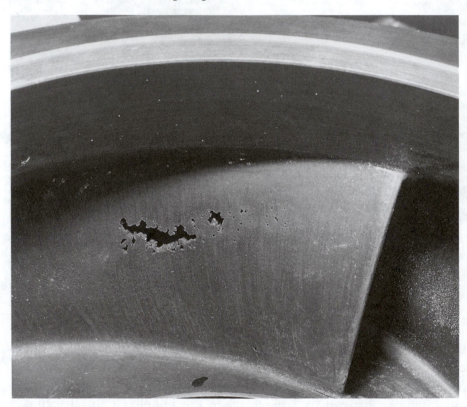

Fig. 9.9   "Pressure" side erosion on a boiler feedwater pump impeller
(viewed from the impeller inlet)

Courtesy of Weir Pumps Ltd

The progressive nature of impeller pressure side erosion allows the cavitation erosion damage it produces to be tolerated in exactly the same way as impeller blade surface erosion. For nearly all impellers, however, the location of the damage makes repair by techniques such as welding impossible so that planned impeller replacement is usually the only viable maintenance policy.

Where, as is typical, the holes produced by cavitation are well away from the leading edge of the impeller blade there is often no immediate risk of mechanical failure. The effect of penetrating the blade can be to relieve the low-pressure region which is causing cavitation. The intensity of cavitation erosion (and the intensity of audible cavitation noise) is then considerably reduced so that the possibility might seem to exist of the impeller being suitable for further service. This possibility is not a real one, especially in plants where operating conditions may change. The beneficial effects of holes in the impeller blades are not predictable and should not be trusted. It is useful to be aware of the presence of this effect as it explains why two impellers with nearly the same amount of pressure side cavitation erosion damage can come from pumps which have been in operational service for very different periods of time.

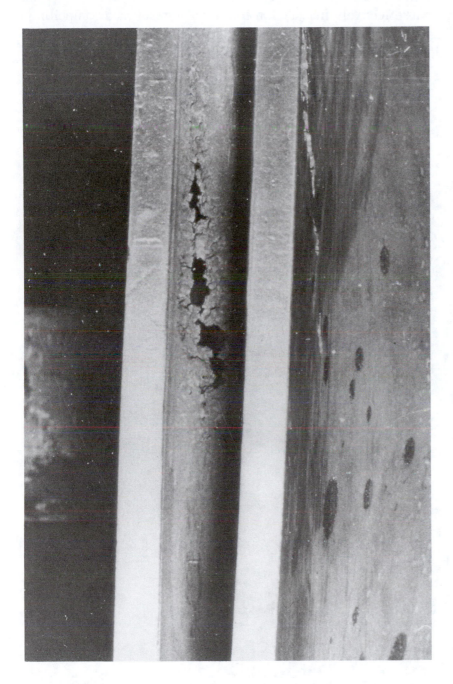

Fig. 9.10 "Pressure" side erosion on a small water pump impeller
(viewed looking down between the shrouds from the impeller outlet)

5. *Impeller outer-blade/shroud erosion*

When NPSH(A) is very low cavitation can extend completely through a centrifugal pump impeller. Flow at the impeller/diffuser or impeller/volute-tongue interface is completely disrupted. In such circumstances cavitation erosion often attacks the outer length of each impeller blade. Usually all the blades become very badly pitted. Damage also often occurs on nearby diffuser blades. Such damage is, of course, accompanied by a significant reduction in generated head.

The impeller shown in Fig. 9.11 was inadvertently operated for a period of about 5000 hours on what was effectively free suction control. The result is a compelling argument for being certain of the operating range and the impeller design suitability before adopting the free suction control method.

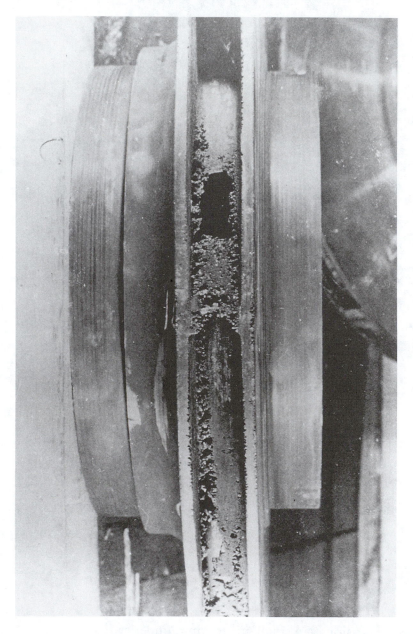

Fig. 9.11   Impeller outer blade/shroud erosion

6. *Impeller blade tip erosion (fan-tail erosion)*

When an impeller blade passes in close proximity to a stationary component such as a diffuser blade a low-pressure pulse occurs. This can reduce the local pressure to that of the vapour pressure. Cavitation erosion can then occur on the blade "land" surface immediately following where the blade ends. A characteristic "fan-tail" erosion pattern (Fig. 9.12) is produced. Opening the impeller-tip/diffuser-blade gap slightly usually cures the problem without penalising pump performance. Opening up the side flow areas to provide relief from local depressurisation is another possibility, but this is often less favoured because of the potential for increased leakage loss. High-speed impellers are more susceptible to fan-tail erosion, although slow-speed, wide impellers having a long blade land (see Fig. 9.1) have been known to exhibit this type of damage.

"Fan-tail" damge pattern

Fig. 9.12 Impeller blade tip (fan-tail) erosion *Courtesy of Weir Pumps Ltd*

The resulting damage is self-limiting and in itself of no significance since the cumulative damage at the impeller/diffuser gap quickly moves the pump to an insignificant cavitation erosion rate. The unacceptable nature of this form of damage arises because it is a very strong indicator of the presence of pressure pulsations that might be of a magnitude that is sufficient, even when relieved slightly, to affect pump shaft dynamics (and hence bearings — external and internal lubricated — and shaft seals). Impeller blade tip erosion indicates a need to investigate immediately whether impeller tip geometry changes are required that are more substantial than those required just to eliminate erosion. On the rare occasions that damage occurs on the diffuser blades or on the volute tongue that defines the gap at the impeller tip the same self-limiting damage pattern emerges.

9.1.2 *Open impellers*

Open impellers, such as those shown in Fig. 9.13 experience the same cavitation erosion damage patterns as the double shrouded impellers. Additionally, cavitation attack is caused by the intense vortex which forms behind the blades as they rotate in close proximity to the stationary pump casing wall. Cavity collapse and the consequent damage is very localised. In this respect open impellers are not well suited to cavitating conditions.

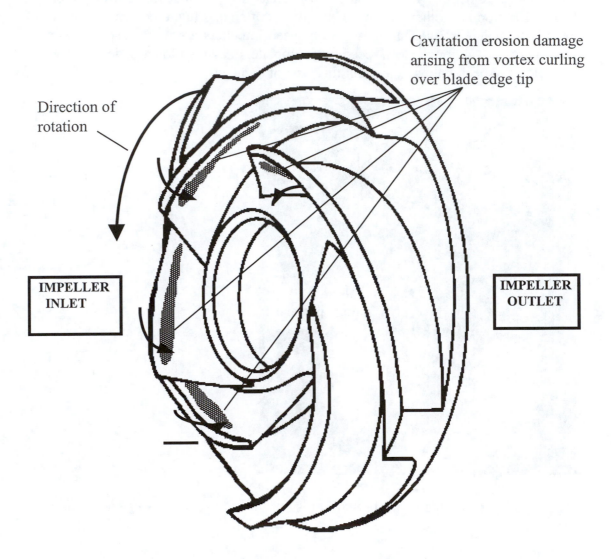

Fig. 9.13   Open impeller — Tip clearance vortex cavitation erosion

9.2 Cavitation Erosion Rates

9.2.1 *The need to know*

It is useful to have an understanding of the rate at which cavitation might produce erosion damage. In an ideal world this would not be needed if the main objective of pump users, to avoid cavitation erosion altogether, could always be met. Unfortunately this is not the case. There are two important circumstances in which cavitation erosion occurs. They are as follows.

(1) Cavitation erosion occurs where a concept of "acceptable" damage on an impeller blade surface has been introduced. This is only appropriate to applications where the impeller design and the pump operating conditions are carefully controlled and the form of damage is "progressive" (see Table 9.1).

(2) Cavitation erosion occurs where an existing plant suffers from the effect of less than adequate design or operating procedures and options such as plant design changes are either prohibitively expensive or, for other reasons, are impractical..

The literature on cavitation is extensive. Cavitation has been the subject of investigations as diverse as its effect on limiting the height of trees and the source of problems in laser surgery in the fluids in the human eye. Even within engineering, publications are numerous. It is important to appreciate this diversity, for although the term "cavitation" originated in mechanical engineering much research work is carried out with other objectives in mind. A large amount of academic research has been devoted to understanding the cavitation process free from the constraints imposed by pump impeller internal pressure contours and flowpaths.

Laboratory research work on the fundamentals of the cavity collapse process that produces damage is far removed from that of the environment in centrifugal pump impellers. In academic circles there is even some uncertainty as to whether the method of cavity collapse bears a true relationship to the mathematical model of cavity collapse. Rolling cavities have been observed by Chizelle et al. (Ref. 9.2) on axisymmetric head-forms immersed in flowing water. These in some ways copy the flow patterns on centrifugal pump impeller blades. Small but intense collapsing vortices have been suggested as the basis of the erosion mechanism rather than the local force generated by cavity collapse. This academic work might one day provide a clear and unambiguous view of the whole process which relates directly to centrifugal pumps. Meanwhile, users of such equipment have to draw conclusions from the information available.

Taking the pragmatic line of the pump user it can be seen from the work of Rayleigh and the simplified approach to cavity dynamics expounded in section 5.3 that cavity collapse can be extremely rapid. This is supported by observation (Fig. 1.1). Simple calculations show the pressure produced at the point of collapse is very large. The local impact force this produces at the point where collapsing cavities impinge upon surface material is very great. This, and the repetitive nature of the process, means that there are no metals that will successfully resist close proximity to cavitation attack for long periods (i.e. over 45,000 hours). Damage to the metals commonly used in pump construction is an inevitable consequence. Those that put up the best resistance are from materials groups that inherently have little ductility. The rotating impeller of a centrifugal pump, especially one running above 1800 rpm, needs ductility to meet all the combinations of centrifugal and hydraulically induced loads which arise over the operating range of the pump.

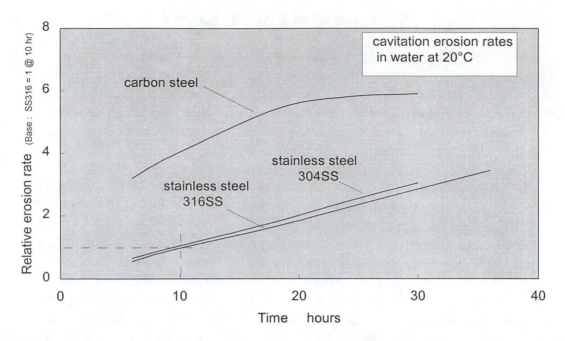

Fig. 9.14 Typical vibratory device cavitation damage results

see Garcia, Hammitt and Nystrom Ref. 9.3

9.2.2 *Comparative erosion rate data*

The literature abounds with papers (Refs. 9.3, 9.4, 9.5, 9.6 and 9.7) on comparative erosion rates. They range from laboratory studies of the erosive effect of very small collapsing cavities (which is not of interest here) to the methodology for improving the outcome with regard to cavitating pump performance (which is). The added complication that results vary with the size and type of test equipment used as well as the test method serves to frustrate the clear definition of practical objectives.

The centrifugal pump user requires to understand the "ranking" of impeller materials to provide a logical basis for dealing with problems which might (and do) arise and a "workshop friendly" method of obtaining data to back this up. The majority of pumps have impellers which are made from materials from one of four groups:

  (1) cast irons,
  (2) gun metals and bronzes,
  (3) mild steels,
  (4) stainless steels.

Within these groups the variety of alloying elements and possible heat treatments create innumerable combinations. All have a different resistance to cavitation erosion attack. Being in one of the common groups is of no particular benefit. The combinations of material, heat treatment, hydraulic operating condition, and impeller design mean each pump is likely to present a unique set of circumstances.

Cavitation erosion laboratory tests aimed at providing data are usually run under one of a number of controlled conditions. Even here difficulties have been reported between test methods. The repeatability using the same method under conditions that outwardly appear to be well controlled has, on occasion, been poor (see Ref. 9.6). Small changes that occur in specimen manufacture, heat treatment, test liquid properties and in test facilities built to a common specification are sufficient to cause measurable variations. Clearly, giving even general guidance is not for the faint hearted. To aid users in putting cavitation erosion into perspective Fig. 9.14 and Tables 9.2 and 9.3 are presented. All the results are taken from vibratory tests using water at 20°C.

Table 9.2 Typical cavitation erosion rate data   *see Hobbs, Ref. 9.7*

Material*	Composition*	Erosion rate range mg/hour	Relative rate (to 316SS)
steel En58J (316SS)	Cr 18; Ni 9, Mo 3, C 0.07	11.3	**1.0 (datum)**
steel En58B (314SS)	Cr 18, Ni 9, C 0.08	11.7	1.1
steel En57	Cr 17, Ni 2, C 0.25(max.)	3.0 to 7.0	0.3 to 0.6
steel BS1603B	Cr13, C 0.2(max.)	11.3	1.0
mild steel	C 0.12, Mn 1.19, S 0.25	50	4.4
aluminium bronze	Al 8.5/10.5, Ni 4.5/6.5 Fe 3.5/5.5, Cu remainder	5.6	0.5
gunmetal	Sn 10. Zn 2, Pb 1.5(max.) Ni 1.0, Cu remainder	34	3.0

*Material and composition descriptions are "as reported" to avoid ambiguity with later standards, etc. For fuller specifications see Refs. 9.4 and 9.7.

Table 9.3 Typical cavitation erosion rate data
see Garcia, Hammitt and Nystrom, Ref. 9.3

Material**	Composition	Erosion rate range mg/hour	Relative rate (to 316SS)
steel 316SS	see Table 9.2	2.81	**1.0 (datum)**
steel 314SS	see Table 9.2	3.04	1.1
carbon steel (mild)	see Table 9.2	7.08	2.5
Plexiglass		6.6	2.4
aluminium 1100-0		28.90	10.2

**Material descriptions are "as reported".

   For fuller specifications see Refs. 9.3 and 9.4.

Notes:

1.   Cavitation damage to cast iron is not readily reparable, particularly on complicated-shaped impellers. No data are quoted.

2.   Plexiglass is added for general interest. Aluminium is sometimes used to conduct cavitation erosion tests on complicated cast shapes where tests on the material proposed for industrial usage would take too long.

3.   Garcia, Hammitt and Nystrom (Ref. 9.3) showed that when these same materials were tested in other liquids (mercury, sodium, etc.) very different cavitation rankings were obtained.

9.2.3 *Evaluating the resistance to cavitation erosion for pump materials*

When, in spite of all precautionary measures, an unacceptable level of cavitation attack occurs it is very useful to have recourse to a cheap, speedy and repeatable method of comparing cavitation erosion rates. This could, for example, be used to compare attack on a sample of impeller material with that on a proposed weld overlay or coating. Pump users (and many pump manufacturers) do not have the time or finance to explore the relative merits of the different test methods. It is easier to choose a good one and get on with the job A most effective way of achieving good results is by means of a Lichtarowicz Cell. This device is detailed in ASTM Standard G134-95 (Ref. 9.8) and its development is described in papers by Momma and Lichtarowicz (Refs. 9.9 and 9.10).

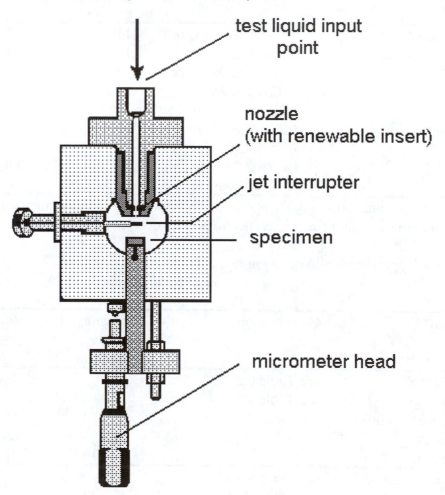

Fig. 9.15   The Lichtarowicz Cell

The outline drawing (Fig. 9.15) shows the simplicity of the device. It lends itself to the fast evaluation of different materials, as a test on even the hardest material will last no more than about 12 hours as cavity intensity can be increased by raising the pressures in the system. The test cell is small and can be made from special materials to accommodate chemically active liquids without great expense. In one example the Lichtarowicz Cell proved invaluable in identifying the magnitude of the changes in cavitation erosion rate that accompanied changes in pumped liquid chemical composition for an application in the nuclear industry. A special stainless steel rig to contain the aggressive test liquid and the test cell was constructed and the comparative tests were completed very quickly. Steller (Ref. 9.6) acknowledged that due to operational advantages jet facilities of the Lichtarowicz Cell type are being used more and more.

A typical set of test data is shown in Fig. 9.16. Like all cavitation erosion tests the Lichtarowicz Cell only partly simulates the impeller environment. The cavities are projected toward the specimen at a high velocity and, following collapse, move radially outward from the collapse surface. So whilst the Lichtarowicz Cell closely approximates to some conditions in a pump impeller it is far from replicating them all. In the centrifugal pump the occurrence of cavitation as streams of bubbles, sheets of vapour in close or distant proximity to the surface being attacked and the pressure gradient under which attack takes place are not simulated — nor could they be. The choice of test apparatus must be a compromise which has adequate validity.

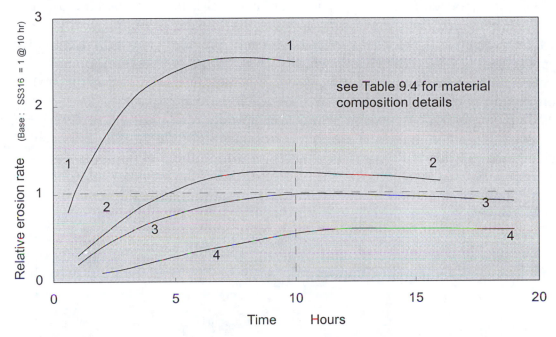

Fig. 9.16   Some comparative cavitation erosion rates obtained
using a Lichtarowicz Cell

Table 9.4 Material compositions relating to Fig. 9.16

Material	Composition
1. corrosion resistant steel	Polish steel 10HNAP $C \le 0.12$, Si $0.25 - 0.60$, Mn $0.40 - 0.90$, P $0.06 - 0.10$, $S \le 0.04$, Cr $0.5 - 1.00$, Al ≈ 0.02
2. nickel aluminium brass	Al $9.9 - 9.5$, Fe $4.0 - 5.0$, Ni $4.5 - 5.5$, Mn $0.75 - 1.3$, Cu remainder
3. steel 316SS	see Table 9.2
4. admiralty bronze	see Table 9.2

9.2.4 *Establishing a basis for cavitation performance assessment in centrifugal pumps*

Actual experience relating to the operation of a particular pump in particular conditions must always take precedence over forecasts based upon "typical" data or even upon works test data. Proof of adequacy by field experience may, and often does, expose the generosity of margins of safety built into methods based upon typical experience resulting from the need of such methods to err on the side of caution when dealing with uncertainties. Operational experience for a pump operating in identical or near-identical conditions should always be sought and should be used instead of using generic methods of evaluating NPSH(A) adequacy. Competent pump manufacturers are able to provide details of such experience when it exists. It is a sad fact that the inconvenience of collecting and collating relevant data means that this most important and potentially most rewarding method of pump performance protection from cavitation erosion is often overlooked.

The vast majority of centrifugal pumps are not at risk from cavitation erosion. This is because most of them pump cold water which is drawn from a nearby free surface where a 1 bar atmospheric pressure provides the equivalent of approximately 10 m NPSH giving, typically, an NPSH(A) of between 10 m and 20 m depending upon the physical arrangement of pumps, pipes and vessels. In small pumps with relatively low shaft speeds — which the majority are — this added NPSH is sufficient to suppress cavitation completely. It is possible to describe pumps which are at risk from cavitation erosion by a defining rule. It is self-evident that if this rule describes conditions where there is an insignificant probability of cavitation occurring then it also describes the boundary up to which there is no need to carry out a cavitation test. Figure 9.17 shows the boundary for small centrifugal pumps.

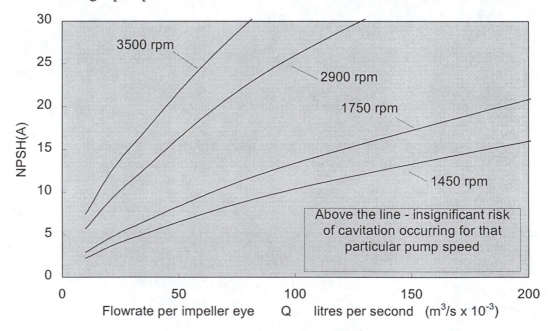

Fig. 9.17   The "no cavitation test" requirement boundary
(operating range 50%Q_{bep} to 120% Q_{bep} — derived in section 9.3.1)

Outside this "no test" group it is necessary to demonstrate adequacy by a cavitation test. For most pumps this entails securing proof of performance by means of the cheap and easy-to-apply NPSH(3pc) test. Where this is inappropriate the more costly NPSH(4mm) test must be undertaken. Pumps that are not able to demonstrate adequacy by either of these test requirements should be rejected. For a new plant this means the specification must be changed (reduced pump speed, increased NPSH(A) are options) and the assessment process repeated until a viable option is found.

Ideally, the aim should be to ensure that no cavitation erosion damage occurs during the lifetime of a pump. This is easily achievable where no cavitation test is required. In cases where pumps running at speeds up to 3600 rpm are used an NPSH(3pc) test is appropriate. The only qualification is that the pumps must always run within the $50\%Q_{bep}$ to $120\%Q_{bep}$ operating range.

For pumps having a running speed of over 3600 rpm it is not practical to be so restrictive. Some damage has to be tolerated if commercially acceptable levels of NPSH(A) are to be used. This tolerance is measured by a definition of "acceptable damage". A definition that is suitable for industrial usage is given by the intent statement that the cavitation damage that occurs should not prevent the pump reaching an operating life of 45,000 hours, again with the proviso that the operating range is within the $50\%Q_{bep}$ to $120\%Q_{bep}$ band.

No guarantee can be given that the guidelines that evolve from the preceding work will produce the described outcome. The guidelines are statements of expectation based upon a considerable amount of pump operating experience. As the basis for the guidelines are explained in section 9.3 it will become clear that they are couched in pessimism so that a better than 99.9% prediction accuracy might reasonably be expected.

It is, of course, possible to operate centrifugal pumps with lesser NPSH(A). For most users the consequences of cavitation erosion attack do not justify taking the greater risk associated with this alternative. In the last analysis, however, taking a greater risk or treating the pumps as expendable is a user choice. The guidelines developed in section 9.4 provide a basis for acceptable long-term operation.

Where no relevant operating experience for a particular pump is available an avoidance strategy has to be developed using generic experience to give confidence in a method of predicting acceptable pump operating conditions.

There are four possibilities when evaluating a particular pump against a given duty and a specified (or actual) NPSH(A):

1. the pump is acceptable without a cavitation test;
2. the pump is acceptable if it passes an NPSH(3pc) test;
3. the pump is acceptable if it passes an NPSH(4mm) test;
4. the pump is unacceptable.

This "layered" approach forms the basis of the cavitation erosion avoidance strategy. The possibilities for acceptance are listed in order of increasing cost.

9.3 Guidelines for Avoiding Unacceptable Cavitation Erosion

9.3.1 *Pumps for which no cavitation test is required*

The need is to define the boundary between a "no testing" and a "testing" requirement. Reviewing the literature (Refs. 9.11 and 9.12) the claim that is made based upon past experience is that the limit of cavitation erosion performance can be expressed by a single value of suction specific speed. Table 9.5 summarises the value of suction specific speed considered appropriate.

Table 9.5

		Suction specific speed n_{ss}		
Source	Cavitating performance limit	SI units	U.K. units	U.S. units
Anderson	Failure of normal pumps	6000	**9000**	-
Anderson	Operation of normal pumps	4200	**6200**	6800
Grist	Operation of normal pumps	4300	-	-
Stepanoff	Operation of normal pumps	4800	7200	**7900**
Anderson	Operation of special pumps*	6000	**9000**	-
Stepanoff	Operation of special pumps*	6400	-	**10500**

Notes: (i) n_{ss} is calculated using values at Q_{bep}, not the duty point.
      (ii) **Bold** type indicates source data value.
*Enlarged impeller eye, double entry.

Using Table 9.5 and the operating experience obtained from U.K. power utility feedwater pumps (Ref. 9.13), a rough guide by which to calculate where cavitation erosion ceases to be a risk encompassing all centrifugal pumps is given in Table 9.6.

Table 9.6

Pump design		Suction specific speed n_{ss}		
basis	Cavitation performance limit	SI units	U.K. units	U.S. units
"Basic"	No risk of cavitation	**2500**	3700	4100
"Best"	No risk of cavitation	**4200**	6300	7000

A pump user does not have (or expect) the design data by which it might be possible to discriminate between "basic" and "best" designs of pump. Proof of acceptability must be by recourse to cavitation test data. To avoid the need for such data a value of $n_{ss} = 2500$ is used to encompass all of the designs.

An NPSH value derived using $n_{ss} = 2500$ is a pessimistically high value for any competently designed pump. However, its use is primarily as a "rough filter". If it can be shown that NPSH(A) in excess of this value is being provided then the centrifugal pump will almost certainly be free from the risk of cavitation erosion over the $50\%Q_{bep}$ to $120\%Q_{bep}$ operational range. There is no engineering reason to undertake a cavitation performance acceptance test in such circumstances.

The minimum NPSH(A) necessary to meet an $n_{ss} = 2500$ requirement and avoid having to prove cavitation performance by test is the basis of the numerical data shown in Fig. 9.17. For many applications centrifugal pumps are driven by electric induction motors with alternating current supply frequencies of either 50 Hz or 60 Hz. With a large number of such pumps having a duty flowrate less than 100 litres per second and NPSH(A) values in the 10 m to 20 m range the reason for not requiring cavitation tests for most centrifugal pumps is evident.

Guideline 1 (Applicable to all pumps)

A centrifugal pump operating in the $50\%Q_{bep}$ to $120\ \%Q_{bep}$ range is likely to be free from cavitation erosion if the suction specific speed is 2500 or less.

A cavitation test is not required for such pumps.

9.3.2 *Pumps for which an NPSH(3pc) test is required*

A cavitation test is essential where it is not practicable to provide NPSH(A) using $n_{ss} = 2500$. The simplest and least costly cavitation test is the NPSH(3pc) test. For pumps operating at speeds less than 3600 rpm it is unlikely that a more expensive form of testing will be justifiable since the possible improvements in NPSH(A) are likely to be very small numerically — a few metres at the most. For such pumps the NPSH(3pc) test is the only viable option. For speeds of 3600 rpm and over the NPSH(4mm) is a practical, albeit expensive, alternative to the NPSH(3pc) test.

The severe consequences arising from cavitation erosion damage necessitate making engineering judgements to ensure all reasonable risk of its occurrence are excluded.

The difficulties in quantifying the extent of damage caused by cavitation erosion from "typical" performance data or works tests arise from two main sources:

1. damage that is unacceptable may take several thousands of hours to appear;
2. small differences in impeller shape, particularly at the blade inlet, may give rise to significant differences in erosion rate.

Making use of published operating experience. Technical papers describing the experience gained from operating a single pump under particular conditions are a useful contribution and there are many documented in the conference proceedings of the professional institutions. However, the greater weight of operational evidence is to be drawn from experience recorded in books by leading pump designers (Refs. 9.11 and 9.12) where the cavitation erosion performance of many hundreds of cavitating pumps has been distilled and general conclusions drawn. Only one small reservation has to be borne in mind in evaluating this most valuable source of information: designers tend to define the boundaries by what is possible for their latest and best design of pumps and occasionally overstep the mark in implying this is applicable to all pumps. Other work on this subject in Ref. 9.1 reflects strongly the user approach of the 1970s. Here pessimism prevails. In the face of unexplained cavitation erosion attack such user-based publications are necessarily over cautious but as pump impeller design methodology has improved in the past decades, especially following the now common use of visual cavitation test loops insisted upon by major users, the views of designers and users are much closer.

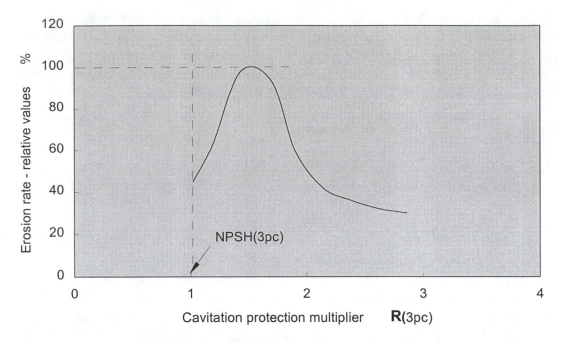

Fig. 9.18   The relative erosion rate — 3% head-drop relationship

The use of published research data.   The relationships between NPSH(3pc) and erosion rate have been established from research work on test pumps. Three findings from this work are of particular relevance.

1. The maximum cavitation erosion rate has been observed to occur above **R**(3pc) = 1 but below **R**(3pc) = 2. The test results presented in Fig. 9.18 show this. This confirms the invalidity of the old assumption that cavitation erosion is not occurring when no head-drop occurs. Whether every pump has such a high maximum **R**(3pc) value is unproven and uncertain. The important finding is that the maximum cavitation erosion rate can occur above NPSH(3pc). It seems likely that the smoother blade contours in well-designed impellers will result in maximum erosion being at a lower **R**(3pc) value, but there is no published data to support this view.

2. Erosion rate at a constant flowrate and pump speed decreases rapidly as NPSH is increased. Floriancic (Ref. 9.14) found that erosion rate varies approximately as the inverse fourth power of NPSH. Whether the fourth-power value is applicable generally is unproven and uncertain. The important finding is that the decay rate is probably rapid as NPSH is increased above the value corresponding to the maximum erosion rate.

3.   Erosion rate at constant NPSH and pump speed is a minimum value at a flowrate at, or close to, that of best efficiency conditions. Erosion rates at lower and higher flowrates increase, those at higher flowrates being more pronounced. Figure 9.19 shows this.

The uncertainties arising in applying this finding from such limited work are large, but a volume of observed operating experience has subsequently served to confirm that the trends are valid and can be applied in a generic sense.

Where provision of the NPSH to meet the impeller life objective with a wide operating range is either physically impossible or only possible at an unacceptably high cost, the limiting of continuous operation to the 80% to 110% best efficiency flowrate range can be seen to confer a significant advantage. In such circumstances operation in the 50% to 80% and the 110% to 120% best efficiency flowrate bands for short periods would be permissible.

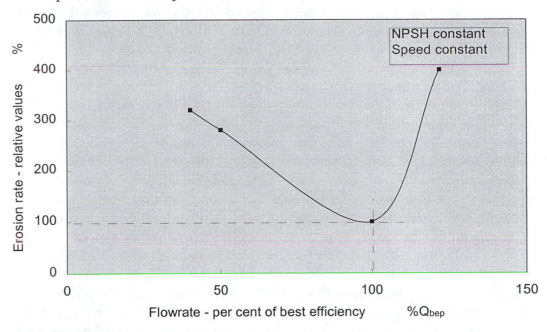

Fig. 9.19   The relative erosion rate/flowrate relationship

Cavitation erosion protection for pumps operating at speeds up to 3600 rpm.

The occurrence of maximum erosion rate has been observed (Fig. 9.18) between $R(3pc) = 1$ and $R(3pc) = 2.0$. A worst-case scenario is taken as maximum erosion rates occurring at $R(3pc) = 2.0$. Given that the erosion rate falls rapidly with the further addition of NPSH it is clear that for a competently designed pump very much reduced cavitation erosion is likely to take place above $R(3pc) = 2.5$. The $R(3pc) = 2.5$ value may be a pessimistically high value, but for pumps below 3600 rpm where NPSH(3pc) values are typically much lower than 10 m, the difference between the added risk of, say, $R(3pc) = 2.0$ and the $R(3pc) = 2.5$ is much less than 5 m. The confidence secured by an $R(3pc) = 2.5$ value makes it an appropriate value for plant design purposes and makes its proof, where applicable, by an NPSH(3pc) test appropriate for contractual acceptance purposes.

Guideline 2   **(Applicability limited to pumps up to 3600 rpm)**

A centrifugal pump operating in the $50\%Q_{bep}$ to $120\%Q_{bep}$ range is likely to be free from cavitation erosion if

$$R(3pc) \geq 2.5$$

     where the NPSH(3pc) value is that at Q_{bep}

     and NPSH(A) is the lowest value in the operating flowrate range.

An NPSH(3pc) test is required if $R(3pc)$ is chosen as the basis for protection.

Cavitation erosion protection for pumps operating at speeds over 3600 rpm.

The establishment of a single suction specific speed value for defining the limits of operation of "normal" pumps described in Refs. 9.11 and 9.12 proves to be most useful in determining minimum NPSH values. Each of these references acknowledged that improved performance associated with "special" pumps could be achieved. As statements of that time — up to 1960 — they were adequate and were taken to apply to all pumps including those running at speeds well over 3600 rpm.

In the period 1960 to 1970 numerous feedwater pumps were installed in large (500 MW and 660 MW turbine-generator set) U.K. power stations. These pumps had running speeds of between 3000 rpm and 8700 rpm. Operation was predominantly continuous at full load. That is, the pumps usually operated at close to maximum speed and best efficiency flowrate. As opportunities arose impellers were examined for cavitation erosion damage. An unexpected trend of damage with increased pump speed was observed even where an increased level of NPSH(3pc) protection was provided. The dotted line in Fig. 9.20 defines the extent of "severe" damage on these early designs of boiler feedwater pump impeller.

From a sample size of over 40 U.K. feedwater pumps only 19 were able to be fully evaluated and, of these, 8 repeated (and confirmed) findings on identical pumps. This is mentioned to demonstrate how difficult it is to obtain information on impeller cavitation erosion rates even in a well-organised survey. Operational constraints imposed by the need for power generation and the programming of maintenance inspections in synchronism with other plant outages mean that the examination of pump impellers in this type of survey has to be opportunistic. Whilst the results of this work presented in Fig. 9.20 may seem few numerically, they represent significant and high-quality observations.

The NPSH(A) to prevent unacceptable cavitation damage to well-designed pump impellers is shown by Fig. 9.20 to be much greater than $\mathbf{R}$(3pc) = 2.5. Chapter 3 showed that NPSH(3pc) varies with the square of the pump speed. It is evident, therefore, that based on an NPSH(3pc) assessment, a very large NPSH will be required to ensure that "damage free" operation is achieved for continuous operation down to 50%Q_{bep}. A guideline which meets such practical needs, i.e. continuous operation anywhere in the 50%Q_{bep} to 120%Q_{bep} range, requires that a compromise be made.

"Damage free" is too onerous. The criterion for "failure" must be reduced from one in which an impeller is expected to remain free of cavitation erosion for its lifetime (approximately 350,000 hours — 40 years continuous operation) to one where it is treated as expendable. When applied to a major capital plant the word expendable means a life in excess of 45,000 hours — five years continuous operation — so that a strategy of biennial inspection can give security to plans based on an impeller change in the fourth year of operation. This allows on-going damage monitoring and assessment of the actual erosion rate which may, of course, turn out to be significantly different from that forecast. In taking what is essentially a riskier strategy the cost reduction associated with a lower NPSH(A) is achievable.

The review of the cavitation erosion damage characteristics in section 9.1 indicates that the expendable impeller strategy has validity when the type of damage is limited to forms of cavitation erosion that can be considered

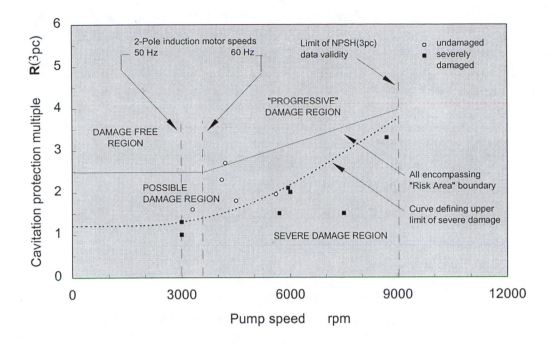

Fig. 9.20 Cavitation erosion data for good designs of boiler feedwater pump (operation in 50%Q_{bep} to 120%Q_{bep}; water temperature 150°C to 185°C; damage inspections at approximately 20,000 hours)

"progressive". This enables a planned approach to maintenance intervals and the component changes which have to be made. Cavitation erosion on either the "suction" or the "pressure" face of impeller blades is regarded as progressive.

The concept of acceptable erosion necessitates the criteria for "failure" to be explicitly linked to a given running period. A definition appropriate to pump selection in the 3600 rpm to 9000 rpm range is as follows:

"Impeller life should be at least 45,000 hours when operating in the 50% to 120% best efficiency flowrate band, the criteria for failure being when a piece of metal of area greater than 1cm^2 becomes detached or a reduction of 5% in generated head occurs."

The boundary for "progressive" damage in Fig. 9.20 enables the minimum allowable NPSH(A) value to be determined for impellers treated as expendable to the above definition. The evidence available supports the validity of providing NPSH(A) given by equation 9.1 up to 9000 rpm.

$$\mathbf{R}(3pc) = 1.5 + \frac{n}{3600} \quad \text{---------------------------------------9.1}$$

Note that $\mathbf{R}(3pc) = 2.5$ at 3600 rpm, and $\mathbf{R}(3pc) = 4.0$ at 9000 rpm.

An important feature of Fig. 9.20 is that the erosion damage data recorded occurred on impellers designed to very exacting standards by major manufacturers. Whilst the cavitation erosion damage reported engendered further improvements to impellers shown in the "severe damage region", it must be appreciated that these were more than just "competently designed". Caution must be exercised in extending these data to a wider range of impeller designs.

The uncertainties in cavity collapse in high velocity flows, the evidence which suggests the greater intensity of attack and the high cost of providing NPSH make the NPSH(3pc) test inappropriate for pump speeds above 9000 rpm. Indeed, the cost of NPSH(A) based on NPSH(3pc) data may on its own make this test inappropriate for pumps in the 3600 rpm to 9000 rpm range. A more costly NPSH(4mm) test to provide the necessary assurance that a lower NPSH(A) is acceptable may become a commercially attractive option.

**Guideline 3   (Applicability limited to pumps in
                  the 3600 rpm to 9000 rpm range)**

A centrifugal pump operating in the $50\%Q_{bep}$ to $120\%Q_{bep}$ range might encounter cavitation erosion which is limited to an acceptable "progressive" form if

$$\mathbf{R}(3pc) \geq 1.5 + \frac{n}{3600}$$

where the NPSH(3pc) value is that at Q_{bep}
and NPSH(A) is the lowest value in the operational flowrate range.

An NPSH(3pc) test is required if $\mathbf{R}(3pc)$ is chosen as the basis for protection.

9.3.3 *Pumps for which an NPSH(4mm) test is required*

For higher speeds the NPSH(3pc) test may well be adequate, but the high NPSH(R) resulting from the pessimism inherent in its use of generic data needs to be evaluated against an NPSH(4mm) test to confirm that it is cost effective.

Studies of the links between NPSH(4mm) and cavitation erosion have been conducted by observation of cavitation within impellers using special test pumps. These have usually consisted of a transparent casing and an impeller made either of transparent materials or with a transparent shroud. The limitations imposed by materials suitable for the pressures contained by such casings generally limit test rig pump speeds to 3600 rpm maximum and the test liquid to water at a temperature in the range 2°C to 50°C. Very special rigs for research purposes have on occasion been designed for higher speeds, hotter water, or other liquids, but for contractual tests, where some scaling up of cavitation performance is in any case inevitable, a visual cavitation test on clean cold deaerated water at speeds well below 3600 rpm is an acceptable norm.

Being able to see cavitation reduces uncertainty. However, it is always difficult to quantify the proximity of cavitating flows to impeller surfaces. This means any link between cavitation and cavitation erosion damage is implied. It is not directly measured. By running damaged impellers it has been visually observed that cavitation erosion damage occurs, not unexpectedly, where the cloud of cavities collapse. A further observation, which is now widely accepted, is that the collapse of bubbly flows produces more damage than smooth sheets of cavitation. It is questionable whether bubbly flows correspond to a cavitating vortex collapse although the damage produced is often very similar.

Applying a coating to the impeller surfaces can help distinguish between where damage is provoked by cavity collapse and where cavities collapse harmlessly some distance away from impeller surfaces thereby not attacking the impeller material. This is often used by manufacturers "in house" to give confidence to supporting design guarantees in service. Poor control of the coating process can easily lead to erroneous results making such tests unacceptable as a basis for a contractual "proof of cavitation erosion" performance test except in very special cases.

Visual tests can be used to distinguish between the presence of cavitation and no cavitation. Against even simple background lighting from stroboscopic flashes the silvery stream of cavities is easily detectable. The assumption that no visible cavitation equates with no risk of unacceptable cavitation erosion clearly applies.

A visual cavitation test can determine the NPSH level that suppresses cavitation completely. The "cavitation-free" zone on the performance characteristic has, as simple logic would deduce, a minimum value close to best efficiency flowrate. At either side of this flowrate value, liquid encounters the fixed impeller blade at a less favourable angle. Vortices generated as the liquid passes the blade create low-pressure regions on the "downstream" side, which is conducive to cavitation and thereby necessitate a higher NPSH to suppress it.

Flowrates up to best efficiency point flowrate are characterised by the effect of a low inlet velocity meeting rotating impeller blades. Even allowing for induced pre-rotation, cavitation occurs on the side of the blade commonly termed the "suction" side in the direction of rotation. Such cavitation results in damage which is easily visible, especially where blades are twisted to better meet the flow entering the impeller eye. Most pumps operate below the best efficiency flowrate as a result of plant designers adding "margins" when calculating the required duty.

Cavitation erosion sometimes occurs on the "pressure" side of impeller blades where a high inlet velocity associated with flowrate greater than that at best efficiency point prevails. Such cavitation and the resultant damage are very difficult to observe. Reflections from polished surfaces and damage inspection of a dismantled pump using mirrors are often employed. In the extreme examples shown in Figs. 9.9 and 9.10 erosion damage has broken through from the pressure side.

The definition of a cavitation-free zone by the first occurrence of observed cavitation is open to a wide variation of interpretations. A practical alternative is to define a cavitation performance characteristic by a measurable amount of cavitation then apply a margin to this to ensure cavitation is suppressed. Chapter 4 explained that a 4 mm cavity measured in the general direction of flow was appropriate.

In 1974 it was concluded (Ref. 9.1) that to avoid cavitation problems at least 1.05 times the visual appearance NPSH(4mm) value is required for pumps running at or below 3000 rpm. Subsequent experience with power industry high-speed pumps strongly suggested that this margin could be safely applied to pumps of all speeds for flowrates in the $50\%Q_{bep}$ to $120\%Q_{bep}$ flowrate range. At the practical observation length of 4 mm the cavitation zone cavities do not have seriously destructive properties and the extra 5% on NPSH(4mm) reduces the chance of damage to being insignificant. The use of a centrifugal pump at NPSH(A) at less than **R**(4mm) = 1.05 is commercially an unacceptable risk with respect to cavitation erosion.

Guideline 4 (Applicability limited to pumps over 3600 rpm)

A centrifugal pump operating in the $50\%Q_{bep}$ to $120\%Q_{bep}$ range is likely to be free from cavitation erosion if

$\mathbf{R}$(4mm) $=$ 1.05 at all operational flowrates.
where NPSH(4mm) and NPSH(A) values are compared throughout the operational flowrate range.

An **NPSH**(4mm) test is required to demonstrate validity if $\mathbf{R}$(4mm) is chosen as the basis for protection.

9.4 Cavitation Erosion — An Avoidance Strategy

9.4.1 *The tabular presentation*

The strategy is presented in tabular form. Table 9.7A lists all the data needed. Table 9.7B details an assessment procedure that is appropriate to most pumps. Those pumps falling outside the conditional statements in the procedure are considered afterward in section 9.4.2.

Table 9.7A   Data collection

Information required to commence an assessment

Plant design data (supplied by plant designer or user)
(i) Q_G and ΔH_G (the duty point)   (see Chapter 6)
(ii) The operating flowrate range
(iii) NPSH(A) throughout the operating range

Pump design data (supplied by pump manufacturer)
(iv) *All pumps* n and Q_{bep} *
(v) *(Stage 2 only)* NPSH(3pc) throughout the operating flowrate range
(vi) *(Stage 3 only)* NPSH(4mm) throughout the operating flowrate range

---

*   For early plant design studies and the preparation of pump specifications an estimate of the pump speed n and flowrate Q_{bep} is required. Chapter 3 provides a basis for doing this. Chapter 12 shows how it can be done.

The assessment process is "layered" in three stages. The stages are based on the four guidelines developed in section 9.3. The objective is to confirm the acceptability of a chosen pump to operate without a risk of unacceptable cavitation erosion damage. Proceed through Table 9.7B until either negligible risk (and any test requirement necessary to prove it) is established or the pump chosen for evaluation is rejected.

Table 9.7B   The assessment procedure

Assessment

NPSH(R) is the minimum value of NPSH(A) for cavitation erosion avoidance.

(See equation 4.1)

Stage 1

1.   All pumps

(a)   Calculate n_{ss} using NPSH(A) at Q_{bep}

(b)   If $n_{ss} \leq 2500$

then cavitation erosion is a negligible risk* $\left[\begin{array}{l}\text{Impeller life: unlimited} \\ \text{Test requirement: none}\end{array}\right]$

(c)   Should n_{ss} be greater than 2500 then proceed to Stage 2

Stage 2

2(A)   For pump speeds up to 3600 rpm

(a)   Calculate **R**(3pc) based on NPSH(3pc) at Q_{bep} and the lowest NPSH(A) in the operating flowrate range

(b)   If **R**(3pc) $\geq$ 2.5 at Q_{bep}

then cavitation erosion is a negligible risk* $\left[\begin{array}{l}\text{Impeller life: unlimited} \\ \text{NPSH(3pc) test required*}\end{array}\right]$

2(B)   For pump speeds between
3600 rpm and 9000 rpm

(a)   Calculate **R**(3pc) based on NPSH(3pc) at Q_{bep} and the lowest NPSH(A) in the operating flowrate range

(b)   If **R**(3pc) $\geq 1.5 + \dfrac{n}{3600}$ at Q_{bep}

then cavitation erosion is a negligible risk* $\left[\begin{array}{l}\text{Impeller life: 45,000 hours min} \\ \text{NPSH(3pc) test required*}\end{array}\right]$

(c)   Should **R**(3pc) be less than the value required in either 2(A) or 2(B) proceed to Stage 3

Stage 3

3   All pumps not cleared by 2(A) or 2(B)

(a)   Calculate **R**(4mm) based on NPSH(4mm) and NPSH(A) at all flowrates throughout the operating range

(b)   If **R**(4mm) ≥ 1.05 x NPSH(4mm) for all flowrates in the operating range then cavitation erosion is a negligible risk* $\left[\begin{array}{l}\text{Impeller life: 45,000 hours min} \\ \text{NPSH(4mm) test required*}\end{array}\right]$

(c)   Pumps failing to meet the above criteria risk being unacceptably damaged by cavitation erosion. They should either be rejected or the risk of a shorter life span accepted

* Applies to flowrates in the $50\%Q_{bep}$ to $120\%Q_{bep}$ operating range.

9.4.2 *Pumps required for operation outside the conditional statements*

Pump operation outside the 50%Q$_{bep}$ to 120%Q$_{bep}$ band. The added risk of cavitation erosion outside the 50%Q$_{bep}$ to 120%Q$_{bep}$ range is shown qualitatively in Fig. 9.19. The conservatism in the **R**(3pc) margin will be lost when operation extends outside this band. Visual tests show that the operation of centrifugal pumps at flowrates less than 50%Q$_{bep}$ is most undesirable. The chaotic and unstable flows shown in Chapter 7 mean that even high levels of NPSH(A) are uncertain to give protection. In practice many of the pumps that are called upon to run at such low flows for any lengthy period of time suffer from non-cavitation related mechanical design problems (axial thrust bearing failures, excessive vibration, etc.) which higher NPSH(A) will not address.

Of course, there are operational circumstances in which pumps must run at low %Q$_{bep}$ values. The high risk of either mechanical damage from cavitation or from other forces acting within the pump must then be accepted. To reduce this risk to a minimum it is important that through the specification for a new pump the pump manufacturer is made aware of the intent to operate under such conditions. More mechanically robust pump designs can often be supplied. As will be shown in Chapter 12, it is sometimes worth considering multiple pumps operating in parallel as a means of raising the %Q$_{bep}$ minimum value for an individual pump.

References

9.1   Grist, E., Nett positive suction head requirements for avoidance of unacceptable cavitation erosion in centrifugal pumps, IMechE Cavitation Conf., pp.153 – 162, 1974.

9.2   Kuhn de Chizelle, Y.P., Brennan, C. E., Ceccio, S. L. and Gowing, S., Scaling experiments on the dynamics and acoustics of travelling bubble cavitation, IMechE Cavitation Conf., pp.165 – 170, 1992.

9.3   Garcia, R., Hammitt, F. G. and Nystrom, R. E., Comprehensive cavitation damage data for water and various liquid metals including correlations with mechanical and fluid properties, ASTM Special Tech. Pub. 408, pp.239 – 279, 1967.

9.4   Knapp, R. T., Daily, J. W. and Hammitt, F. G., Cavitation, Chapter 9 - Evaluating Resistance of Materials to Cavitation Damage, pp.362 – 443, McGraw-Hill, New York, 1970.

9.5   Lush, P. A. and Ewunkern, A. E., Cavitation erosion of engineering materials IMechE Cavitation Conf., pp.95 – 102, 1992.

9.6   Steller, J. K., International cavitation erosion test - summary of results, IMechE Cavitation Conf., pp.121–132, 1992.

9.7   Hobbs, J. M., Practical aspects of cavitation, Phil. Trans., pp.267 – 275, 1966. 9.8ASTM Standard G 134-95, Standard Test Method for Erosion of Solid Materials by a Cavitating Liquid Jet, 1995.

9.9   Momma, T. and Lichtarowicz, A., Some experiments on cavitation damage produced by a submerged jet, ASME Nuclear Engg. Conf., Vol. 2, pp.877 – 884, 1993.

9.10   Momma, T. and Lichtarowicz, A., A study of pressures and erosion produced by collapsing cavitation, Wear, Vol. 186 and 187, pp.425 – 436, 1995.

9.11   Stepanoff, A. J., Centrifugal and Axial Flow Pumps, Second edition, p. 255, Wiley, New York, 1957.

9.12   Anderson, H. H., <u>Centrifugal Pumps</u>, p 26,
       Trade & Technical Press, Sutton, U.K., 1962.

9.13   Grist, E., <u>Advanced Class Boiler Feed Pumps – Operational Experience</u>,
       BPMA Tech. Conf., pp.1 – 21, 1981.

9.14   Floriancic, D., <u>Experimentalle untersuchungen an einer pumpe zur
       feststellung anderung der saugfahigkeit durch oberflashenraughigkeit und
       durch heisswasserforderung</u>, Thesis 4406, E T H Zurich.

Chapter 10

Centrifugal Pump Low-Flow Protection

10.1 Functional Requirements

Every centrifugal pump must have low-flow protection. The primary purpose of this protection is to prevent low-flow conditions arising which cause vapour locking. This is achieved by ensuring that either (i) the pump is shut down safely before it can happen, or (ii) a flow route back to the inlet side of the pump is opened up which prevents the flowrate through the pump falling below a prescribed level. The minimum continuous flowrate Q_v necessary to prevent vapour locking is typically much less than $5\%Q_{bep}$.

Where a flow route back to the pump inlet is provided this is described as a leak-off system. A typical leak-off system is shown in Fig. 10.1. In other technical literature it is sometimes called a recirculation system or a bypass system.

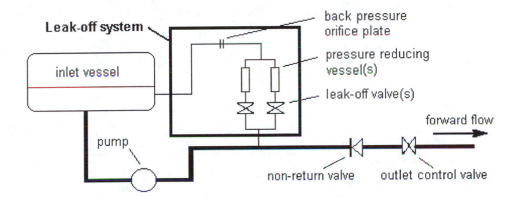

Fig. 10.1   A typical leak-off system for a high-energy centrifugal pump

Where a leak-off system is part of the plant design it is often convenient to use it to provide protection against other unacceptable low flowrate related operating conditions such as

(i) cavitation surging — hydrodynamically induced (see Chapter 7),

(ii) cavitation surging — thermodynamically induced (see Chapter 8),

(iii) cavitation erosion damage (see Chapter 9).

The minimum continuous flowrate to prevent (i), (ii) and (iii) is typically between $20\%Q_{bep}$ and $50\%Q_{bep}$. The leak-off system design specification should take a value for the minimum continuous flowrate Q_{lo} based on the greatest of these secondary protection requirements.

181

Many years ago the need for low-flow protection was perceived as being to prevent "overheating". Whilst at that time some pumps included components vulnerable to a rise in temperature above the normal liquid temperature this was not usually so. However, the outward sign of distress to a pump close to or at zero flowrate was the increased temperature of the pump casing. Prevention was seen to be effected by stopping the pump before a predetermined temperature "high" was reached. The term "overheating" became used generally. This legacy of the past is now almost always a euphemism for vapour locking.

It is prudent to include at an early stage of design a transitional flowrate band of up to $10\%Q_{bep}$ as illustrated in Fig. 10.2. The lower limit of this band defines the flowrate Q_{tr}. Its function is to cater for the short period of operation that occurs whilst the leak-off valve is in the process of opening. Prior allocation of this band allows preliminary leak-off system design calculations to proceed knowing the minimum permissible continuous flowrate in the pump operating range is always Q_{lo}.

In determining the flowrate values for secondary protection requirements the cavitation surge limit — the point where undesirable surges can be triggered — should relate to the minimum transient flowrate Q_{tr}. The cavitation erosion value relates to the minimum continuous flowrate Q_{lo} since short periods below this value will produce insignificant cavitation erosion damage.

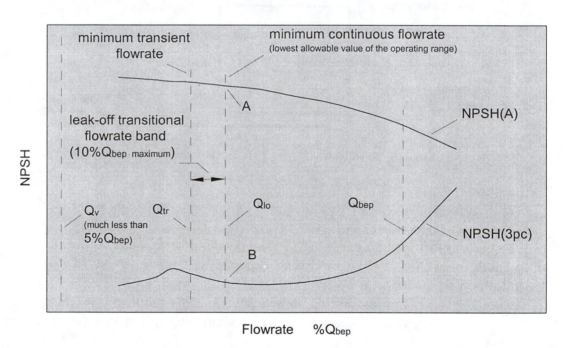

Fig. 10.2 Quantifying NPSH(A) and NPSH(3pc) values at Q_{lo}

Historically a wide range of Q_{lo} values have been tried. Pump damage was the consequence in too many cases where values less than $20\%Q_{bep}$ were chosen and long periods of running at leak-off or very low forward flow occurred. Such unacceptable operational experience led to a consensus of opinion that the value of Q_{lo} should be increased. It is now considered prudent to ensure that, irrespective of all calculated requirements, it is never less than $20\%Q_{bep}$.

Choosing a leak-off system which ensures that the flowrate never falls below a minimum value much higher than that necessary to prevent vapour locking also has an additional benefit. It ensures that the mechanical integrity of the centrifugal pump is less likely to be challenged by the large hydraulically induced radial loads on the impeller which often accompany very low flowrate operation.

The secondary purposes are clearly very important but they are only secondary. Prevention of vapour locking is the main aim.

10.2   Choosing an Appropriate Low-flow Protection System

10.2.1   *Quantifying the options*

Defining the problem.   The propensity of a centrifugal pump to vapour lock needs to be expressed in measurable terms to enable quantitative guidelines to be drawn up that relate to low-flow protection system choice. As stated earlier, the essential requirement of low-flow protection is that it must either allow the pump to be shut down safely or establish a secondary flowpath within a given time. It is the time available for action that determines whether pump shut down is a practical possibility or, if not, which types of leak-off system are appropriate. Cost considerations are then used to make the final choice where competing leak-off systems are available. Additional security can always be obtained by paying more. The need is to understand the risk so that an appropriate level of protection is provided commensurate with minimum expenditure.

Four low-flow protection system categories are described in section 10.2.2. These categories derive from operational experience and a simple analysis of essential factors. More complicated systems can be developed from them. However, as might be expected, operational experience shows it is advantageous to keep the design simple wherever possible.

Low-flow protection basics.   The minimum acceptable time for low-flow protective action to be taken is that which just avoids the risk of vapour locking. This is the time taken for the NPSH to fall from NPSH(A), the value at the outset, to the NPSH necessary to protect against vapour locking. Equation 6.1 defines the minimum value of NPSH for vapour lock protection. For a given flowrate and pump speed this gives

maximum permissible ΔNPSH = NPSH(A) - 1.3 NPSH(3pc)   ----10.1
reduction resulting from
inlet pipe transients

The maximum permissible change in vapour pressure Δp_{vap} can be calculated using equation 10.2:

$$\Delta p_{vap} = \Delta NPSH \; \rho_L \; g$$
$$\Delta p_{vap} = \{NPSH(A) - 1.3 \; NPSH(3pc)\} \; \rho_L \; g \quad -----10.2$$

The values of NPSH(A) and NPSH(3pc) appropriate to leak-off protection system design are those at the minimum continuous flowrate Q_{lo}. They are quantified by Points A and B in Fig. 10.2. Chapter 4 shows how to put a value to NPSH(A). The NPSH(3pc) curve is obtained from test data.

Calculating trip temperature T_2. The temperature T_2 of the liquid at the pump inlet which brings about the change in Δp_{vap} is obtained using thermodynamic property tables (e.g. steam tables — Ref. 10.1). This is shown graphically as curve C-D in Fig. 10.3. Point C is fixed by the normal pump liquid temperature T_0. Adding Δp_{vap} to the value of p_{vap} at point C fixes point D. The maximum permissible temperature T_2 can then be read off the graph. Where a pump trip on "temperature high" is provided the trip temperature is T_2.

Important note: The use of a temperature trip should not be the sole means of protection. It is hydraulically very crude. Industrial systems based on temperature sensors alone have often proved unreliable. The main deficiency is the assumption that the NPSH(A) value is always as calculated. Should an unexpected inlet pipe depressurisation occur — say, due to a blockage — the maximum ΔNPSH value from equation 10.1 is considerably reduced. Just at the time it is needed the trip setting temperature is invalid.

A temperature trip can be a practical, relatively low-cost, but occasionally undependable, "catch-all". It should be relegated to protection support. It should not be used as the sole means of protection except where the pump itself can be considered expendable.

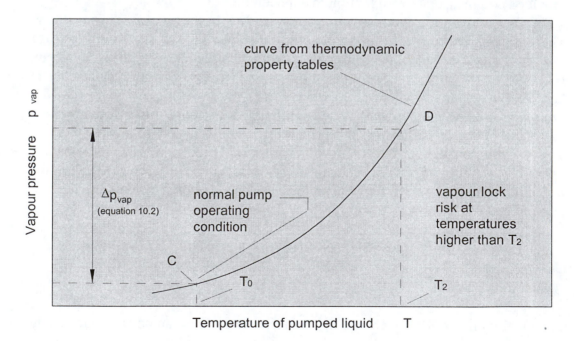

Fig. 10.3   Determination of the maximum permissible temperature T_2

Another problem with using temperature values as a basis for setting NPSH protection was highlighted in Chapter 5 and is evident in Fig. 10.3. The oversensitive nature of the signal is very evident when it is used to provide protection for a pump when the liquid is one such as very hot water where vapour pressure changes rapidly with temperature. Too low a setting by 0.1°C can overprotect: too high by 0.1°C can put the pump at risk.

Calculating the maximum permissible leak-off opening time t_0 . The time t_0 taken to heat the liquid to temperature T_2 by the pump input power loss is the maximum time within which the leak-off system must operate. At the simplest level a good approximation of the heat balance in a centrifugal pump at zero flowrate and constant pump speed is given by the following:

heat from pump   = change in heat content of
input power in time t   the pumped liquid

$$P_{AO} t = m \, C_p \, (T - T_0)$$

The time t_2 to reach the limiting temperature T_2 is given by

$$t_2 = \frac{m \, C_p \, (T_2 - T_0)}{P_{AO}} \qquad \text{------10.3}$$

where   m = the mass of liquid churned by the impeller
at zero flowrate ($\rho_L V_c$)

10.2.2 *The categories of low-flow protection system*

In normal practice a pump is required to respond promptly to a rapid decline in demand for forward flow. Low-flow protection systems can be broadly categorised by a measure of the time it takes to do this. Rather than try to describe each system individually (and risk getting it wrong) it is prudent to seek a "worst case" that will encompass all possibilities. This is achieved by assuming the pump is at zero flowrate and all the input power goes into heating the liquid churned around by the impeller. Leakage losses, thermal radiation to the surrounding environment, and heat carried away by liquid as the outlet valve reduces flowrate from Q_{lo} to zero are all ignored.

At zero flowrate the ratio **p**ump **i**nput **p**ower to **i**mpeller **chu**rned **m**ass, for convenience named here the **pipichum** value, is an appropriate all-encompassing value.

As should be expected this power/mass ratio is a measure of the time it takes the zero flowrate value of input power P_{AO} to reach the maximum allowable temperature T_2 after a pump is "dead ended". For a particular liquid the pipichum value can be used, in broad terms, to determine if a leak-off system is appropriate for a particular pump by quantifying whether it can respond in time:

pipichum value φ = $\dfrac{\text{pump input power}}{\text{impeller churned mass}}$ = $\dfrac{P_{AO}}{m}$ = $\dfrac{C_p \, (T_2 - T_0)}{t}$ ------10.4

Table 10.1 The four categories of low-flow protection system

Category Pipichum value φ kW/kg		System design			Comments
		Design basis	Criterion for operation	Method	
Manual systems					
A	below 0.1	Operator intervention	Time to vapour lock after single* valve closure	Operator intervention required before time limit is exceeded	Subject to human error
B	0.1 up to 1	Operator intervention	Time to vapour lock after single* valve closure	Operator intervention required before time limit is exceeded	Subject to human error
		Back up temperature alarm set at T_2	Temperature at NPSH(A) minimum acceptable value	Failure to intervene results in a pump trip at T_2	Back up only
Leak-off systems					
C	1 up to 100	Leak-off system always maintains a minimum flowrate	Minimum allowable flowrate through impeller(s)	Open leak-off line at Q_{lo}	Automatic leak-off protection
		Back up temperature alarm set at T_2.	Temperature at NPSH(A) minimum acceptable value	Failure to intervene results in a pump trip at T_2	Back up only
D	100 plus (high energy pumps)	Leak-off system always maintains a minimum flowrate	Minimum allowable flowrate through impeller(s)	Open leak-off line at Q_{lo}** (Duplicate to achieve reliability if necessary)	Automatic leak-off protection

* Simultaneous isolation at inlet and outlet with a running pump is not allowed
     (see Chapter 8).
** A pump trip provides ineffective protection (vapour lock and consequential damage will probably occur before run-down is complete).

Note: As the temperature rises the pump input power and the liquid density of the churned mass both decrease in the same proportion; the pipichum value of a non-cavitating pump remains the same.

Table 10.1 requires pump users to determine a pipichum value. Accuracy is not essential. The objective is to distinguish between groups that are at least one order of magnitude apart. A way of doing this that is easily applied by pump users *at the specification production stage* is given by:

$$\text{pipichum } \varphi = \frac{\text{pump input power}}{\text{impeller churned mass}} = \frac{P_{AO}}{\rho_L \tau (\pi D b)} \qquad \text{----------------10.5}$$

where P_{AO} = pump input power at zero flowrate,
τ = churned mass multiplier,
D = impeller outer diameter,
b = impeller flow passage width at outlet.

Equation 10.5 should only be used as a guide. If necessary a better figure for the impeller churned mass can be substituted when the geometry of the proposed pump is known. Prior to this equation 10.5 can be used with the following:

(i)   The power at zero flowrate is shown in Fig. 3.3(b) to be between 25% and 75% of the value at Q_{bep}. A mean value of 50% is chosen:
$$P_{AO} = 0.5 P_A \text{ at } Q_{bep}$$
(ii)   Impeller flow passage width at outlet = b:
$$b = \frac{Q_{bep}}{\pi D v_1}$$

The velocity of liquid in the inlet and outlet pipework of a centrifugal pump impeller is typically less than 3 m/s. Higher values lead to excessive pipe friction losses. Impellers are designed so that the liquid passes through without any abrupt change in velocity. The radial velocity of liquid being ejected from the impeller, sometimes referred to as the meridional velocity, is near to 3 m/s. Higher values are found in special pump designs. An all-encompassing value of b can be obtained for a large number of pumps by assuming the radial velocity is equal to the inlet pipe velocity and is 4 m/s. A value of $v_1 = 4$ m/s is chosen.

(iii) The impeller churned mass is quantified as a multiple of the volume swept by the outlet blade edge. The liquid within the impeller and that in the pump casing/impeller space are clearly disturbed. Also, some liquid is drawn into the impeller at its inlet and outlet by local circulations. It is difficult to put a value on the total mass involved but fortunately a rough approximation is adequate for the purpose at hand. Examination of a number of pump construction arrangements shows that two times the area swept by the impeller πDb is probably representative. A value of $\tau = 2$ is chosen.

Combining (i), (ii) and (iii) gives: $\text{pipichum } \varphi = \dfrac{P_A}{\rho_L Q_{bep}}$ ------------10.6

Pipichum values are not dimensionless. It is only appropriate to use the pipichum values given in Table 10.1 when the measurement units are P_A in kW, ρ_L in kg/m^3, and Q_{bep} in litres per second.

10.3 Leak-off System Design

A review of features that must be considered in a leak-off system design is presented. The aim is to provide guidance on the design issues that are important to avoid cavitation vapour locking and where, with applied diligence, improved performance might be obtained which could be demonstrated by test.

The three flow paths of interest are shown in Figs. 10.4 and 10.5. They are

     (i)   the pump flow path   (containing flowrate Q),

     (ii) the leak-off flow path (containing flowrate q_{lo}),

     (iii) the forward flow path (containing flowrate q_{fo}).

To avoid ambiguity care is taken to refer to the descriptors (i), (ii) and (iii) for flow paths and flowrates as appropriate when describing leak-off system design.

Leak-off provision is described by the flowrate value Q_{lo}. It is convenient to express this as a percentage of best efficiency pump flowrate. Thus a "30% leak-off system" describes a system with a flowrate $Q_{lo} = 30\%Q_{bep}$ when the leak-off alone is open.

10.3.1 *The basics*

Having established a value for the minimum continuous flowrate Q_{lo} taking account of Chapters 7 and 9 consideration should be given to recognising the following general points before proceeding to more particular aspects of leak-off system design.

Opening and closing time requirements. The purpose of the leak-off valve is to establish flowrate Q_{lo} as a secondary flow path in the leak-off pipework by opening promptly. It is required to operate and establish an adequate secondary flow path within the minimum time limit set by equation 10.3. Opening too slowly risks vapour locking.

Conversely closing a leak-off valve very quickly can present a risk of water hammer. However, unlike opening there is normally no functional need for rapid closure. A pump builds up to its full flowrate in a time scale determined by the speed pickup of the prime mover. For electric motor driven pumps this is usually very quick — a matter of a few seconds at the most. The total volume of liquid lost by recirculation back to the inlet vessel due to inclusion of a leisurely leak-off valve closing time measured in many tens of seconds is insignificantly small.

For leak-off system design the functional ideal is open quickly, close slowly.

Fail-safe protection. If the pump flowrate falls below Q_{lo} whilst the leak-off valve is closing it should always immediately go to the fully open position again.

For leak-off system design the safety function ideal is always fail open.

Leak-off valve position protecting a stationary pump. The position of the leak-off valve when the pump is at rest needs to be considered. For single pump systems there is no problem. Here, as the pump flowrate is reduced below Q_{lo} either by controlled flowrate reduction or by stopping the pump, the valve opens and remains open until called upon to close again by a flowrate dependent signal. However, where the pump has an installed spare (usually referred to as a standby pump) and the incoming liquid is supplied by a booster pump the "stand open" position for the leak-off valve on a stationary pump is very wasteful. A considerable amount of power is lost by the booster pump recirculating liquid through the leak-off line of the stationary pump back

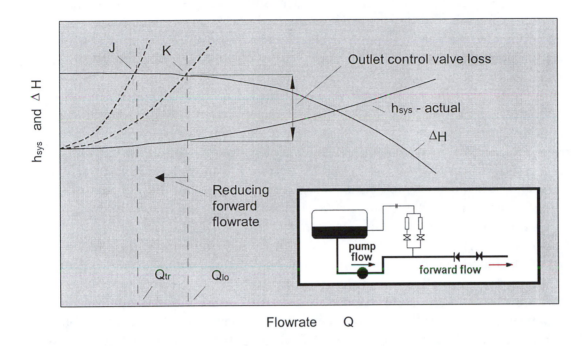

Fig. 10.4 Extreme of leak-off functional requirement — forward flow only

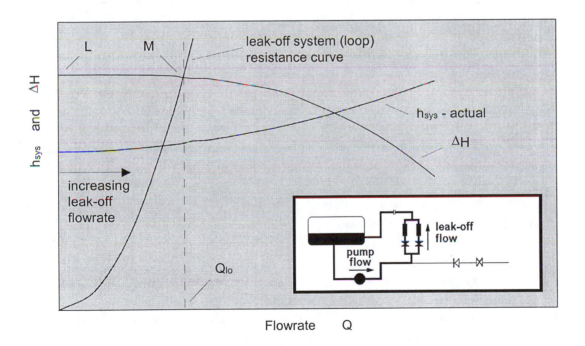

Fig. 10.5 Extreme of leak-off functional requirement — leak-off flow only

to the inlet vessel. In this case it is necessary to add to the operating procedures and, where a control system is used, to the logic diagram, a requirement for the leak-off line to be closed on "pump shaft stationary" and reopened prior to a pump start. Failure to open the valve on pump start will, of course, lead to immediate vapour locking so very reliable procedures are required.

Maintenance procedures. It is very easy to forget that the leak-off system isolating valve must be re-opened following maintenance on the pump. Running of an unprotected pump soon results in vapour lock and pump seizure. Lockable (open) isolating valves and/or enforceable maintenance procedures are essential.

Leak-off valve seat design. Nearly all the service life of a leak-off valve is spent in the closed position. Its seat forms a seal against a pressure drop approximating to the pump generated pressure. A seat design that resists wire drawing and cavitation erosion attack is a functional need.

Essential redundancy. The major consequences of pump vapour locking are the cost of repair and the cost of lost production. The latter can be significant where no standby pump is installed. Quantifying the reliability of a leak-off system — the hardware, the software (computer control program), and the contribution made by human error — to give a sound basis for choice is commercially impracticable in all but the most exotic applications. The amount of system redundancy — items/subsystems that are automatically replaced when a fault occurs so that continuous operation is maintained — has to be decided by a more pragmatic route. The question posed has to be turned round from "How can absolute performance be estimated?" to "What level of plant failure can be tolerated?" The cost of failure gives the answer, and an approximate value for it must be established. Added security costs more. The degree of security that is justified has to be measured against the additional cost incurred should the existing protection be ineffective.

Given that a decision to install a leak-off system has been made and that a single failure is likely to be intolerable, a double train system (e.g. Fig. 10.1) needs to be costed and, if justifiable, added to the plant design. Included in this decision must be consideration of the need to provide independent duplication of (i) flow signal pressure switches, (ii) flow signal measurement take-off points, and (iii) leak-off actuator power supply.

It is common practice for leak-off valves to be opened on receipt of a signal generated by closure of a single pressure switch when it senses low flowrate. The protection provided is totally dependent upon this signal being reliable. Duplication may be necessary to provide protection against inadequate signal — due to blockages in the signal lines, etc. and electronic signal acquisition failure — spurious or zero value. It is worth considering two independent locations for signal acquisition (e.g. orifice tappings 180 degrees to each other) and electronic interrogation which compares the two readings and raises an alarm when there is a disparity greater than, say, 10%. A single signal provides adequate protection for most pumps. Adding a second signal is usually a low-cost method of improving security where this is not the case.

10.3.2 *Aggregating the flowrate*

As the leak-off system is brought into use the transition to be made is from Fig. 10.4 to Fig. 10.5. During this transition from "forward flow only" to "leak-off flow only" the pump flowrate is only a little less than the sum of forward flowrate and leak-off flowrate. This is because fortuitously the generated head for a centrifugal pump in both the transitional range (Fig. 10.4 J to K) and in the leak-off range (Fig. 10.5 L to M) is almost constant. To a first approximation little error is incurred by adding the flowrates in the two systems together. This is because the generated head driving the liquid through the two systems at the total pump flowrate is reduced by only a small amount, usually much less than $15\%\Delta H_{bep}$.

A major benefit that accrues from the above is that both (i) the time to reduce forward flowrate from Q_{lo} to Q_{tr} with the leak-off shut and (ii) the time for the leak-off flowrate to become Q_{lo} during which time forward flowrate is approximately Q_{tr} can be demonstrated by a relatively simple test without putting the mechanical integrity of the pump at risk. Design and/or manufacturing errors often occur which result in a leak-off system operating at a much lower flowrate than intended. When an error is suspected it is useful for all concerned to know that site tests can eliminate this possibility by confirming that the leak-off flowrate is the "as designed" value.

10.3.3 *Setting the set points*

Background considerations. There are a great number of possible ways to operate a leak-off system. As might be expected the less complicated systems tend to be the most reliable. This section describes systems based on simple on/off power operated designs of leak-off valve. Modulating leak-off valves are dealt with in section 10.3.7. "Automatic" leak-off valves whose seat position is determined by a mechanical link to outlet non-return valve position are described in section 10.4.1(c).

The "set points", the pump flowrate values at which leak-off opening and closing are initiated, determine the operating characteristics of a leak-off system. The opening set point for at least one valve is always Q_{lo}.

The set point value for the majority of leak-off systems can be fixed using the **flowrate overlap** criterion described in section 10.3.4. Implicit in the use of this criterion is that, as a pump is taken out of service, the rate of forward flowrate reduction is slow compared with the rate at which full leak-off flowrate is established. This is commonly the case in leak-off systems where pipichum values less than 100 (i.e. Table 10.1, category C) pertain, particularly those using the relatively fast opening electrically operated solenoid or relay operated globe valves. The flowrate overlap method is easily applied and the conditional timing requirement is usually confirmed as being at least an order of magnitude different. This makes a more detailed evaluation unnecessary.

For leak-off systems where meeting the flowrate overlap criterion is in doubt or is impossible the **valve timing** criterion described in section 10.3.4 is used. This criterion provides an essential check for high-energy pumps where the time to vaporise a trapped or slow moving volume of liquid is very short. All leak-off systems with pipichum values of over 100 (i.e. Table 10.1, category D) should be checked. Using the valve timing criterion entails a detailed look at forward flowrate control and leak-off valve opening times. The chosen values should be the subject of confirmatory site tests.

Idealised leak-off system characteristics.   The presentation convention for centrifugal pump operation during the leak-off operation is shown in Figs. 10.6 and 10.7. The idealised system characteristic describes the general intent of the leak-off system design in a readily understandable format. It is a plot of pump flowrate Q against forward flowrate q_{fo}. Both flowrates are expressed as $\%Q_{bep}$ values.

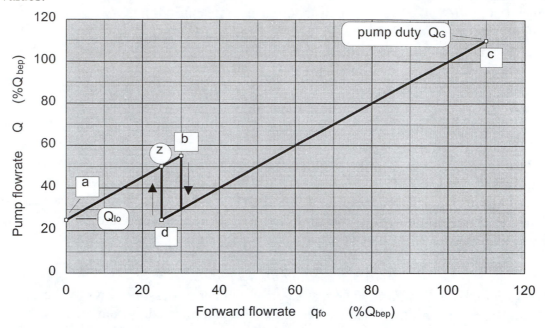

Fig. 10.6   An idealised "25% leak-off system" (one 25% valve) — Example 1

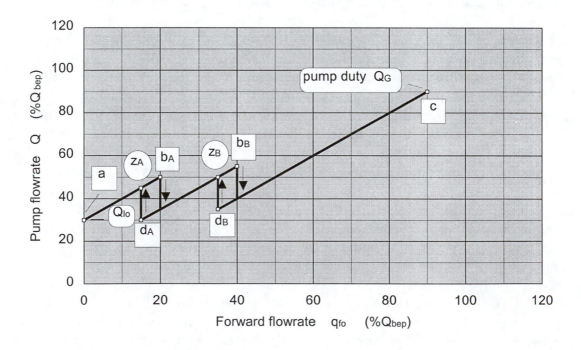

Fig. 10.7   An idealised "30% leak-off system" (two 15% valves) — Example 2

10.3.4 *Set point values*

The flowrate overlap criterion. The leak-off valve opening set point for at least one valve must be Q_{lo}. As the forward flow and leak-off flow can occur simultaneously the pump flowrate could reach a maximum of twice this value. This worst-case scenario would occur if the leak-off valve were opened fully before a particularly tardy forward flowrate control system had a chance to reduce forward flow significantly below Q_{lo}. It is common practice to fix the closing set point by assuming this worst case and adding on a further small margin to avoid hunting.

For a single leak-off valve system Brown, Muntz and Reil (Ref. 10.2), reporting the experience of a major pump manufacturer, state that to avoid operational difficulties the opening and closing set points should be fixed at 1 and 2.2 times the value of Q_{lo}, respectively.

For a two leak-off valve system in which equally rated valves work in parallel (i.e. not as an installed standby) Brown, Muntz and Reil point to the need for separation of the valve set points to avoid hunting between them. A $5\%Q_{bep}$ margin is described as sufficient.

Such a well-established practice is a sound basis for future designs.

Notes:

(i)   In a two valve system where each valve is rated at half the Q_{lo} value failure of one valve leaves the pump safe from vapour lock but possibly at risk from other secondary effects. Such a single valve failure must initiate a controlled shutdown of the pump.

(ii)   To allow "on-line" maintenance so that production is not disrupted two full size ($2 \times Q_{lo}$ rating) leak-off valves should be installed so that one can operate as standby to the other.

Idealised leak-off characteristics are presented in Figs. 10.6 and 10.7 for the following examples.

Example 1 — A 25% leak-off system based on a single valve

> $Q_{lo} = 25\%Q_{bep}$
> Leak-off valve rating: $25\%Q_{bep}$
> Set point "open" : at $Q = 25\%Q_{bep}$ on falling flowrate
> Set point "close" : at $Q = 55\%Q_{bep}$ on rising flowrate

The pump duty is arbitrarily chosen in this example as being $110\%Q_{bep}$ and is denoted as point "c". All Q_{bep} values are pump flowrates unless otherwise stated. The operating intent described by Fig. 10.6 is as follows.

At pump start the leak-off valves are open and the pump runs up speed to operate at point "a" ($25\%Q_{bep}$). On commencement of forward flow demand the pump flowrate increases until at point "b" ($55\%Q_{bep}$) when the leak-off valve closes and pump flowrate drops back to $30\%Q_{bep}$. The pump flowrate then equals the forward flowrate demand of $30\%Q_{bep}$. The pump then operates only in the forward flow mode over the range "c" to "d". Normal demand should lead it to operate at the pump duty point "c".

As demand falls and the pump control system reduces flowrate below point "c" the leak-off valve remains closed until point "d" is reached. The leak-off valve then opens and pump flowrate rises to point "z". As forward flowrate continues to fall so does the pump flowrate. Finally, when point "a" is reached forward flow ceases and all the pump flowrate passes through the leak-off system. The pump can then be safely stopped. Tripping the pump (i.e. removing the driving power supply) at any forward flowrate results in the same close-down sequence.

Notes (i) Point "d" has the value Q_{lo};
     (ii) The individual leak-off valve rating is "z" minus "d";
     (iii) "b" minus "z" is 5% of pump best efficiency flowrate.

Example 2 — A 30% leak-off system based on two parallel operating valves

$Q_{lo} = 30\%Q_{bep}$
Leak-off valve rating: $15\%Q_{bep}$ for each of two valves
Set point "open" Valve A: at $Q = 35\%Q_{bep}$ on falling flowrate
Set point "open" Valve B: at $Q = 30\%Q_{bep}$ on falling flowrate
Set point "close" Valve A: at $Q = 50\%Q_{bep}$ on rising flowrate
Set point "close" Valve B: at $Q = 55\%Q_{bep}$ on rising flowrate

The pump duty is arbitrarily chosen in this example as being $90\%Q_{bep}$ and is denoted as point "c". All Q_{bep} values are pump flowrates unless otherwise stated. The operating intent described by Fig. 10.7 is as follows.

At pump start both leak-off valves A and B are open and the pump runs up to speed to operate at point "a" ($30\%Q_{bep}$). On commencement of forward flow demand the pump flowrate increases until at point "b_A" ($50\%Q_{bep}$) is reached. Leak-off valve A then closes and the pump flowrate drops back to $35\%Q_{bep}$. Further increase in forward flowrate demand leads to pump flowrate increasing to point "b_B" ($55\%Q_{bep}$). Leak-off valve B then closes and pump flowrate falls back to $40\%Q_{bep}$. The pump flowrate then equals the forward flowrate demand of $40\%Q_{bep}$. The pump then operates only in the forward flow mode over the range "c" to "d_B". Normal demand should lead it to operate at the pump duty point "c".

As demand falls and the pump control system reduces flowrate below point "c" both the leak-off valves remain closed until point "d_B" is reached. Leak-off valve B then opens and pump flowrate rises to point "z_B". Further fall in demand produces a similar pattern as pump flowrate passes first through point "d_A", then to point "z_A", and on to point "a". Finally, when point "a" is reached forward flow ceases and all the pump flowrate passes through the leak-off system. The pump can then be safely stopped. Tripping the pump (i.e. removing the driving power supply) at any forward flowrate value results in the same close-down sequence.

Notes: (i) Point "d_A" has the value Q_{lo};
     (ii) the individual leak-off "A" valve rating is "z_A" minus "d_A";
     (iii) "b_A" minus "z_A" is 5% of pump best efficiency flowrate.
     Similarly for the "B" valve except the values "d_B", "z_B" and "b_B" all have an additional 5% of pump best efficiency flowrate.

Pump operation is possible at all forward flowrates. The leak-off system provides protection for the pump from the unacceptable aspects of low-flow performance described in earlier chapters. The operation of the leak-off system, however, can be improved by two further measures.

1. Repeated cycling of the leak-off valves is prevented if the lower flowrate of the pump operating range Q_L is limited to that given by the forward flowrate at point "d".

2. Energy wastage is prevented if, by an overriding signal*, the leak-off valve is closed after running above point "z" for a period of over one minute. This prevents long periods of operation with the leak-off open at forward flowrates between points "z" and "b" (or "z_B" and "b_B").

Adopting the two measures places a lower limit upon the forward flowrate. For Examples 1 and 2 it can be seen from Figs. 10.6 and 10 7 that the minimum forward flowrates (i.e. the flowrate to the process) would be $25\%Q_{bep}$ and $35\%Q_{bep}$, respectively. This restriction is often considered acceptable.

* This signal should not have priority over the fail-safe requirement of the leak-off to go to the open position whenever the pump flowrate falls below Q_{lo}.

The valve timing criterion. The flowrate overlap approach implies that for all operating conditions the pump flowrate never falls below Q_{tr}. Ideally to maintain pump flowrate above the minimum acceptable value during the transition from forward flow to leak-off flow the sum of the two component flows must always equal or exceed Q_{tr}:

$$\text{i.e. } q_{fo} + q_{lo} \geq Q_{tr} \qquad \text{--------------------10.7}$$

In practice it is unrealistic to expect to be able to obtain reliable curves giving q_{fo}/time and q_{lo}/time data for all plant operating conditions. This makes it impossible to demonstrate that the minimum requirement of equation 10.7 has been met. An alternative method of assessment is required which can be verified by test.

It is clear that the pump flowrate will never fall below Q_{tr} if the time for the leak-off flowrate to rise from zero to Q_{lo} is at least equal to, but preferably much less than, the time for forward flowrate to fall from Q_{lo} to Q_{tr}:

$$\text{i.e. } t_{lo} \leq t_{fo} \qquad \text{--------------------10.8}$$

The time to fully open the leak-off valve, including any dead band, is made the fixed point for assessment. This avoids the need to understand, initially at least, the vagaries of the valve characteristic and provides a value that is demonstrable by calculation or test. A margin against uncertainties of measurement, etc. must be included. A practical interpretation of this valve timing criterion is that the time t_{fo} is less than the time t_{lo} by at least 3 seconds.

Essential design changes. Where on evaluation it is found that the valve timing criterion is not met then the plant control system must be changed. Adjustment must be made to either (a) decrease the rate at which the flowrate falls in the transient band or (b) increase the rate at which the leak-off valve(s) can open. There is a balance to be struck between the forward flowrate change time t_{fo} and leak-off opening time t_{lo}. In marginal situations significant improvements in leak-off system reliability can often be made if care is taken to optimise the two. Competent pump manufacturers ensure this is done and leave just a final "trimming" by small adjustments to the set points to be made during new pump commissioning to take into account the differences between design and actual plant operating conditions.

The constraints put on the forward flowrate change time t_{fo} and leak-off opening time t_{lo} at the outset of design studies should not be regarded as unchangeable. The rate of reduction in forward flowrate at and below Q_{lo} is often irrelevant to plant performance requirements and can usually be slowed down without penalty. Leak-off valve opening rate, and hence the rate of increase in leak-off flowrate, can sometimes be speeded up.

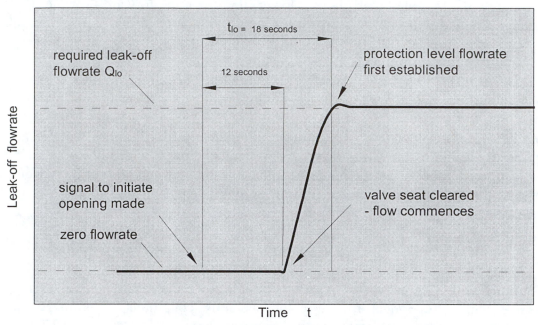

Fig. 10.8   Opening characteristic for a leak-off valve (test data)
Courtesy of Western Power Corporation

10.3.5 *Opening and closing times*

Leak-off opening time t_{lo}. The time taken by a leak-off valve to allow the required flowrate to be reached is crucial to its acceptability. The many different designs of valve and valve actuator provide a wide range of opening characteristics. For all of them the operating time is the sum of

      (i)   the time for the control system to initiate movement by the actuator,
      (ii)   the additional time to clear the valve seat (the "dead band"),
      (iii)   the additional time to establish the protection level flowrate in the leak-off line.

The results of opening time tests on a very slow leak-off valve are shown in Fig. 10.8. Notice the long time, some 12 seconds, taken to initiate a flow following a signal to open. Many of the simpler designs of leak-off valve and actuator have similar characteristics. A good part of this is sometimes taken up by "reset" logic in the computer control system. The time for a particular valve to clear the dead band and establish leak-off flowrate at Q_{lo} depends to some extent upon the seat design but mainly upon actuation method.

From calculation or test a value must be obtained for the leak-off opening time t_{lo}.

Notes:
1. Pneumatic actuation is the preferred choice for leak-off valves, particularly when a fast response time (less than say 5 seconds) is required. With suitable spring assistance a prompt fail open is also achieved on loss of air supply.
2. Electric motor operated actuators are less able to respond reliably on a very short time scale. Where a drive shaft thread dictates the opening time, then usually reversal of shaft rotation means it also fixes the closure time at the same rate. This is often unnecessarily fast. Standard motor driven actuators also suffer from an inability to fail to open in the event of actuator power supply failure.

Forward flowrate change time t_{fo}. An accurate value for the forward flowrate change time t_{fo} must be established. Leak-off systems respond to changes in forward flowrate. The time for forward flow to fall through the transitional flowrate range (Q_{lo} to Q_{tr}) must be determined. Confirmation of the adequacy of the leak-off system is crucially dependent upon knowing the correct value.

(a) *Outlet control valve closure.* Many leak-off systems respond to a reduction in forward flow caused by closure of an outlet control valve. Where this is electrically or pneumatically actuated the rate of closure through the transitional range is determined by the outlet control valve characteristic (available from the valve manufacturer) and the actuator operating speed.

(b) *Pump speed.* For variable speed pumps the rate of reduction in pump speed is usually the controlling factor. Automated systems allow a value t_{fo} to be readily quantified by prior calculation and verified by test. Systems that are either manual or have a manual override need to be carefully assessed. Examples can be found of systems in which the operator can reduce pump speed and hence forward flowrate much faster than the leak-off valve can clear its seat. If it can be done, it will be done; the result is vapour locking.

(c) *Static pressure.* There are a few applications in which the rate of flowrate reduction is produced by a rapid change in the static pressure in the inlet and/or outlet vessels. An accurate assessment of the rate at which this can occur and how this affects pump flowrate is then essential. Such applications are exceptional.

From an evaluation of (a), (b) or (c) a value must be obtained for the forward flowrate change time t_{fo}.

Reducing set point spacing. In many systems the leak-off valve response can be considered instantaneous compared with the rate at which pump flowrate value changes (i.e. the transitional flowrate band is negligibly small). In these systems the set points can be chosen closer to Q_{lo} than shown in Examples 1 and 2. It is essential to obtain and evaluate t_{fo} and t_{lo} values for all plant operating conditions before doing this.

10.3.6 *Flowrate measurement signals*

Flowrate measurement can be made at one (or more) of three locations:

    (i) in the pump outlet pipe before the leak-off take-off point,

    (ii) in the pump inlet pipe,

    (iii) in the pump outlet pipe after the leak-off take-off point.

The preferred location of the device producing the flow measurement signal is (i) between the pump outlet and the leak-off take-off point. Pump flowrate is the parameter that relates directly to vapour lock risk and to the secondary risk factors. Using it avoids any ambiguity.

Location (ii) is not preferred. In pumps where inlet conditions at low flowrate may pose a risk of cavitation the introduction of the additional pressure drop associated with commonly used flow measurement devices (orifice plates, venturi meters, etc.) is not acceptable.

Location (iii), a forward flowrate measurement, is not preferred. This view takes account of the plant control system argument that the use of forward flowrate is more progressive (i.e. without the "swingback" associated with opening and closing leak-off valves) and the probability that forward flowrate is likely to be a reliable plant measurement made for other purposes.

10.3.7 *Modulating leak-off valves*

A modulating leak-off valve is sometimes used. This is not a preferred method of operation. The leak-off system is primarily a plant safety system and an added complication of this kind introduces further potential failure modes. The transient nature of pump start-up and shut-down benefits from clean and positive on/off function with decisive valve sealing.

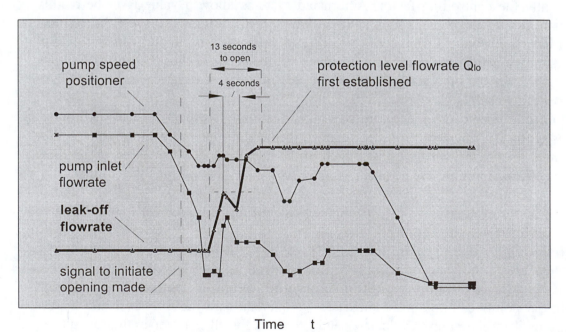

Fig. 10.9   Hunting between a modulating leak-off valve and the pump speed control

Courtesy of Western Power Corporation

Given the undesirability of running for long periods at low $\%Q_{bep}$ values explained in Chapters 7 and 8, the arguments sometimes put forward related to energy saving benefits derived from installing modulating leak-off systems are misplaced. Modulation often confers the advantage of a smoother transition, but as the transition usually takes less than one minute it is a very doubtful advantage since it is usually more than offset by reliability considerations.

The way many modulating leak-off valves operate gives additional risk of cavitation damage across the valve seat. Being proportional to a function of flowrate or a parameter related to flowrate there is usually a small band of flowrates where the valve is nearly but not quite closed, i.e. simmering. The potential for extensive damage exists where a large generated head is present. At such very low flowrates any following pressure reducing device is ineffective. More practical and commercial difficulties that arise when attempting to justify the use of modulating control on leak-off valves are described by Brown, Muntz and Reil (Ref. 10.2).

An unnecessary delay in the valve opening sequence can be introduced by a modulating leak-off valve. An example of this is shown in Fig. 10.9. Here a 4 second period of "hunting" between the leak-off and pump speed control systems is evident which, in an opening time of 13 seconds, invalidated assumptions made on the rate at which the protection level could be established.

10.3.8   *The leak-off system as a heat sink*

The leak-off system provides a means of dissipating the heat generated from the energy supplied to the pump impeller. It prevents dead-heading and with it the immediate consequence — a vapour lock. However, this is not the end of the story and for most systems accommodating the dissipated heat still places limitations on the system design.

In essence the churned mass of the impeller in equation 10.3 is replaced by the mass of liquid circulated by the leak-off system. That is, the leak-off system just provides a larger mass and delays the eventual vapour lock of the pump. Using the liquid storage capacity of the inlet vessel (or its equivalent) and the leak-off system pipework, equation 10.3 can again be used to calculate the time to reach an unacceptable value of temperature T_2.

In practice the limiting temperature may not be that which produces pump vapour locking. The generation of steam from the free surface of an inlet vessel which is open to the environment may prove unacceptable long before vapour lock occurs. The power/mass is again crucial. This can have unexpected results. For example, the following may occur.

1. A small domestic 0.1 kW circulator will run practically indefinitely without coming near to vapour locking when sufficient heat is lost through system piping that is not insulated.

2. A 1 kW pump recirculating into an open 5000 litre water tank will fill a workshop with steam if left on overnight.

3. A power station boiler feedwater pump running at leak-off absorbing 4000 kW and discharging into a 100 tonne deaerator storage vessel (the pump inlet vessel) could result in the water temperature climbing at about 1°C per minute unless further secondary systems are brought into operation.

10.4 Leak-off System Hardware

A detailed evaluation of leak-off system hardware is outside the scope of this book. However, as it is important to understand and to be able to evaluate the relative merits of the options available an overview of equipment commonly found in use is presented. The closely matched options — proved by their continuing presence in the marketplace — inevitably mean that an element of personal preference dictates final choice.

The most heavily protected pumps are high-energy pumps. These have a high capital value and their failure due to the mechanical damage caused by a vapour lock gives rise to a severe financial penalty. The leak-off system for these machines may comprise up to three hardware items as shown in Fig. 10.1, namely, (1) leak-off valve, (2) pressure reducing device, and (3) back pressure orifice plate.

Examining the function of these items provides guidance as to where they are essential, where they may be safely omitted, and the robustness of mechanical design which ensures reliable operation either as a single train or, as in Fig. 10.1, as a double train.

10.4.1 *Leak-off valves*

There are three main groups to be considered
> (a) valves that provide leak-off control only,
> (b) valves that combine leak-off and pressure reduction functions,
> (c) valves that combine leak-off, pressure reduction and outlet
> non-return valve functions.

All leak-off valve groups have to open with the full pump outlet pressure on one side and a pressure approximately that of the pump inlet on the other (i.e. approximately the full generated pressure). Where this pressure breakdown is over 10 bar, it can lead to considerable cavitation erosion damage or wire drawing occurring in the leak-off valve, usually across the valve seat. Examples of this are shown in Fig. 10.10. In these circumstances valves specially designed for leak-off applications are essential. To minimise the risk of damage a pressure reducing device is often incorporated into the valve or added to the leak-off system separately directly after the valve.

Where the pressure breakdown is 10 bar or less, then a standard plug valve without a pressure reducing vessel is usually adequate.

(a) Cavitation erosion   (b) Wire-drawing
see Light, Ref. 10.3   *see Chemoul, Ref. 10.4*

Fig. 10.10   Damage to leak-off valves

a) *Valves that provide leak-off control only.* A multitude of valve designs are offered for this application. Some valves are not suitable so care has to be taken.

The parallel slide valve (Fig. 10.11) meets all the requirements for high-energy applications. It incorporates a wide seating face that gives the opportunity to withstand cavitation and wire-drawing brought about by the associated high pressure drops.

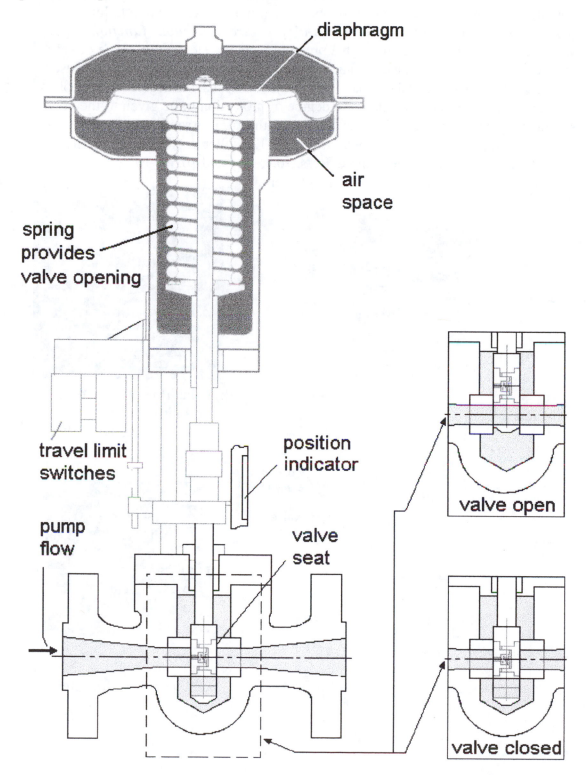

Fig. 10.11 Parallel slide leak-off valve　　　　*Courtesy of Hopkinsons Ltd*

Fitted with a pneumatic actuator incorporating a spring to open should the air supply fail this design has proved eminently suitable for protecting power utility boiler feedwater pumps. The parallel slide design is also inherently good at accommodating thermal transients. As the sudden passage of hot liquid takes place through the valve differential expansion of valve body and internals takes place. Some types of valves (e.g. plug valves) are less able to operate reliably under such conditions. A leak-off valve which sticks is a common cause of pump failure.

(b) *Valves that combine leak-off and pressure reduction functions.* By withdrawing the valve stem through a pressure reducing element this type of valve progressively brings resistance channels into play as leak-off flowrate is increased. Being able to vary the leakage path geometry gives a useful degree of control to the valve characteristic and enables a degree of protection to be given to the valve seat at very low flowrates. Two examples of the pack assembly controlling the pressure reducing feature are shown in Fig. 10.12. Fuller details of this "cascade" type of leak-off valve can be found in Refs. 10.3, 10.4 and 10.5.

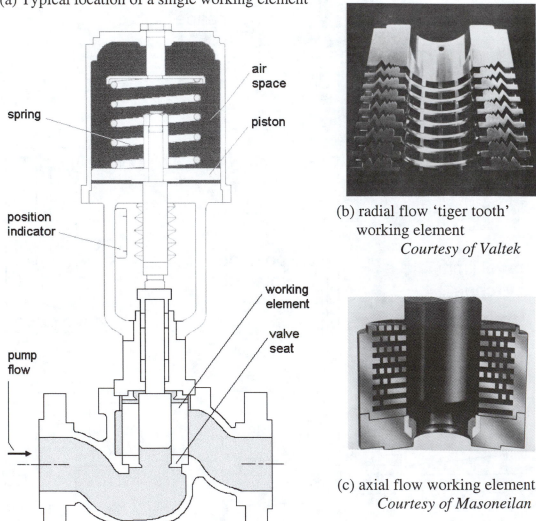

(a) Typical location of a single working element

(b) radial flow 'tiger tooth' working element
Courtesy of Valtek

(c) axial flow working element
Courtesy of Masoneilan

Fig. 10.12   Examples of "cascade" leak-off valve working elements

(c) ***Valves that combine leak-off, pressure reduction and outlet non-return valve functions.*** In all leak-off systems low-flow protection is provided by the leak-off flowpath being opened up as forward flowrate is reduced and the outlet non-return valve closes. Several valve designs are available which combine, by a mechanical linkage, the leak-off line sealing and pressure reduction functions with the non-return valve function. All three functions are incorporated into the non-return valve. The principle of operation is shown in Fig. 10.13. The advantage of an inclusive leak-off function is the elimination of the external control system necessary for the valves described previously. This has led to the valve sometimes being described as "automatic". The disadvantages are (i) the linkage may be subject to a considerable

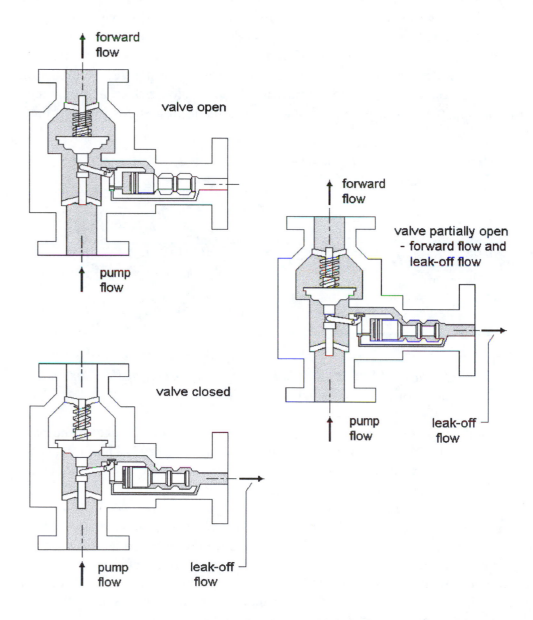

Fig. 10.13. Automatic recirculation valve   *Courtesy of Yarway Inc.*
(Combines leak-off, pressure reduction, and outlet non-return valve functions)

buffeting by the main forward flow of liquid, (ii) maintenance online is not a practical option (a very costly 2 x Q_G rated valves and piping are needed to provide a standby facility), and (iii) maintenance necessitates the removal of the valve casing from the main outlet pipe; this is not practical where large pipe forces and moments are present. Additionally this type of valve does not give any external indication of whether or not it has operated nor does it give anyone the opportunity to open it in an emergency. Fuller details of this "automatic" type of leak-off valve are in Ref. 10.6.

10.4.2 *Pressure reducing devices*

A clear line back to the inlet vessel would mean that the leak-off valve has to break down the pump generated pressure rise for as long as the pump was operating on leak-off. Should it be held at leak-off waiting to go on load — a common practice with some high-energy pumps in process and power generation plant — serious levels of cavitation damage would occur. To avoid such damage a pressure reducing device — a means of destroying surplus energy — is fitted on the outlet side of the leak-off valve. This clearly eases the design requirements of the leak-off valve and transfers long-term cavitation erosion risk to a component with no moving parts. The pressure breakdown is made in a series of controlled stages.

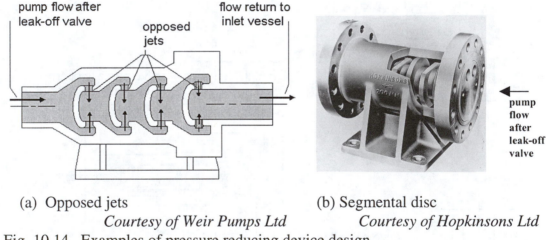

(a) Opposed jets (b) Segmental disc
Courtesy of Weir Pumps Ltd   *Courtesy of Hopkinsons Ltd*
Fig. 10.14   Examples of pressure reducing device design

Pressure reducing device is the description used in this book. Usage in the past has often referred to it as the pressure reducing vessel or "prv". This has led to frequent confusion by plant operators and valve manufacturers on a number of occasions. In plant documents and working procedures the pressure relief valve more commonly uses the abbreviation prv.

There are several "stand alone" designs of pressure reducing devices. Examples of designs in common use shown in Fig. 10.14 are (a) the opposed jets and (b) the segmental disc. With a pressure reducing device located shortly after each leak-off valve these items are usually situated between the isolating valves provided to facilitate leak-off maintenance. Where a standby leak-off valve is installed this enables maintenance to be undertaken without stopping the pump.

In an excellent survey Brown, Muntz and Reil (Ref. 10.2) reviewed the reliability records of many designs of the pressure reducing devices used to protect boiler feedwater pump leak-off valves. A maximum period of 1000 hours operation between inspections is recommended for the opposed jets design. This reflects the arduous task which, averaging the records, was based on destroying 850 m generated head for each of the four stages.

10.4.3 *Back pressure orifice plates*

A back pressure orifice plate, sometimes called a trimming orifice, is often included in the leak-off system design as shown in Fig. 10.1. In high-energy pumps the pressure reducing device or its outlet pipework may itself be subject to cavitation erosion attack as high velocity jets emerge at its downstream end. Crossland (Ref. 10.7) suggests a safe pressure drop is 85% of the difference between the inlet pressure and its vapour pressure. The plant design has to provide the 15% remainder as a back pressure by a combination of static height and back pressure orifice. Irrespective of this, it is often prudent to impose a modest back pressure. The orifice plate also enables on-site pressure drop "adjustments" to be carried out readily without having to change to the pressure reducing device internal design.

The back pressure orifice plate should be located at or lower than the leak-off valve if possible to prevent cavitation erosion damage. It should be sized and provided with tapping points to enable temporary instrumentation to be fitted that enable a rudimentary check on Q_{lo} to be made either during commissioning or later should problems arise.

10.5 In Summary

Table 10.2 is presented as an aide-memoire and quick reference guide to the preceding sections.

The boundaries of viable operation for the various low-flow protection systems given in Table 10.1 are not precise. The effect of changes in some thermodynamic properties are assumed insignificant compared with the order of magnitude change in t. This is not always the case. In practice there is also a considerable overlap due to the wide range of quality and response times for commercial grade hardware. The boundaries enumerated are only for preliminary guidance as to the conditions in which a particular system is likely to be appropriate. This guidance should not be considered prescriptive.

Table 10.2   A guide to evaluating low-flow protection

The steps to take	Action	Comments
Part 1 Low-flow protection requirements		
1. Data required to start evaluation	1a. Q_{lo} — leak-off flowrate	Minimum flowrate for cavitation erosion protection probably fixes this (see Chapter 9). Must not be less than $20\%Q_{bep}$.
	1b. t_{lo} — Leak-off valve opening time	⎞
	1c. t_{fo} — Forward flow reduction time through $(Q_{lo} - Q_{tr})$ band	⎟ see section 10.3.5 ⎟ ⎠
	1d. Estimated cost of a vapour-lock incident arising from: - pump repair - loss of production	A +/- 20% estimate is usually sufficient
2. Choose appropriate level of low-flow protection	2a. Calculate pipichum value	Equation 10.6
	2b. Take guidance from Table 10.1	If category A or B, consider if operator intervention (with temperature alarm if necessary) is acceptable. If OK end evaluation
		Otherwise a leak-off system is probably required — see below
Part 2 Leak-off system evaluation		
3. Evaluate options in leak-off system design	3a. Ensure the basics are covered	See section 10.3.1
	3b. Estimate the set points	See section 10.3.4
	3c. Compare t_{fo} and t_{lo}	See section 10.3.4
4. Evaluate hardware	Compare alternatives	Take guidance from section 10.4
5. Essential site tests	Tests to confirm: Q_{lo} t_{lo} t_{fo}	t_{lo} and t_{fo} are not applicable to valves which include the non-return function See section 10.4.1(c)

References

10.1 <u>UK Steam Tables in SI Units</u>, Pub. Edward Arnold, London, 1970.

10.2 Brown, R. W., Muntz, N. R. A. and Reil, B.,
<u>Boiler feed pump leak-off systems</u>,
IMechE Conf., Centrifugal Pump Low-Flow Protection, pp.1–16, 1991.

10.3 Light, G., <u>The practicalities of feedpump recirculation valve design</u>,
IMechE Conf., Centrifugal Pump Low-Flow Protection, pp.37–45, 1991.

10.4 Chemoul, J., <u>Control valve problems and solutions</u>,
IMechE Conf., Centrifugal Pump Low-Flow Protection, pp.47–57, 1991.

10.5 Mott, W. M. and Badowski, D. F., <u>Feedpump leak-off control systems</u>,
IMechE Conf., Centrifugal Pump Low-Flow Protection, pp.59–70, 1991.

10.6 Eckford, N. D., <u>Integral leak-off/non-return valve design</u>,
IMechE Conf., Centrifugal Pump Low-Flow Protection, pp.29–35, 1991.

10.7 Crossland, A. W., <u>Pump leak-off valve and pressure reducing vessel
development</u>, IMechE Conf., Centrifugal Pump Low-Flow Protection,
pp.17–28, 1991.

Chapter 11

Cavitation Tests for Centrifugal Pumps

11.1 Test Specifications

11.1.1 *Test types*

Industrial users demand that centrifugal pumps are tested to demonstrate cavitation performance where there is a risk of unacceptable cavitation occurring. Chapters 6, 7, 8 and 9 show that for most industrial applications the risks from cavitation can be expressed by an NPSH(3pc) value. Chapter 9 shows that the risk can be expressed more precisely by an NPSH(4mm) value. Tests to determine the NPSH(4mm) value are shown to be justifiable only when the added cost of this test is more than offset by the savings obtained by the reduced level of NPSH(A) brought about by the increased certainty of where cavitation performance limits are. Test methods that are robust in both an engineering and a contractual sense are available to demonstrate pump performance as measured by either NPSH(3pc) or NPSH(4mm).

More specialised tests are sometimes carried out by pump manufacturers to provide additional information on cavitation performance. The inclusion of a rudimentary check for the onset risk of cavitation surging is to be welcomed where pump and test loop geometry makes this a practical possibility. By providing test conditions conducive to its onset and carrying out tests over the full operating range a limited but worthwhile assurance is obtained. Another example of specialised testing is the paint erosion test where clues to the location of cavitation erosion damage are obtained.

On rare occasions special tests are included in test specifications. Normally the difficulty of defining appropriate boundaries for contractual acceptance makes inclusion in specifications impractical. The lack of contractual clarity does not, of course, mean that such tests have no value to pump users. Quite the contrary. The additional design specific information obtained helps to give a clearer understanding of cavitation performance than by recourse to generic data analysis alone.

11.1.2 *National and international standards*

The majority of pump users prefer to specify centrifugal pump test requirements based upon an available test standard or test code. Typical examples of these are Refs. 11.1 to 11.5. By relying on the collective expertise of those charged with producing such documents a great deal of time and effort can be saved. Sadly, the standards and codes for centrifugal pump cavitation testing often prove incomplete or unsuitable.

It is self-evident that a good test standard is one that is (a) appropriate in quality, (b) easily understood by both parties to a contract, (c) easily applied. In an engineering community in which user adoption of a test standard is essentially voluntary the failure of a particular standard can be measured by the extent of pump users reluctance to use it. The evidence of unsatisfactory national and international standards is exemplified by a proliferation of user-company-based amendment sheets, "special" test requirements, and the diversity of "industry-based" test specifications in areas where a good basic standard should suffice. The centrifugal pump industry world wide has always had an unjustifiably large number of such documents. Efforts to consolidate the best of these in international standards have achieved only limited success. This makes it necessary for pump users to understand the basic requirements of cavitation testing, test loop design and test methodology.

This chapter is not written as an alternative test standard or test code. Its aim is to identify important areas for pump user consideration so that when a pump specification is prepared, with or without recourse to an existing pump standard or code, due regard is given to technical issues that must be addressed if deficiencies are to be avoided.

11.1.3 *Test quality*

The use to which cavitation test data are put varies greatly. For most centrifugal pump applications a rudimentary test will often prove sufficient to demonstrate that the risk from cavitation is minimal. For a few applications more precise methodologies are necessary. To distinguish between them the levels of testing that are available for consideration are described in this book, in descending order of complexity and cost, as quality levels A, B and C. Table 11.1 describes in outline the choice of test method.

Table 11.1 Test quality levels

Quality level	Description		Contractual guarantee
C	Cheap and simple	Rudimentary — adequate for most industrial applications where the objective is to confirm a pump is clearly not at risk	NPSH(3pc)
B	Better	Well engineered — where data must be capable of withstanding vigorous questioning as to validity sufficient to meet commercial needs arising from the contractual use of test results	NPSH(3pc) NPSH(4mm)
A	Advanced	Research and development — where high quality and repeatability are essential for measuring performance in absolute terms	Not applicable

11.2 Test Basics

11.2.1 *The test liquid*

Commercial reality dictates that cavitation tests are, almost without exception, carried out on "clean cold water" at the pump manufacturer's works. In practice the water used is likely to be drawn from a coarsely filtered industrial supply. This filtering is usually sufficient to reduce to an acceptable level the effects of debris, rust, etc. which might otherwise occur in holding tanks or in the test loop pipework.

The property values described by Table 11.2 will suffice for tests where the description clean cold water is in some doubt and where a contractual basis for testing is deemed necessary.

Table 11.2 "Clean cold water" — A definition

Pump liquid property	Acceptable limits	
Temperature	10°C to 40°C	
Density	990 kg/m³ to 1050 kg/m³	(see Note 1)
Kinematic viscosity	up to 1.75 x 10⁻⁶m²/s	
Free solids content	up to 2.5 kg/m³	

Note 1: The maximum density of pure water in the 10°C to 40°C range is between approximately 990 kg/m³ and 1000 kg/m³. The figures in the table allow for a small amount of dissolved particulate matter.

11.2.2 *The water temperature limit*

Chapter 5 showed that the accurate measurement of NPSH is practically unaffected by changes in the thermodynamic properties of water temperature in the range 0°C to 40°C. So whilst the vapour pressure appropriate to the test water temperature must be used in calculations, the value of NPSH obtained remains essentially constant.

Chapter 3 (Fig.3.8) showed that the kinematic viscosity of water could affect head-drop measurements below 10°C. It follows that requirements which restrict cavitation tests on water to between 10°C and 40°C will be less likely to be susceptible to error.

It is sometimes tempting to carry out industrial cavitation tests using water above the 40°C limit given in Table 11.2. Hard won experience shows the potential folly of doing this where NPSH is calculated using a vapour pressure determined by the measured temperature and thermodynamic property tables. Even in research quality testing the results can be seriously invalidated by errors in temperature measurement. Many workers in this field have encountered the problem of erroneous NPSH values. The error only really becomes apparent at temperatures above about 100°C and many inexperienced users will probably fail to appreciate a false reading particularly where the temperature is not much greater than 40°C.

Tests in which the liquid is boiling at a large free surface and for which the NPSH is calculated using primarily the physical height of the liquid above the pump inlet is reliable above 40°C. The extra cost of a "very hot" water test where NPSH is based upon control of liquid level above the pump inlet usually makes this prohibitive.

11.2.3 *NPSH measurement reference plane*

The NPSH reference plane for a centrifugal pump defines the datum for measuring and comparing cavitation performance. Usually the comparison made is between test performance and plant design values. In principle as long as the same reference plane is used for the data then the validity of comparative conclusions is not compromised irrespective of where the datum is. It is worth remembering this general point when access to pump shaft/impeller geometry is difficult. Exceptional measures can then be applied to advantage. An example of this is a pump mounted within a tank where it may prove expedient to use a feature other than the pump to define the datum for NPSH values. A point on the tank outlet flange is a popular choice.

In practice it is normally prudent to follow the established conventions of the centrifugal pump industry. This ensures that data recorded during a test can be compared with data from other sources. The following two mutually exclusive definitions of the reference plane for NPSH measurement are in use.

(i) NPSH values for pumps with the impeller(s) mounted on a horizontal shaft are measured from the shaft centreline.

(ii) NPSH values for all other pumps are measured from the leading edge of the impeller blade at the outer diameter of the impeller eye. For multistage pumps this relates to the first stage impeller.

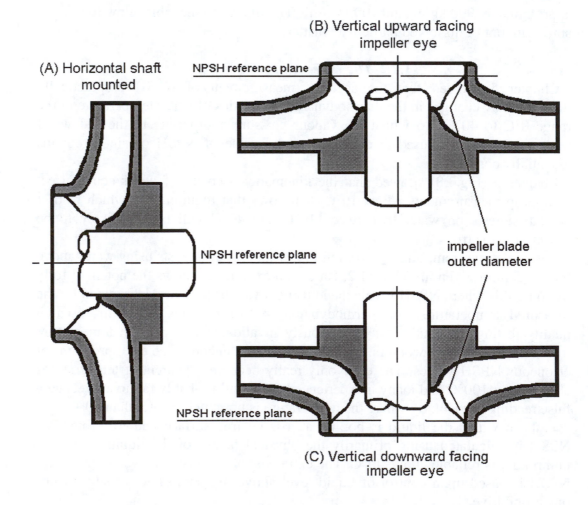

Fig. 11.1 Pictorial definition of the NPSH measurement reference plane

Figure 11.1 illustrates how the definitions of NPSH reference plane are interpreted. The two definitions are inconsistent as can be seen by turning an end-suction impeller on its side. In a strict sense data from one cannot be used for the other. However, for many pumps the difference is usually only a few centimetres and this is usually regarded as insignificant where NPSH values of interest are a metre or more.

Common practice is to group NPSH data together irrespective of whether determined by definition (i) or (ii) when the generic analysis of pumps is undertaken. The exception is very large eye diameter, slow-speed impellers (e.g. power utility cooling water pumps) where considerable error would be introduced by using the definitions inconsistently.

11.2.4 *Inlet pipework configuration*

Test installation pipework geometry is defined in standards in a way that facilitates reliable and repeatable performance measurement. It also provides a common basis for performance comparisons between pumps. Usually the inlet pipework configuration where the pump is installed is very different from that on test. Consideration needs to be given to deciding whether this difference matters.

The method of calculating ΔH_G and NPSH(A) described in Chapter 6 caters for differences between test and site pipework layout when measured as a head-loss. Provided that common sense "good practice" is applied to inlet pipework design (e.g. avoidance of a bend immediately before the pump inlet and of pockets where vapour/air can accumulate); then test data and plant design data that refer to the same measurement reference plane will have validity in predicting plant ΔH and NPSH values.

Pump performance measurement by standard methods does not extend to cavitation surging. Chapter 7 describes the NPSH and flowrate conditions that are necessary for hydrodynamically induced surging to occur. The guidelines given in this book avoid these conditions. However, if a pump is, for some unexpected reason, expected to function in this risk area then a cavitation test may be needed to give some assurance that hydrodynamically induced surging is unlikely to be a problem. Most standards recommend that total inlet head measurement (used as the base for both ΔH and NPSH calculations) is made at a distance of two diameters of straight pipe upstream of the pump inlet flange position and that an additional minimum of four diameters of straight inlet pipe is added to secure repeatable measurements of ΔH_G, NPSH(3pc) and NPSH(4mm). Hydrodynamically induced surging is shown in Chapter 7 to have a greater chance of developing if a longer straight inlet pipe is used. The provision of test conditions that encourage surging is considered worthwhile. Although not a basis for conclusive proof of the absence of cavitation surging, the provision of a minimum inlet pipe straight length to the pump flange of 15 diameters will give such an inlet instability every chance to present itself when cavitation tests extend to the region where surging is a risk. It has to be recognised that the physical design of some pumps makes such an inlet configuration impossible. However a 15 diameter straight pipe requirement should be included in test specifications wherever it is reasonable to do so.

11.2.5 *The speed of the test pump*

The head generating capability of a cavitating centrifugal pump and the NPSH vary with approximately the square of the pump speed. Chapters 3 and 5 provide confirmatory test and theoretical explanations for this. Pump characteristics, including NPSH curves, are usually presented based on a nominal speed. To provide a sound contractual basis for pump specifications it is important to establish the actual pump speed and how this is arrived at. Where NPSH test data are being recalculated for a higher speed users should consider using the square value to avoid underestimating NPSH requirements.

Induction electric motor driven pumps. A large number of centrifugal pumps are driven by electric motors taking power from an alternating current supply. Induction motors exhibit a small percentage "slip" in speed compared with the synchronous frequency. This slip increases as the power transmitted by the motor to the pump increases. For such pumps it is important to recognise that this slip speed is an integral part of the final plant. Performance should be achieved with whatever speed pertains at the guaranteed flowrate Q_G. A correction is made by the pump manufacturer for the reduction in speed associated with slip when the pump is selected. The pump user has no need to know the precise speed: a test to demonstrate the ability of the pump to meet the specified duty is all that is needed.

Conversely, these same machines change speed in direct proportion to the frequency of the electrical supply. The supply frequency during a pump test is normally outside the control of the pump manufacturer so in a cavitation test specification it is reasonable to (i) nominate a supply frequency*, (ii) measure the frequency on test, (iii) correct test data back to the nominated frequency value.

* Where a pumping plant is required to function during conditions of low supply frequency an appropriate low value must be specified. An example of this is power utility boiler feedwater pumps where the pump duty is required to be maintained when the supply frequency falls to 5% below the nominal value.

Changes in electrical supply voltage within the band permitted by the motor design and given on the motor nameplate are usually assumed to have a negligible effect on pump NPSH(3pc) performance.

Variable speed pumps. Where a variable speed drive is provided the cavitation test speed must be controlled to give the specified value.

11.2.6 *Effects made insignificant by appropriate test methodology*

Hysteresis. Cavitation characteristics often exhibit hysteresis. The ΔH/NPSH curve for decreasing values of NPSH, that is, as cavitation becomes more extensive, is different from that when the reverse occurs as NPSH values are rising. Remembering that the data are used to determine what will happen when NPSH falls and approaches a level higher than the NPSH datum it is clear that all tests should be carried out with NPSH progressively falling from non-cavitating conditions. Recovery from the very much lower arbitrary reference NPSH value is not of practical interest.

To secure reliable and unambiguous results that are appropriate to pump usage NPSH should always be progressively reduced during cavitation tests. The test specification should always state this as a requirement.

Dissolved air. Chapter 3 showed that NPSH(3pc) and NPSH(4mm) values correspond to well-developed cavitation on the impeller blades. Such cavitation has been shown (Refs. 11.6, 11.7 and 11.8) to be influenced very little by air content. Le Fur, David and Pecot (Ref. 11.9) conclude that ".... the influence of dissolved oxygen content depends upon the type of pump: it may be considered negligible for centrifugal pumps....".

Where the ability to observe the water entering the pump is available (usually only the case for Class A tests) the effect of dissolved air is immediately apparent. As water enters the low pressure regions near the throttle valve or the pump inlet the dissolved air immediately comes out of solution. This makes the water appear milky. The air is released in the form of very small bubbles. These congregate, coalesce, and rise to points where they can be vented off. In a closed loop system this milkiness disappears in a matter of minutes. Cavitation performance as measured by NPSH(3pc) during and after this "milky" stage shows no change confirming that even relatively large amounts of air in and released from the pumped liquid have no measurable effect on NPSH values of commercial interest. This is reassuring since academic research has shown that the inception of very small cavities (at conditions much less than those at NPSH(0pc)) can be affected by both dissolved air and free air.

After reducing the dissolved air content to levels where the water passing through the low pressure regions remains transparent, the conditions in the pumped liquid become relatively stable. It takes many hours (and a large free surface area) for water to return to being saturated with dissolved air.

Where air bubbles are not vented off and remain recirculating in a loop they can markedly affect the hydraulic performance of a centrifugal pump. Most machines suffer an almost total drop off in generated head at gas void fractions of about 10%. Special pump designs produced for the off-shore petroleum industry (Ref. 11.10) are capable of handling a gas void fraction much greater than 50%. The testing of such vapour/liquid mixtures is outside a definition of cavitation relating to vapour phase growth and collapse within the local vicinity of the impeller. However, it is relevant to note the risk to test result validity posed by inadequate venting of a test loop.

As a precaution and to secure reliable and unambiguous results the test pump should be run immediately prior to cavitation tests for a period of 10 minutes or ten times the nominal recirculation time of the test liquid in the test loop whichever is the greater. During this time the venting of free air or other vapour should be undertaken. The test specification should always state this as a requirement.

Barometric pressure. Where open sump or open tank tests are carried out the value of NPSH is calculated using the barometric pressure. Accurate measurement of barometric pressure, which is readily obtained, needs to be recorded during tests — at least every hour. Although usually slow and progressive, the value of barometric pressure does change very rapidly on occasion making it potentially the largest source of error in calculating NPSH. Quantifying the change in barometric pressure as a change in NPSH value shows it to be as much as 0.5 m in an hour. It is prudent to take regular barometer readings and, when a pump is to be tested for which the NPSH values are expected to be very low, keep an eye on the weather!

11.2.7 *The limits of measurement uncertainty*

Appropriate values for principal cavitation performance parameters measured to quality levels C and B are:

flowrate measurement by orifice plate	+/-2%,
head measurement by standard test gauges	+/-0.5%,
water temperature by thermometer or thermocouple	+/-0.5°C,
pump speed by tachometer or digital revolution counter	+/-0.5%,
cavity length	+/-1 mm.

These combine to give rise to an overall limit of uncertainty in the value of NPSH(3pc) and NPSH(4mm) of less than +/-10%.

Note: Table 4.1 assumes and includes the effect of a 10% uncertainty.

The limits of uncertainty of interest during a pump test are those pertaining to the uncertainty in the measurement process at the point on the instrument scale where the readings are taken. Values for measurement instruments are often quoted at full scale deflection or maximum instrument range value. Care has to be taken not to confuse these two figures.

The main source of error in cavitation test data is flow measurement. It is good practice to change the measurement device when, as is usual, tests are carried out over a wide range of flowrate values. Typically this might mean switching from a single to a multiple combination in a bank of orifice plates or vice versa.

Similarly, changing pressure gauges during a test in which the uncertainty in the measured value becomes unacceptably poor must be undertaken. It is prudent to establish where changeover points are before a test commences. Many measurement instruments are not suitable for use below about 40% of the full scale reading.

All contractual tests should be firmly linked to the pump manufacturer's Quality Assurance system. This should provide documented evidence that all measurement instruments have a proof of calibration based on a route back to a certified primary standard.

11.3 NPSH Test Hardware

11.3.1 *Essential requirements*

The essential requirements for cavitation testing a centrifugal pump are

    (i) an outlet valve for flowrate control,

    (ii) a flowrate measurement device, usually an orifice plate or venturi meter,

    (iii) a means of controlling the NPSH.

Where tests are carried out for a lengthy period of time and the volume of test liquid is small a means of controlling the water temperature has to be provided.

Provision of the essentials makes NPSH(3pc) tests possible. To facilitate viewing cavitation for NPSH(4mm) tests additional modifications to the test pump and sometimes the pump impeller have, of course, to be undertaken.

11.3.2 *Quality level C — NPSH(3pc) tests*

In tests on a small centrifugal pump Rütschi (Ref. 11.11) showed that the quality level C test methods can lead to errors in NPSH measurement at low $\%Q_{bep}$ values. This is illustrated in Fig. 11.2. At flowrates near to Q_{bep} the test results for quality level C and B test loops are shown to be much the same. Whilst these particular tests may exaggerate the differences that arise they are useful in drawing attention to potential deficiencies. The values of NPSH(3pc) at flowrates at and around Q_{bep} are likely to be the values of practical interest. Quality level C tests therefore provide a relatively cheap way of obtaining reliable NPSH data in the region of Q_{bep}.

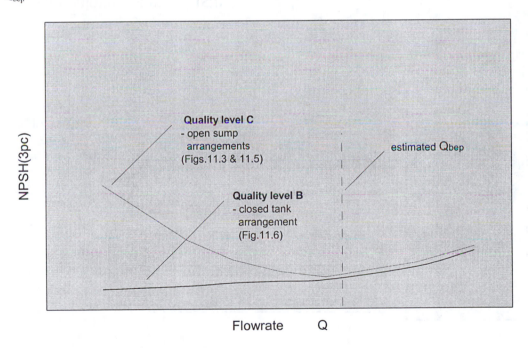

Fig. 11.2 Test NPSH values affected by test loop design   *see Rütschi, Ref. 11.11*

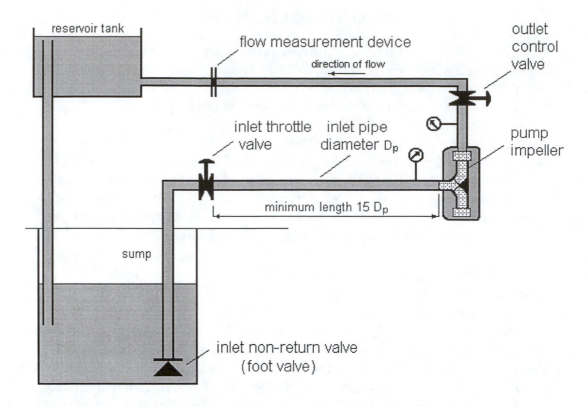

Fig. 11.3   Cavitation test layout; quality level C — open sump arrangement
(NPSH control by throttle valve)

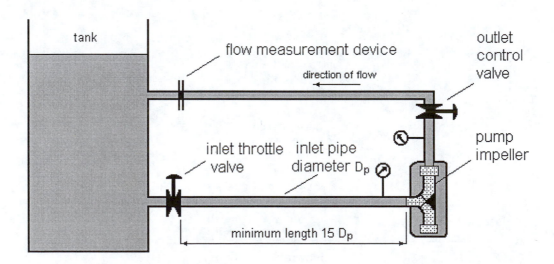

Fig. 11.4   Cavitation test layout; quality level C — open tank arrangement
(NPSH control by throttle valve)

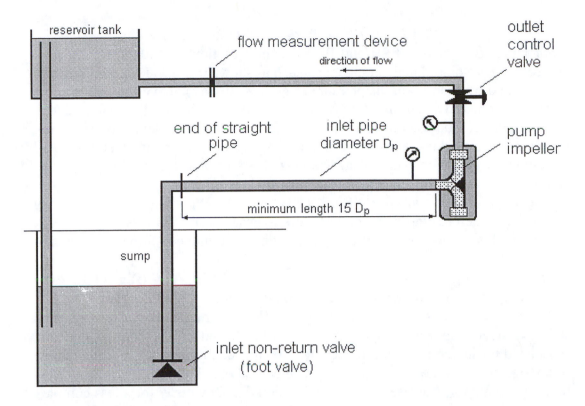

Fig. 11.5 Cavitation test layout; quality level C — open sump arrangement
(NPSH control by sump level)

Variations on the simplest of cavitation test loop design are illustrated in Figs. 11.3, 11.4 and 11.5. These are all quality level C. These loops produce surprisingly good results when used to evaluate pump performance where cavitation is well developed such as is the case for NPSH(3pc) tests.

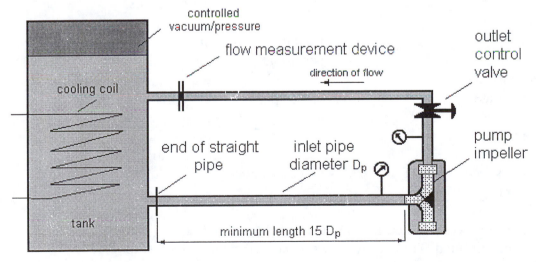

Fig. 11.6 Cavitation test layout; quality level B — closed tank arrangement
(NPSH control by tank pressure)

11.3.3 *Quality level B — NPSH(3pc) tests*

For better control of NPSH and the avoidance of flow disturbance from inlet control valves a closed tank arrangement such as shown in Fig. 11.6 is used. NPSH is controlled by changes in the pressure of gas (usually air) above the liquid in the tank. Tests on cold water are often carried out under a vacuum. Provision of such low pressure can be used to control dissolved air to very low levels if this is felt necessary. A closed test loop also has the advantage of isolating the test liquid and its vapour from its immediate surroundings. Tests at high (or low) temperatures and on liquids other than water are readily carried out.

Test loops come in all sizes (Refs. 11.12 and 11.13). It is prudent to review requirements before embarking on test work to ensure that loop capabilities match the need. This is best done by listing the limiting operating conditions as shown in Table 11.3.

An example of a quality level B test loop is shown in Fig. 11.7 (Ref. 11.12). The loop incorporates a stainless steel bellows to eliminate air/liquid interfaces. A bank of three orifice plates gives a wide range of flowrate measurement options. The stainless steel loop was designed so that, with the aid of the cooling coil and associated refrigeration plant provided, tests on a wide range of water temperatures or tests on liquids other than water could be carried out. The facility, designed and built by the author, was used for university research. Obviously for a less complex industrial testing some of the more "exotic" options could be discarded.

Table 11.3 Test loop operating range

Operating variable	Range min.	max.	Limiting operating condition At minimum	At maximum	Measurement instrument accuracy
Temperature °C	-20	150	refrigeration plant capacity	insulation losses	+/- 0.1
Pressure at outlet bar (absolute)	0.5.	10	air pump capacity	loop design hydraulic test	+/- 0.5%
Pressure at inlet bar (absolute)	0.1	10	vacuum pump performance	pump mechanical seal	+/- 0.5%
Flowrate litres per second	2	30	orifice plate accuracy	high pressure drop maximum manometer differential	+/- 2%
pump speed rpm	900	2850	Fixed camplate hydraulic motor		+/- 1 rpm

Note: The power measurement arrangements (necessary if Q_{bep} is to be quantified) have been omitted. These were very special to the particular test loop.

(a) Test facility photograph

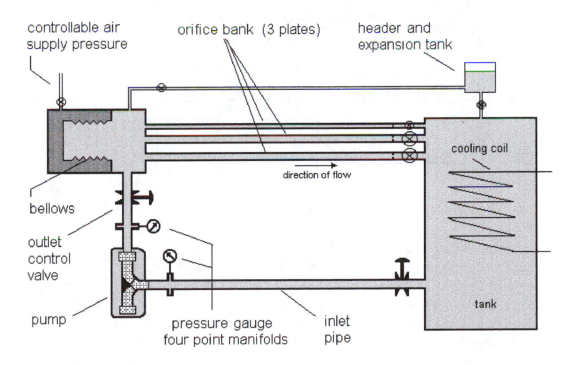

(b) General arrangement

Fig. 11.7 Cavitation test facility; quality level B — closed tank arrangement
(NPSH control by bellows pressure)

Flowrate measurement is the source of greatest uncertainty. The standards covering measurement devices give the conditions necessary to achieve a given accuracy at a given flowrate. Troskolanski (Ref. 11.14) gives additional data that are useful in quantifying the changes in device coefficient over the wide range of conditions over which a test loop may be required to work.

11.3.4 *Quality level B — NPSH(4mm) test layout*

Visual cavitation tests offer an opportunity to understand the conditions under which erosive cavitation thrives. Transparent observation windows do however introduce several obvious disadvantages. Whilst hydraulic contours can be maintained (sometimes at considerable expense) there is no easy way of accommodating high inlet pressures or high inlet temperatures. A compromise has to be sought.

In sections 3.7 and 5.10 it was evident that scaling NPSH upward with speed using a square law and ignoring the thermodynamic effect led to conservative NPSH performance forecasts. That is, the added uncertainties introduced by testing on "clean cold water" at a lower speed all err on the side of caution. This approach, favoured by users, has for many years been the basis of cavitation tests where NPSH(4mm) data form part of contractual guarantees.

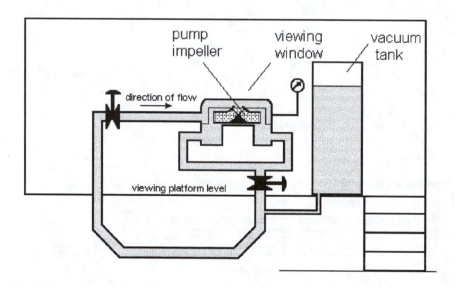

Fig. 11.8 Visual test facility — general arrangement   *Courtesy of Weir Pumps Ltd*

An example of a visual test facility suitable for the cavitation testing of centrifugal pumps is shown in Figs. 11.8 and 11.9. Such a test loop is suitable for contractual testing to ascertain quality level B NPSH(4mm) data and for carrying out quality level A research and development testing work. The test loop shown accommodates a full size boiler feedwater pump impeller and simulates the associated impeller inlet and outlet passages. By building the test loop with the pump shaft mounted vertically a clear view of the cavitating impeller is obtained. Typical of the output that the facility can produce is (i) photographic evidence of cavitation (Fig. 1.1) and (ii) NPSH(4mm) data (Fig. 11.12).

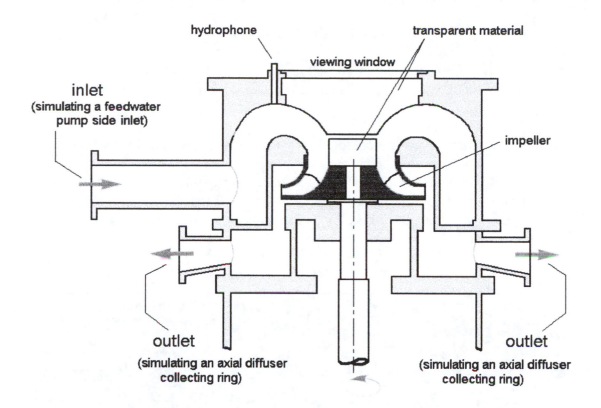

(a) Sectional arrangement

(b) During a test

Fig. 11.9   Visual test facility — Observation of impeller cavitation

Courtesy of Weir Pumps Ltd

11.4 NPSH(3pc) and NPSH(4mm) Test Procedures

11.4.1 *NPSH(3pc) tests*

The objective is to produce an NPSH(3pc)/Q curve. From this the guaranteed value is established. Cavitation tests should be carried out by varying NPSH with the following held constant:

> flowrate,
> pump speed (except for electric motors where section 11.2.4 applies),
> test water temperature.

The alternative "constant NPSH" method of testing described in section 3.6 is not preferred; the onset of cavitation is not clear and this leads to the probability that the pump will be put at risk by being subjected to excessive amounts of cavitation whilst under test.

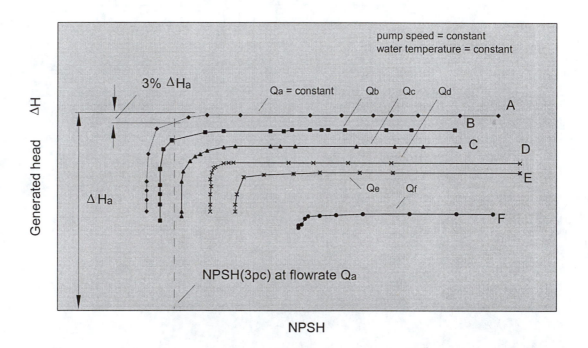

Fig. 11.10 ΔH/NPSH test curves — constant flowrate method

Starting from non-cavitating conditions progressive reduction in NPSH produces a ΔH/NPSH characteristic typically as curve "A" in Fig. 11.10. By keeping to a near constant flowrate control is maintained easily since the system resistance changes very little. A small reduction in NPSH is produced by closing the inlet throttle valve a little. Flowrate falls slightly. Importantly, this is to a condition of lesser cavitation risk. The procedure is then to counteract this by opening the outlet control valve. Using this method it is very noticeable when the "knee" of the curve is reached. Beyond this point cavitation develops extensively within the impeller causing the NPSH value to remain practically constant as the generated head is further reduced. Test observation confirms this as the inlet pressure reading changes very little in value as the generated head is reduced. The 3% deviation in generated head, the NPSH(3pc) value, is easy to observe and the test can be completed without the risk of unknowingly subjecting the pump to unacceptable levels of cavitation. Repeating the procedure at flowrates Q_a, Q_b, to Q_f produces curves "A", "B", to "F". From them the corresponding NPSH(3pc) values are obtained. Transferring the data produces the NPSH(3pc)/Q curve shown in Fig. 11.11. The Q_{bep} can be added to put performance into perspective and, for designers, to assist in confirming the suction specific speed value. The further addition of the guarantee flowrate value Q_G enables the NPSH(3pc) guarantee to be evaluated.

The method requires good control of NPSH at the pump inlet. This is helped considerably if loop designers pay special attention to choosing an inlet throttle valve with good pressure control characteristics. However, it has to be said that, where pump power is small (less than 50 kW) reasonable results can be obtained using a simple gate valve.

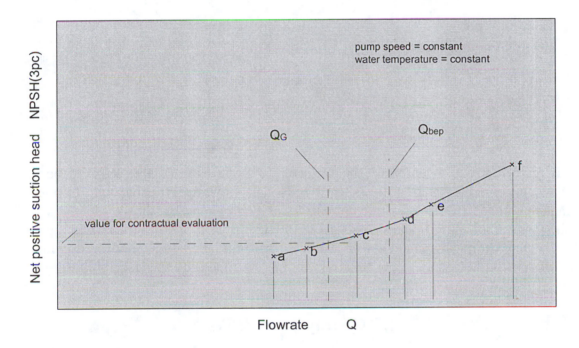

Fig. 11.11 NPSH(3pc)/Q curve

11.4.2 *NPSH(4mm) tests*

The objective is to produce an NPSH(4mm)/Q curve like that shown in Fig. 11.12. From it the guaranteed value at Q_G is established. The test procedure is similar to that for the NPSH(3pc) test — indeed it is common practice for the two to be carried out at the same time. Starting at the non-cavitating condition flowrate is held constant whilst NPSH is progressively reduced. The NPSH corresponding to the first visible cavity is noted. This is close to the value associated with the detection of cavities by hydrophone. At a lower NPSH the value for a 4 mm cavity length is noted. To reduce uncertainty in measuring cavity length it is helpful if, prior to the test, a marker pen line is drawn on the impeller blade 4 mm from the inlet edge. The procedure is repeated at a number of flowrates throughout the operating range. The Q_{bep} can be added to Fig. 11.12 to put performance into perspective and, for designers, to assist in confirming the suction specific speed value. The further addition of the guarantee flowrate value Q_G enables the NPSH(4mm) guarantee to be evaluated.

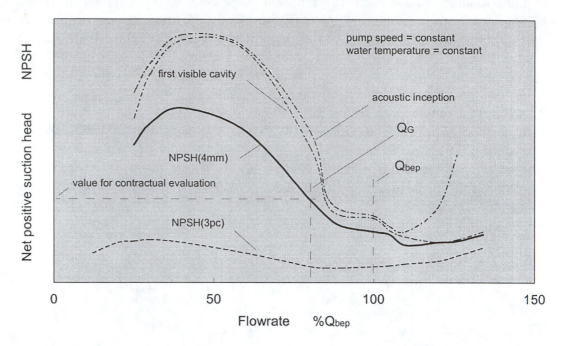

Fig. 11.12 NPSH(4mm)/Q curve   *Courtesy of Weir Pumps Ltd*

In carrying out visual tests the observer is impressed by the stability and repeatability of cavitating flows as NPSH and flowrate are changed. In the apparent chaos of flows within a centrifugal pump impeller shown clearly by Simpson, Cinnamond and Wood (Ref. 11.15) this is both remarkable and reassuring. It means that tests to ascribe particular cavitating conditions with particular NPSH values are unlikely to be swamped by unacceptably large perturbations.

11.5 Quality Level A Tests

These tests do not have a basis on which fair and rigorous contractual arrangements can be defined. Their use is in very special circumstances and is one of qualitative support. Quality level A tests usually fall within the generic description of pump manufacturer research and development. To obtain repeatability the techniques described depend heavily upon in-house knowledge and quality control experience, matters that cannot be expressed satisfactorily in contract specifications.

Evidence from quality level A tests often has significance for users. Where data relates directly to a pump being considered for application within a particular plant it can complement the generic data on which selection is made and give reassurance that the assumed secondary effects remain insignificant. Equally valuable but unquantifiable is the added confidence that accrues from dealing with a pump manufacturer that is (i) demonstrating an understanding of how to minimise risk and (ii) showing in-house competence in controlling difficult techniques.

11.5.1 *Transparent impeller tests*

Transparent plastic impellers have been used to provide an understanding of the cavitating flows on the "pressure" side of impeller blades. This is normally obscured from view at flowrates greater than Q_{bep} during an NPSH(4mm) test. These impellers are both costly and difficult to make. An example is shown in Fig. 11.13.

Fig. 11.13   Transparent plastic impeller

Courtesy of Weir Pumps Ltd

Transparent materials have poor strength characteristics and, where these are plastic, are much more ductile than the metal impeller material being copied. The deflections produced by (i) the forces resulting from impeller rotation and (ii) the hydraulic pressure differences across impeller blades and impeller shrouds have to be such that the validity of results is not impaired. To achieve this, low speed testing is usually inevitable. Even then for impellers of light construction the blade-to-flow angles may not be truly representative of the metal original. A metal centre boss is often found to be necessary to transmit power and facilitate centring the impeller with the shaft axis. Taking the overall view, the use of a fully transparent impeller is not favoured generally on grounds of cost, manufacturing difficulty and poor replication of the operating geometry.

11.5.2 *Paint erosion tests*

Cavitation erosion is the most prevalent of the unacceptable forms of cavitation affecting centrifugal pump performance. Chapter 9 describes generically based rules for avoiding it in many cases and containing it in others. The NPSH(4mm) test described in section 11.4.2 gives a good indication of where cavitation is occurring within an impeller. It does not provide any reliable information as to whether any cavitation observed is likely to produce unacceptable damage in the long term. The question arises as to how satisfactory pump performance is to be demonstrated when cavitation is seen on the impeller blades. Paint erosion tests provide the answer in many cases.

The difficulty in choosing a paint that is sufficiently adhesive that it withstands the high velocity non-cavitating flows and yet will respond to cavitation erosion attack by being removed is evident. The quality control associated with applying the paint to ensure consistency of test data is also a problem. Manufacturers who regularly use the technique, usually to confirm the acceptability of changes in impeller design, develop a great deal of in-house experience. Competence is hard won and is the result of many trials ranging from paints that are "flaky-washable" to those that are "superglue" quality. It is clear that a rigorous contractual basis for such tests is a practical impossibility. The paint erosion test is a test which provides users with assurance but with little proof.

The objective of paint erosion tests is to establish in a short period of time the propensity of cavitating flows to cause erosion. If an impeller surface is attacked within a few hours then a question arises as to its likely long-term survival in service in the same conditions. Such tests can be carried out at full speed, although it is often expedient and instructive to combine this test with visual observation tests where a reduced speed may be necessary. The results from a paint erosion test using a loop with a visual facility are shown in Fig. 11.14. These results are compared with metal erosion test results obtained on the same impeller running at the same %Qbep. The photographs show clearly that the same locations have been attacked by cavitation during both paint removal and metal removal tests. The white paint in Fig. 11.14(a) has been lifted to reveal the darker underlying metal and has produced an irregular removal pattern at the edge of the area under attack. The metal removal pattern in Fig. 11.14(b) is in the identical region but is continuous with a graduated increase in roughness and has a more clearly defined boundary. It is evident that a properly conducted paint erosion test can give a good indication of whether there is a risk from cavitation erosion attack.

The NPSH below which paint erosion takes place has been investigated. Figure 11.15 shows typically where this is. To put it into context the acoustic, NPSH(6mm), NPSH(25mm), and NPSH(3pc) values are also shown. It has to be recognised that the scouring action of different velocity non-cavitating flows and the form/intensity of cavitation implosions at the metal surface makes analysis of results somewhat subjective. Experienced research workers can recognise the signs when paint removal is not due to cavitation alone. This is clearly not an area suitable for securing contractual proof.

(a) Paint removal by cavitation   (b) Metal removal by cavitation

Fig. 11.14 Cavitation erosion test results —
paint and metal removal on the same impeller compared

Courtesy of Weir Pumps Ltd

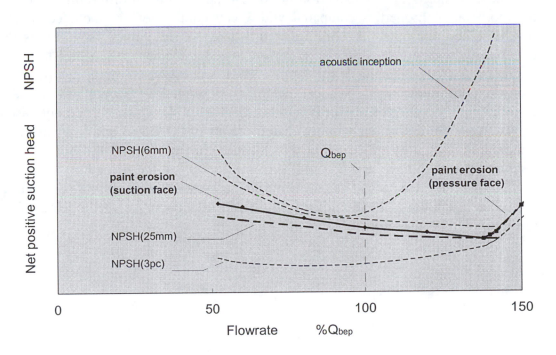

Fig. 11.15 NPSH and paint removal for a particular impeller

Courtesy of Weir Pumps Ltd

11.6 Special NPSH(3pc) and NPSH(4mm) Tests

It is rarely possible to carry out quality level B cavitation tests on pumps that operate at a high speed or pump a liquid other than "clean cold water". The reason is usually one of the following:

(i)   the power absorbed is beyond the capability of the test facility,

(ii) the properties of the liquid make it too dangerous to test in a works environment,

(iii) the cost of testing is prohibitive.

11.6.1 *Tests at reduced speed on clean cold water*

NPSH(3pc) tests. These should be carried out using the procedure described in section 11.4.1. No special requirements are necessary. The results should be scaled using the affinity laws for cavitating pumps. Section 3.7.2 showed that scaling test results that have been obtained on clean cold water to a higher speed is likely to overestimate NPSH(3pc) slightly. This may give users an extra small margin of protection.

NPSH(4mm) tests. These should be carried out using the procedure described in section 11.4.2. It is important to use the actual impeller wherever possible, not a model. Small differences in cast contours (fillet radii, etc.) and surface blemishes may materially alter the cavitation performance of an impeller. The consequence of this is the introduction of a source of error which invalidates the use of **R**(4mm) = 1.05 to achieve acceptable operation. Moving to a higher value of **R**(4mm) significantly reduces the potential economic benefits of visual testing.

11.6.2 *NPSH(3pc) tests on liquids other than clean cold water*

The principal advantage of a test using the liquid for which the pump is supplied is that the benefits of any reduction in NPSH(3pc) arising from the difference in thermodynamic properties compared with clean cold water are included in the results. This advantage is only secured where NPSH values have limits of uncertainty comparable with the +/-10% allowed for clean cold water tests. A potential source of major error from NPSH values based upon temperature measurement rather than physical height is described in section 11.2.5. An evaluation of the sources of NPSH measurement error should be carried out based on the vapour pressure/temperature characteristics of the liquid and the particular temperature measurement device (and its location) proposed.

Apart from the above, tests should be carried out using the procedure described in section 11.4.1. No special requirements are necessary other than those pertaining to the safe handling of the pumped liquid and establishing a methodology that ensures NPSH data are of sufficient accuracy.

11.7 Cavitation Surge Testing

11.7.1 *Identifying the need for a test*

Hydrodynamically induced surge testing should be avoided unless a real risk of its occurrence is perceived. Previous chapters provide direction as to the operating conditions under which risk is likely to be acceptable. Chapter 7 showed that the boundaries of a region within which cavitation surging is possible are loosely described by (i) the flowrate $50\%Q_{bep}$, (ii) the lower value of the operating flowrate range Q_L (where this is less than $50\%Q_{bep}$), (iii) the cavitation head-drop inception curve NPSH(0pc), and (iv) the NPSH(3pc) curve. Chapter 6 showed that for all forms of hydraulic performance the NPSH provided should never be less than 1.3 times the highest value of NPSH(3pc) in the operating range — point A in Fig. 11.16. By following this simple rule the evidence presented earlier suggests that conditions promoting surging be removed. Conversely, the risk of hydrodynamically induced cavitation surging can be said to exist if operation (including transient operation) enters the region CDEF on Fig. 11.16. If the pump is to operate in the region CDEF then a test should be considered.

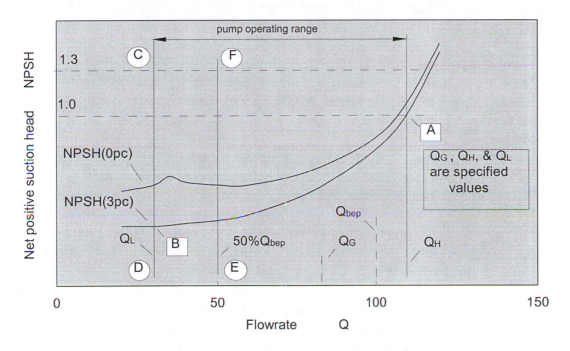

Fig. 11.16 The region of cavitation surge risk (CDEF)

11.7.2 *Identifying the limitations of practical tests*

It is not practical or economical to seek to determine quantitative data relating to hydrodynamically induced cavitation surging by test. Rudimentary qualitative tests are possible.

Specifying a qualitative test raises the question of its value. For although the risk area has been identified, a value measurement cannot be placed on operating performance parameters (i.e. the magnitudes of inlet pressure swing, frequency; axial thrust) to define the limits of acceptability. However, it is prudent to provide for a simple qualitative test when the need is identified and the opportunity arises. All that is required is a minimum 15 diameters of straight inlet pipework, a suitable pressure transducer, and a short period of additional test time.

Cavitation surging is typically a 1 Hz to 10 Hz phenomenon and can be measured by sudden changes to either impeller axial thrust or inlet pipe pressure. For such a rudimentary test the inlet pressure measurement point is normally the only practical option. Tapping into the source of this pressure measurement with a suitable device for measuring dynamic pressure (e.g. an electrical transducer) is the only addition necessary to the instrumentation provided for an NPSH(3pc) test.

11.7.3 *Cavitation surge test procedure and the criterion for concern*

The cavitation surge test should be carried out at two flowrates:

(i) the minimum flowrate in the operating range Q_L,

(ii) 50% Q_{bep}.

The procedure for carrying out the test is simple. At each of the two flowrates the inlet pressure should be progressively reduced from the 1.3 x NPSH(3pc) value associated with point A to the lowest NPSH(3pc) in the pump operating range — point B on Fig. 11.16. The pressure pulsations at the inlet head measurement point should be noted.

Choosing a criterion on which surge test results are contractually guaranteed is impractical. The unpredictable nature of cavitation surging and the limitations of the qualitative test method mean that the results can only be used to indicate an area of concern. The criterion for being concerned that pump performance is inadequate is when the amplitude of pressure pulsations in the 1 Hz to 10 Hz band is more than 10% of the inlet pipe pressure — a value well outside that permitted by many test standards (see Refs. 11.1, 11.2 and 11.3).

11.7.4 *Making use of the test results*

Test results will either (a) provide a limited amount of assurance at little extra cost or (b) demonstrate that pressure perturbations exist that are a cause for concern. Only the latter needs to be addressed here. Evidence of large pressure perturbations during a cavitation surge test lead to a need for pump user/pump manufacturer dialogue. The questions that have to be addressed are:

(i)   Is the risk of operating at the NPSH values at which cavitation surging is observed real?

(ii)   Does cavitation surging pose a threat to the mechanical integrity of the pump? (Are the axial thrust bearings and any mechanical seals able to cope?)

(iii)   Is the risk of operating at the flowrates at which surging occurs real? (Can the operating range be changed to avoid the surge range?) (Can the impeller be changed to modify the region or intensity of surging?)

(iv)   What effect will differences between test and plant inlet pipework geometry make?

The results of the test described do not provide conclusive proof of hydrodynamically induced cavitation surging. They provide evidence for opening a user/pump manufacturer dialogue on assessing the risk of an unacceptable form of cavitation occurring when, for whatever reason, pump operation at low NPSH levels is required. It is worth recalling that if the guidelines given in Chapters 6 and 7 can be adhered to the need for a cavitation surge test is not necessary.

References

11.1   ISO Standard 2548: 1973, Acceptance Tests for Centrifugal, Mixed Flow and Axial Pumps, Part 1, Class C tests, (also published as BS 5316: Part 1: 1976).

11.2   ISO Standard 3555: 1977, Acceptance Tests for Centrifugal, Mixed Flow and Axial Pumps, Part 2, Class B tests, (also published as BS 5316: Part 2: 1977).

11.3   ISO Standard 5198: 1987, Acceptance Tests for Centrifugal, Mixed Flow and Axial Pumps, Part 1, Precision class tests, (also published as BS 5316: Part 3: 1988).

11.4   Centrifugal Pumps – 1.6, Hydraulic Institute Test Standard, 1988.

11.5   API 610, 8th edition, Centrifugal Pumps for Petroleum, Heavy Duty Chemical, and Gas Industry Services, American Petroleum Institute Standard 1995.

11.6   Kanellopoulos, E. V., The influence of total gas content on the inception of cavitation in a small centrifugal pump, Dept. of Aeronautics and Fluid Mechanics, University of Glasgow, April 1959.

11.7   Vuskovic, I., Investigations on the influence of air content on cavitation and corrosion, Escher Wyss News, Vol. XIII, p. 83, 1940.

11.8   Shmuglyakov, L. S., The relation between the cavitation coefficient and the content of air dissolved in water of a hydraulic turbine, Energomashinostroenie, No. 5, p. 11, 1956.

11.9   Le Fur, B., David, J. F. and Pecot, D., The influence of occluded or dissolved air contained in water on pump cavitation, IMechE Conf., Part Load Pumping, pp.97 – 104, 1988.

11.10 Cohen, D. J., Dallas, J. and Knight, F. I., <u>Development of a gas handling hydraulic submersible pump and planning a field trial, Captain field</u>, Offshore Technology Conf., Houston, TX, pp.175–193, 1997.

11.11 Rütschi, K. Von, <u>Messung und drehzahlemrechung des NPSH – wertes bei kreiselpumpen</u>, Schweizer Ingenieur und Architekt, No. 59, pp.971 – 974, 1980.

11.12 Grist, E., <u>The correlation of cavitation characteristics of centrifugal pumps</u>, University College, Cardiff, 1965 (unpublished).

11.13 Paterson, I. and Grist, E., <u>Designing a reactor coolant pump test facility</u>, Nuclear Engineering International journal, pp.22 – 23, Oct,. 1986.

11.14 Troskolanski, A. T., <u>Hydrometry</u>, pp.326 – 376 and Fig. 29, Pergamon Press, Elmsford, NY, 1960.

11.15 Simpson, H. C., Cinnamond, C. and Wood, F. J., <u>The quantitative study of 3D flow patterns in centrifugal pumps</u>, International Assoc. for Hydraulic Research, 10th Congress, London, 1963.

Chapter 12

New Centrifugal Pump Specifications —
Pump Selection and Cavitation

12.1 An Introduction to the Selection Process

This chapter describes a pump selection process and the principal considerations that have to be taken into account when writing a specification if the unacceptable effects of cavitation are to be avoided. It is a "clean paper" approach. That is, a free choice is assumed for all the options. It is a demonstration of how to implement the logic developed in preceding chapters. The broad objective is summed up as being to provide guidance in producing a specification that will secure competitive capital prices for a pump that has the least running cost whilst being free from the risk of unacceptable cavitation.

The objectives and the methodology used in pump selection are described in some detail. This is followed by examples that demonstrate how the selection process can be developed in particular circumstances. A selection based on generic information readily identifies an "optimum" combination described by (a) a pump construction, (b) a prime mover shaft speed, and (c) pump performance curves. Where appropriate a range of options close to this optimum also emerges from the same calculations. Armed with such knowledge a specification can then be produced that anticipates the engineering offers that are likely to be made. A reminder is given that no specification should be written so as to exclude the possibility of a novel (and better) approach being taken. Also due consideration should be given to offers supported by "Application Lists" demonstrating good performance from machines that are identical or nearly identical to that offered. All specifications should request the latter and state that when the data provided have direct relevance the assessment of technical advantages will be biased toward it. Particular data are always better than generic data. Finally, it should be expected that the offers made will be different from the optimum generic choice. Pump manufacturers are, quite rightly, strongly influenced by the "least-cost-to-make" option and use existing proven designs wherever possible. This makes it unlikely that the guaranteed flowrate will be at the best efficiency flowrate even for pumps with a single operational flowrate.

12.2 Specific Objectives Identified

The principal objective is to select an energy efficient pump (or set of pumps) that has a minimal risk of unacceptable cavitation occurring when the plant is running continuously at any flowrate within its operating flowrate range. Additionally the aim is

235

to produce data which assist in defining the contents of a specification for the chosen pump(s) and their low-flow protection requirements. The specific objectives reached by the output from the calculational process preceded by the input data necessary to reach them are listed in sections 12.2.1 and 12.2.2.

12.2.1 *The input: Information provided by the pump user*

The requirements that the contractual guarantees are based upon and which the user must quantify are

- (i)   the pumped liquid (e.g. clean water at between 20°C and 150°C),
- (ii) plant flowrate Q_p = a value (litres per second),
- (iii) plant generated head ΔH_p = a value (metres) at Q_p,
- (iv) estimated annualised plant running by flowrate band = a bar chart*
- (v)   either NPSH(A) = a value at Q_p (metres),

  or   NPSH(R) = an implied request to be advised of the minimum acceptable NPSH(A) for economic operation.

Additionally, any restrictions on the choice of prime mover imposed by the available energy source (e.g. electrical supply, steam quality, etc.) should also be stated.

For variable speed pumps the equation of the system resistance curve

$$h_{sys} = h_s + k_2 Q^2 \quad \text{(equation 4.8)} \quad \text{should also be provided.}$$

* A bar chart, such as shown in Fig. 12.1, is often better than a table in that errors in estimates are more apparent. Where the plant running pattern is particularly complex, however, such as where summer, winter, stand-by and start-up modes exist, a tabular presentation is sometimes more manageable.

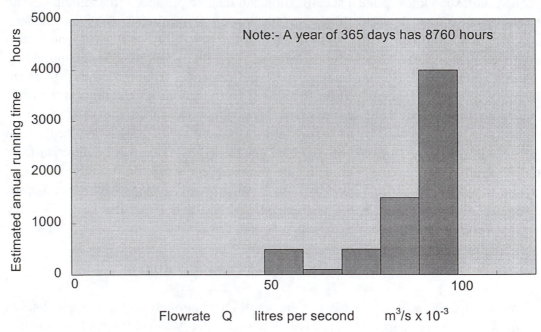

Fig. 12.1 Typical annualised plant running estimate by flowrate band

12.2.2 *The output: Data which will be produced by the selection process*
(i)   The number and rating of pump(s) required;
          the pump duty: guaranteed generated head ΔH_G;
                            guaranteed flowrate Q_G;
          the probable pump speed n;
          the NPSH(R) to ensure risk from cavitation is minimal
          (where NPSH(A) is not given);
(ii)   estimated $\Delta H/Q$ and P_A/Q performance curves;
(iii)   estimate of yearly energy consumption = a value (kilowatt hours);
(iv)   low-flow protection = guidance on probable needs.

12.3 The Methodology

12.3.1 *The plant duty and the plant operating range*

The plant duty $(\Delta H_p, Q_p)$ is set by the design of the plant and is regarded as a fixed point in the pump selection process. Very few pumping applications have a single operating point and even where this is the case consideration usually has to be given to accommodating the differences in on-load running and start-up as described in section 4.2.5. Most pumps operate over a range of flowrates. This range must be identified so that pumps can be chosen to give the most economic operating option. An estimate needs to be made of the plant running patterns. An annual basis is usually sufficient. Longer term differences can be included in a specification if there is a need to include provision for plant upgrading at a later date. Annualised totals can be usefully presented in the form shown in Fig. 12.1.

12.3.2 *Selecting the pump duty and the pump operating range —*
Basic rules

Choosing a pump (or set of pumps) to avoid the unacceptable effects of cavitation during continuous service is achieved with a very high degree of certainty by selecting each pump to have

   (i)   an operating range lower limit Q_L of greater than $50\%Q_{bep}$
        (cavitation surge Chapter 7 and cavitation erosion Chapter 9),
  (ii)   an operating range upper limit Q_H of less than $120\%Q_{bep}$
        (cavitation erosion Chapter 9),
 (iii)   a value for NPSH(A) and a pump suction specific speed n_{ss} of less than 2500
        (summary of international experience — Table 9.2).

Combining the above means that a pump selection based on minimum annualised operating power consumption achieves the principal objective.

The above means that continuous operation should ideally be limited to the $50\%Q_{bep}$ to $120\%Q_{bep}$ flowrate band. Least capital cost is obtained when this can be achieved using a single fixed-speed pump. If the required operating range is greater than that allowed for such a pump then the choice is to

                1. select more than one pump and run them in parallel or
                2. select a variable speed pump.

12.3.3 *Single fixed-speed pumps*

A first estimate of the pump guaranteed duty point (ΔH_G, Q_G) and the operating range can now be made. This is done by assuming single fixed-speed pump operation. Non-cavitating performance is first evaluated to determine the least power absorbed option. Cavitating performance is then examined. If cavitation is a problem then a change is made to conditions that are less efficient but are compatible with reduced cavitation risk. The pump choice is changed to one with a different speed or physical configuration and the whole process repeated. If the change proves insufficient or impracticable then altering the inlet configuration to provide additional NPSH(A) can be evaluated. At each stage an estimated annualised energy cost can be calculated so that, with the change in capital cost attributable to plant design change, a comparative cost can be ascertained. Obviously changes that give rise to better cavitation performance for a trivial cost increase are readily justified. Where a cost change is significant then a rudimentary added-cost/reduced-risk analysis has to be considered.

It is self-evident that the most economic mode of running is likely to be achieved with a single pump operating at the flowrate with the highest cumulative running hours per year. This proves to be true for a great many applications. Only a plant with a very large flowrate spread requirement or a high-energy absorbing plant with unusual operating patterns (e.g. twin flowrate usage peaks, seasonal variations, etc.) is likely to justify changing this view.

Selecting the pump constructional configuration. The available impeller configurations are illustrated in Fig. 2.2. The pumps they are used in are based on

(a) a single-entry impeller,
(b) a double-entry impeller,
(c) a multistage (series mounted impellers) design in which (a) or (b), or combinations of (a) and (b), are used.

More complex pump designs are sometimes used. An example is a multistage pump where part of the flow is "tapped off" at an interstage point. Up to the tapping point full flowrate rated impellers are used; afterward, residual flowrate designs are provided. Whatever the arrangement the selection logic presented using (a), (b) and (c) applies.

Non-cavitation performance. The selection process starts by assuming a single-entry impeller configuration and choosing a pump speed. Calculation of the specific speed enables the best efficiency value at the plant flowrate to be determined. If the specific speed is outside the centrifugal pump range 300 to 2500 then a change is made to another speed. Of course, this has to be within the capability of the chosen prime mover. Too low a specific speed requires the pump speed to be increased and vice versa. As the specific speed value is directly proportional to the pump speed so the available possibilities are soon investigated. Having exhausted all the single impeller options (likely to be the least costly) the more complex double-entry and multistage options are then considered. At the end of this process the most favourable options are listed. It is prudent to list at least four. Some will probably have to be discarded when cavitating conditions are considered.

With experience this evaluation process can be completed as a "back-of-an-envelope" exercise in less than fifteen minutes. An orderly and documented evaluation by an inexperienced person might take up to one hour.

Sufficient data are now available to construct $\Delta H/Q$ and P_A/Q curves. Combining information from the latter with annualised plant running by flowrate band data the estimated annual energy consumption can be calculated. This provides the basis for assessing the change in energy cost when quantifying the disadvantages of changing to pumps that are likely to provide better cavitation protection whilst being inherently less efficient. Such information is a valuable guide. However, it must be remembered that it is only a guide since it is based upon all-encompassing generic statistics that will not reflect the exact performance of pumps tendered against the specification.

Cavitation performance. The evaluation process has so far concentrated on pump choice to give economic operation over a $50\%Q_{bep}$ to $120\%Q_{bep}$ flowrate range. Designs that risk unacceptable cavitation now need to be eliminated.

Table 9.2 gave the suction specific speed values that experienced observers had noted always to be consistent with "no risk of cavitation". The values were presented in two groups to distinguish between pumps in which the design practice might be described as "basic" and those in which it might be described as "best". This former group allows users to make a generic evaluation based on a large degree of certainty – the cautious approach. The second provides a means of quantifying the potential offered by more technically advanced designs. Certainty is given by a suction specific speed value of 2500; possible risk is given by a suction specific speed of 4200. The prudent user reviews plant design at an early stage using these two to determine what difference in NPSH(A) is implied by the two values. Very often the cost and inconvenience of providing NPSH(A) to meet the requirements of $n_{ss} = 2500$ is minimal if identified at this early stage. This contrasts with the near "fait accompli" situation if consideration is left until pump tenders are received.

Note: The evaluation produces a minimum NPSH(A) for continuous pump operation. The lower NPSH values during transients (Chapter 6) require separate consideration.

12.3.4 *Pumps in parallel*

Pumps in parallel are identified by a rating percentage. This is a percentage of the plant flowrate Q_P. For example, two identical pumps sharing the plant duty are described as two 50% pumps (i.e. 2 x $50\%Q_P$ pumps). For low flowrate demand one pump can be stopped thus giving a flowrate spread of $25\%Q_P$ (50% of 50%) to $120\%Q_P$. The flowrate at which changeover occurs has to be above $50\%Q_{bep}$ of the pump that remains in operation. The actual value is determined by comparing and minimising the power absorbed during two-pump and one-pump operation in the overlap region.

Pumps in parallel can give rise to problems. Uneven load sharing or hunting sometimes occurs. Any measures needed to prevent this must be detailed in the offer.

Pumps in parallel do not have to be identical. However, they usually are since such a selection reaps considerable benefits from a reduced spare inventory and a reduced need for supporting paperwork (operation and maintenance manuals etc.). Using a large pump with a small pump adds very little to the plant duty spread. The minimum continuous flowrate for the small pump is proportionally less than for the large pump. In applications where an installed standby pump is necessary to ensure plant flowrate is maintained when pump failure or pump maintenance occurs, the advantages of the 2 x 50% operating mode increase. A 3 x 50% selection is strongly recommended where requirements are met by a flowrate spread of 25%Q_p to 120%Q_p; a minimum spares inventory is an important consideration and an installed spare provides standby security. Regularly changing the pumps chosen to be the operating pair improves standby availability, a feature not easily possible with other combinations. Too often deterioration in a standing pump goes unnoticed and results in it not being available when required.

The choice of 2 x 50% pumps has one obvious disadvantage. Avoidance of operation below 50%Q_{bep} for each individual pump has to be enforced. It is too easy to leave both pumps running and reduce flowrate to below 50%Q_p by putting less than 50%Q_{bep} through each. A strategy for ensuring safe operating practice is required. This can be either by automatic control logic or by operator action. Consideration as to how this is achieved should be reflected in the specification for such pumps.

12.3.5 *Booster and main pump sets*

The NPSH(A) characteristic for a single pump shows a gradual fall with increasing flowrate. This is brought about by the increasing effect of pipe losses incurred in the inlet pipework. Where the plant duty (ΔH_p, Q_p) is met by more than one pump arranged in series the pattern changes for the second pump. This is usually the pump at risk. The extent of the change depends upon the pump set arrangement chosen. Important considerations arise at flowrates on either side of Q_{bep}.

The NPSH(A) and NPSH(3pc) characteristics presented in Fig. 12.2 correspond to the pump set arrangements shown in Fig. 12.3. The characteristics are compared at the same protection level for the plant design flowrate Q_p. There are many other forms of variable speed prime movers and speed changing devices than those named in the illustrations. The principles applied to them are the same. Of all the variants the one designated "pump set arrangement 3" in Fig. 12.3 is by far the most popular.

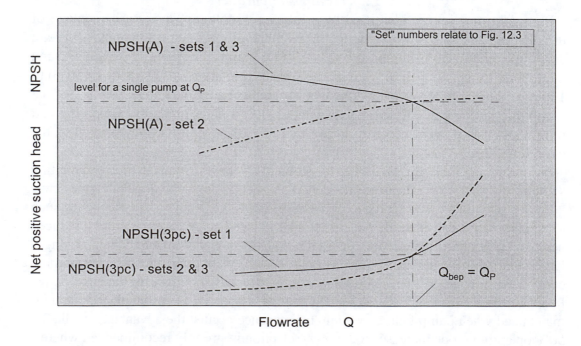

Fig. 12.2 Pump set cavitation protection for main pumps at off-Q_{bep} flowrates

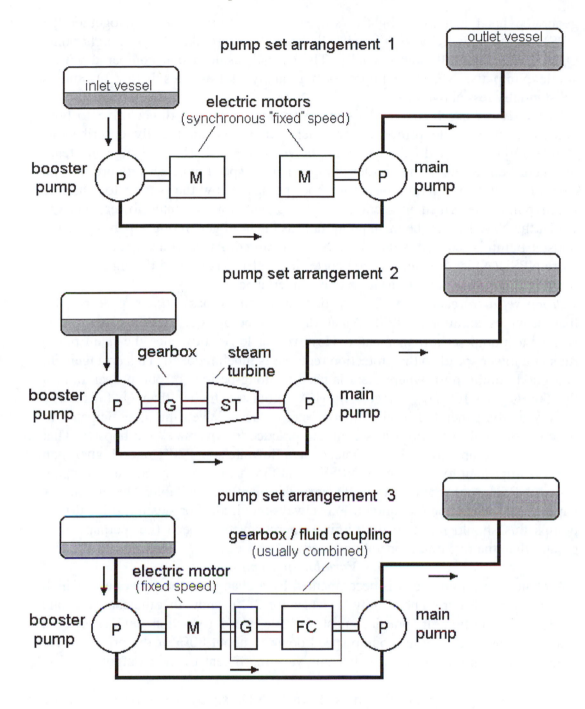

Fig. 12.3 Typical booster and main pump set combinations

The cavitation level of interest is that at the main pump inlet. This is where the greatest risk from cavitation exists. The protection level at the booster inlet is set by the single pump logic discussed earlier. For the purposes of the illustrations it is assumed that $Q_{bep} = Q_p$.

***Pump set arrangement* 1.** The NPSH(A) at the main pump inlet is that at the booster inlet plus the booster pump generated head. As flowrate reduces below Q_p two effects improve matters: the inlet pipework losses are reduced and the booster generated head increases slightly (see Fig. 3.3a). Taking the two together the NPSH(A) rises by perhaps 10% as it falls from Q_p to zero. At flowrates higher than Q_p the NPSH(A) falls away rapidly. The overall result is that compared with a single pump the NPSH(A) improves marginally at flowrates below Q_p and is substantially less above it.

Pump set arrangement 2. This pump set incorporates a direct drive to both pumps from a single prime mover. Such pump sets are usually described as "tandem" pump sets. In the example shown in Fig. 12.3(b) a change in steam turbine speed results in the booster pump changing speed in direct proportion. The cavitation protection requirements for both pumps follow the pattern for variable speed pumps described in section 12.3.6, so that, for flowrates lower than Q_p, whilst the NPSH(A) at the main pump falls as the head generating capability of the booster pump is reduced, so does the NPSH required. At flowrates greater than Q_p more NPSH(A) is provided by the booster just when it is needed if long periods of running under such conditions have to be undertaken.

Pump set arrangement 3. This tandem arrangement has a fixed speed booster that provides additional NPSH(A) at flowrates below Q_p. In this respect it is similar to pump set arrangement 1. However, at lower flowrates the main pump runs at a lower speed so the protection required is much reduced. This arrangement is a good compromise where there is need for long periods of pump operation in the $50\%Q_{bep}$ to $100\%Q_{bep}$ range. NPSH(A) provision higher than Q_p falls away rapidly as the generated head of the booster reduces. When this is anticipated to pose operational difficulties it is common practice to "oversize" the booster. That is, choose a pump with its best efficiency at a flowrate some 20% or so higher than Q_p. It is important to evaluate the NPSH(A) at the maximum flowrate and compare it with NPSH(3pc) when data for the particular plant are available. The margin for vapour lock protection (equation 6.1) is always required. For steam turbine driven pumps this should relate to the turbine emergency trip speed (i.e. perhaps 10% greater than the maximum operating speed).

12.3.6 *Variable speed pumps*

A change in pump speed is accompanied by a change in the flowrate at which best efficiency occurs. The affinity laws (section 3.2.1) show that the change is one of direct proportion. In practice most variable speed reduction devices offer a usable means of changing a pump speed down to about 80% of the prime mover speed. Below this many devices become very inefficient or their output speed is unstable.

A variable speed pump that runs down to 80% speed has a best efficiency flowrate at 80% of the full speed value. By the guidelines given earlier operation is permitted down to 50% of this value so that a flowrate spread of $40\%Q_p$ (50% of 80%) to $120\%Q_p$ is achieved.

Using a variable speed device has the advantage of reducing NPSH(3pc) and NPSH(4mm) where pump operation is at flowrates less than Q_{bep} since, as shown in section 3.7.2, these both vary with the square of the pump speed. For a given plant NPSH(A) the protection at lower flowrates is increased.

The principal disadvantages of variable speed devices are usually a significantly higher capital cost and added control system complexity. Also, at flowrates greater than Q_{bep} the effectiveness of a given level of NPSH(A) protection is significantly reduced.

12.3.7 *Progressing through the evaluation process*

The majority of applications are met by a single pump and most of these pumps are free from any risk from cavitation. The evaluation starts by a simple check of the options available to see if this is the case. A single pump of the most basic design (single-stage, single-entry impeller) is assumed. If no risk is identified the evaluation stops. Where the single pump choice is at risk from cavitation a number of options are considered at speeds that are available from the chosen prime mover. If marginal conditions arise then a change to a double-entry impeller can be considered. The options are expanded to include other variants in pump and pump set design until favoured options are established.

12.3.8 *Low-flowrate protection*

The requirements for specifying appropriate protection measures are discussed in Chapter 10. To evaluate the pump manufacturer's proposals it is necessary to request (a) the value of pump input power P_{AO} and (b) the volume of liquid enclosed by the pump casing V_C.

12.4 Learning by Example

12.4.1 *The range of examples*

Simple examples show how the guidelines are implemented. The procedure leads to choosing a hydraulic design that can be used as the basis of a technical specification. It does not cover mechanical engineering design aspects.

Six examples are given. These are drawn from typical applications. Examples 1 and 2 are representative of general water service pump applications, Example 3 of a chemical process duty, and Examples 4 and 5 of boiler feedwater duties. Example 6 is the duty of a domestic central heating pump. The data used are listed in Table 12.1.

Each example shows how, through the consideration of a number of options, a pump choice is arrived at. The calculations are simple. Examples 1, 2 and 3 are explained in some detail including the rather wordy pump configuration descriptions. As the complexity of examples increases this is abbreviated. An explanation of the abbreviations precedes Example 4.

The following is a reminder of the essentials:

NPSH(A) = available NPSH

NPSH(R) = minimum permissible NPSH(A)   equation 4.1

specific speed $n_s = \dfrac{n\,Q^{1/2}}{\Delta H^{3/4}}$ equation 3.4

suction specific speed $n_{ss} = \dfrac{n\,Q^{1/2}}{NPSH(A)^{3/4}}$ equation 3.11

where ΔH = head per stage at Q_{bep}

Q = flowrate per eye at Q_{bep} section 2.4.1

Table 12.1 Data used in examples

Example no.	1	2	3	4	5	6
Pumped liquid	water @ 20°C	water @ 20°C	ammonia @ 20°C	water @ 150°C	water @ 20°C to 150°C	water @ 80°C
Flowrate Q_P	100 dm³/s	100 dm³/s	10 dm³/s	100 dm³/s	500 dm³/s	0.5 dm³/s
Generated head ΔH_P	50 m	5 m	200 m	2000 m	2000 m	3.5 m
Annual plant running hours	see Fig. 12.1	see Fig. 12.1	see Fig. 12.1	see Fig. 12.8	see Fig. 12.10	5000 @ 0.5 dm³/s
NPSH(A)	not given*	10 m	not given**	50 m	50 m***	0.5 m
Prime mover details (abridged)	electric motor 50 Hz	electric motor 60 Hz	electric motor 60 Hz	electric motor 50 Hz	steam turbine 8000 rpm	electric motor 50 Hz

*   Open vessel – surface at atmospheric pressure
**   Closed vessel – surface at vapour pressure of liquid
*** Additionally evaluate a low level (10 m) inlet vessel option

Example 1

PART A — INPUT
Information provided by the pump user

   Pumped liquid: Water at 20°C
   Plant flowrate Q_P: 100 litres per second
   Plant generated head ΔH_P: 50 m (at 100 litres per second)
   Estimated annualised plant running by flowrate band = see Fig. 12.1
   NPSH(A): Not given; advise NPSH(R), i.e. minimum permissible NPSH(A)
         Inlet taken from a vessel which is open to the atmosphere.
   Prime mover: Electric motor running on 50 Hz supply

PART B — OUTPUT
(i) *The evaluation of pump design options*
***Option* 1A**

Step 1. Assume $Q_G = Q_P$ Spread is easily within the range
   (i.e. pump duty Q_G, $\Delta H_G = Q_P$, ΔH_P) $50\%Q_{bep}$ to $120\%Q_{bep}$

Step 2. Assume pump duty is at Q_{bep}

Step 3. Choose a pump speed   Approximate motor speed (rpm)
   1450 rpm (4-pole) selected   = [60 x number of pairs of poles x
                                                       supply frequency Hz]
                                                    less a small per cent "slip"
                                                    e.g. 2950, 1450, 960 and 720 rpm
                                                    at 50 Hz
                                                 n.b. The higher the speed the smaller
                                                 the pump but the more the risk from
                                                 cavitation.

Step 4. Choose to evaluate single – stage,   This is the basic design —
single-entry impeller pump   see section 2.4.1
Calculate specific speed

$$n_s = \frac{1450\,(100)^{1/2}}{(50)^{3/4}}$$

$$\approx 770$$

Estimate a best efficiency value   A rough guide is required — use Fig. 3.4
$\eta = 79\ \%$ at Q_{bep}

Step 5. Calculate NPSH(R)   (No NPSH(A) is given)
Choose a "no-risk" 2500 suction
specific speed value

$$\text{NPSH(R)} = \left[\frac{1450\,(100)^{1/2}}{2500}\right]^{4/3} \qquad \text{NPSH(R) = minimum NPSH(A) value}$$

$$= 10.2\ \text{m}$$

Assume barometric pressure = 1 bar   [1 x 10.2 = 10.2 m]
Assume inlet pipe velocity is 4 m/s: for this velocity $v_1^2/2g$ is less than 1 m.
Assume the friction loss in the inlet pipe h_{F1} is negligible.
P_{vap} for water at 20°C = 0.02 bar absolute (approx.)   [0.02 x 10.2 ≈ 0.2 m]
From equation 4.10, $\text{NPSH(A)} = h_{S1} - h_{F1} + \dfrac{p_{v1}}{\rho_L g} + \dfrac{v_1^2}{2g} - \dfrac{p_{vap}}{\rho_L g}$

$$10.2 = h_{S1} - 0 + 10.2 + 1 - 0.2$$

$$h_{S1} = -0.8\ \text{m}$$

The selection indicates that a pump could be chosen that draws from a tank with a free surface 0.8 m below the pump shaft centreline as shown in Fig. 12.4. This negative value of h_{S1} is often referred to as "suction lift".

A tank mounted at the same level as the pump baseplate is likely to provide more than adequate NPSH with respect to cavitation risk over a 50%Q_{bep} to 120%Q_{bep} flowrate range. The more mundane considerations necessary to ensure trouble-free performance, measures to avoid drawing air into the inlet pipe through surface vortices, or the ingestion of debris if the pump takes suction from the bottom of a tank need to be considered.

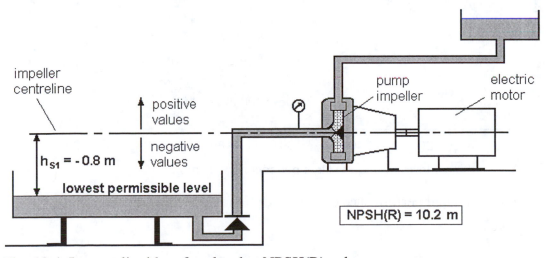

Fig. 12.4 Lowest liquid surface level at NPSH(R) value

***Option* 1B**

Choose pump speed = 2900 rpm

specific speed = $\dfrac{2900 \times 770}{1450}$ = 1540

best efficiency value = 84%   Use Fig. 3.4 again

$$NPSH(R) = 10.2 \times \left[\dfrac{2900}{1450}\right]^{4/3}$$

$$= 25.5 \text{ m}$$

The calculation shows that compared with Option 1A an *increase* in NPSH(R) of 25.5 - 10.2 = 15.3 m is necessary to provide the same level of protection against cavitation. If the added minimum free surface height can be accommodated then the higher speed (smaller) pump with a higher efficiency will be the more commercially advantageous option. If not then a specification based on Option 1A is appropriate.

***Option* 1C**

Choose pump speed = 1450 rpm again

Change to an n_{ss} = 4200 base assessment   See Table 9.2

This gives a best efficiency value of 79 % at Q_{bep} and a minimum NPSH(R) of 5.2 m.

The calculation shows that compared with Option 1A a *decrease* in NPSH(R) of 10.2 - 5.2 = 5.0 m in the level of protection is necessary for the "best" pump designs. The ability to draw down to a lower level is of no advantage for a floor mounted tank application. As a new 4-pole motor will cost more than a new 2-pole motor then a specification based upon Option 1A is appropriate.

THE CHOSEN OPTION — Option 1A.   A single-stage, single-entry 1450 rpm
pump with NPSH(R) = 10.2 m

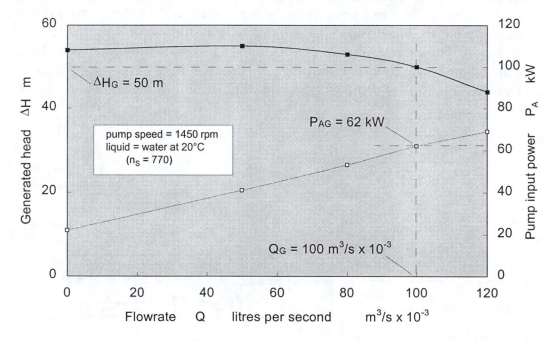

Fig. 12.5 Estimated performance curves — Option 1A

(ii) *Estimated performance curves — Option* **1A**

For the chosen pump ($n_s = 770$) $\Delta H/Q$ curve can be constructed by interpolating between curves in Fig. 3.3(a). The only prior knowledge required is that the pump duty flowrate $Q_G = 100$ litres per second is at $\%Q_{bep} = 100$. Additionally, the P_A/Q curve can be constructed using a value of best efficiency of $\eta = 79\%$ from Fig. 3.4 and interpolating between curves in Fig. 3.3(b). The result is Fig. 12.5.

$$\text{Power absorbed (at } Q_{bep}) \; P_A = \frac{\rho_L \, g \, Q \, \Delta H}{\eta} \qquad \text{(equation 2.5)}$$

$$= \frac{1000 \times 9.81 \times 100 \times 50}{0.79}$$

$$= 62 \text{ kW}$$

Table 12.2 $\Delta H/Q$ and P_A/Q data

$\%Q_{bep}$	%	0	50	80	100	120	specified flowrate ($\%Q_{bep}$)
Q	dm³/s	0	50	80	**100**	120	calculated value at $\%Q_{bep}$
$\%\Delta H$	%	108	110	106	100	88	from Chapter 3, Fig. 3.3(a)
ΔH	m	54	55	53	**50**		calculated value at $\%\Delta H$
$\%P_A$	%	35	63	86	100	112	from Chapter 3, Fig. 3.3(b)
P_A	kW	22	41	53	62	69	calculated value at $\%P_A$

The pump duty is given in **bold**.

A lack of precision in the calculations should be noted. A second significant figure approach is usually adequate. Greater calculational effort produces smoother curves — that is all. Figure 3.5 provides a reminder that considerable variation occurs in commercially available designs. The calculated curves provide users with an indication of performance not a certainty.

(iii) *Estimated annual energy consumption — Option* **1A**

Table 12.3 Power consumed by flowrate band

$\%Q_p$ band $(Q_p = Q_G = Q_{bep})$	50 to 60	60 to 70	70 to 80	80 to 90	90 to 100	100 to 120
Annual hours run (Fig. 12.1)	500	100	500	1500	4000	0
Power absorbed (Fig. 12.5) kW	34	41	48	54	60	0
Yearly energy used in band kWh	17,000	4100	24,000	81,000	240,000	nil

Estimated annual energy consumption = sum of energy used in each band
= 366,100 kWh

(iv) *Low flowrate protection requirements*

Calculate the pipichum value:

pipichum $\varphi = \dfrac{P_A}{\rho_L \, Q_{bep}} \approx 0.6$ (Table 10.1, Category B)

$P_A = 62$ kW at Q_{bep}, $\rho_L = 1000$ kg/m^3, $Q_{bep} = 100$ m^3/s x 10^{-3}

(v) *Observations on cavitation risk to be taken into account when preparing a specification*

Ask for an Applications List showing where one or more plant similar to that offered is in operation.

A 4-pole (1450 rpm) motor driven pump is likely to be a very safe option. It may be worth looking at whether NPSH(A) in excess of 25 m (h_{S1} greater than about +15 m) can be provided to take advantage of cheaper 2-pole (2900 rpm) motor driven pump.

Low flowrate protection by means of a leak-off system is probably unnecessary. However, the following two tier system is worth considering if the pump control system permits its use.

    (i) Trip the pump on outlet vessel "level high".

    (ii) Trip the pump on liquid in the pump reaching "temperature high".

Example 2

PART A — INPUT

Information provided by the pump user

    Pumped liquid: Water at 20°C

    Plant flowrate Q_p: 100 litres per second

    Plant generated head ΔH_p: 5 m (at 100 litres per second)

    Estimated annualised plant running by flowrate band = see Fig. 12.1

    NPSH(A): 10 m

    Prime mover: Electric motor running on 60 Hz supply

PART B — OUTPUT

 (i) *The evaluation of pump design options*

Option 2A

Step 1. Assume $Q_G = Q_P$ Spread is easily within the range
       (i.e. pump duty Q_G, $\Delta H_G = Q_P$, ΔH_p)   50%Q_{bep} to 120%Q_{bep}

Step 2. Assume pump duty is at Q_{bep}

Step 3. Choose a pump speed   e.g. 3550, 1750, 1150, 875 rpm
       1750 rpm (4-pole) selected

Step 4. Choose to evaluate single-stage,   The basic design — see section 2.4.1
       single-entry impeller pump
       Calculate specific speed

$$n_s = \dfrac{1750 \, (100)^{1/2}}{(5)^{3/4}}$$

       ≈ 5200 Too high for a centrifugal pump

Option 2B

Step 3. Choose a pump speed
  875 rpm (8-pole) selected

Step 4. Choose to evaluate single-stage, The basic design — see section 2.4.1
  single-entry impeller pump
  Calculate specific speed
  $n_s = \dfrac{875 \times 5200}{1750}$

  ≈ 2600 Top end of allowable specific speed range

  Estimate a best efficiency value
  $\eta = 82\%$ at Q_{bep} A rough guide obtained from Fig. 3.4

Step 5. Assess the adequacy of NPSH(A)
  Calculate suction specific speed
  $n_{ss} = \dfrac{875 \, (100)^{1/2}}{10^{3/4}}$

  $= 1560$ Satisfactory (less than $n_{ss} = 2500$)

It is instructive to quantify how sensitive the conclusion is by comparing the specified provision of NPSH(A) with that given by a "no risk" calculation.

Step 5A. Recalculate NPSH(A)
  Choose "no-risk" suction specific speed value 2500
  $NPSH(A) = \left[\dfrac{875 \, (100)^{1/2}}{2500} \right]^{4/3}$

  $= 5.3 \text{ m}$

It can be seen from the result that the cover over the minimum that is acceptable using generic data is 10.0 - 5.3 = 4.7 m. A substantial extra margin in this case.

  Power absorbed (at Q_{bep}) $P_A = \dfrac{\rho_L \, g \, Q \, \Delta H}{\eta}$ (equation 2.5)

  $= \dfrac{1000 \times 9.8 \times 100 \times 5}{0.82}$

  $\approx 6.0 \text{ kW}$

An estimate of the shapes of the $\Delta H/Q$ and P_A/Q performance curves and the estimated power consumption is obtained using the procedure described in Example 1. This is presented in Fig. 12.6.

Option 2C

Calculations for Options 2A and 2B indicate a possibility that the duty may be better met by a pump with a specific speed outside the centrifugal range. The high flowrate/low generated head combination characterises that for an axial flow machine. Such a machine may be an alternative to Option 2B.

High specific speed pumps (e.g. axial flow pumps) are outside the scope of this book. However, it may be necessary to compare the two types of machine where borderline conditions arise such as those described by this example.

The trends shown in Fig. 3.3 continue in a more exaggerated fashion for specific speeds greater than 2500. High specific speed pumps are characterised by

(i)   a very high generated head at zero flowrate (may necessitate a higher pressure rating for the outlet pipework);

(ii)   a high absorbed power at zero flowrate, often more than that at the duty point (may necessitate a higher rating for the prime mover — e.g. a larger electric motor);

(iii) an efficiency curve which does not maintain high values over a wide band below or above Q_{bep} (less suited to extensive usage over a wide range).

A fuller description of the trends for axial flow pumps is given by A. J. Stepanoff in Ref. 12.1.

For this example an axial flow pump is assumed unsuitable. Option 2B is chosen.

THE CHOSEN PUMP — Option 2B. A single-stage, single-entry 875 rpm pump
with NPSH(A) =10 m.

(ii)   *Estimated performance curves — Option* **2B**

Table 12.4 $\Delta H/Q$ and P_A/Q data

$\%Q_{bep}$	%	0	50	80	100	120	specified flowrate $\%Q_{bep}$
Q	dm³/s	0	50	80	**100**	120	calculated value at $\%Q_{bep}$
$\%\Delta H$	%	150	135	115	100	75	from Chapter 3, Fig. 3.3(a)
ΔH	m	7.5	6.8	5.8	**5**	3.8	calculated value at $\%\Delta H$
$\%P_A$	%	70	82	90	100	102	from Chapter 3, Fig. 3.3(b)
P_A	kW	4.2	4.9	5.6	6	6.3	calculated value for 20°C water at $\%P_A$

The pump duty is given in **bold**.

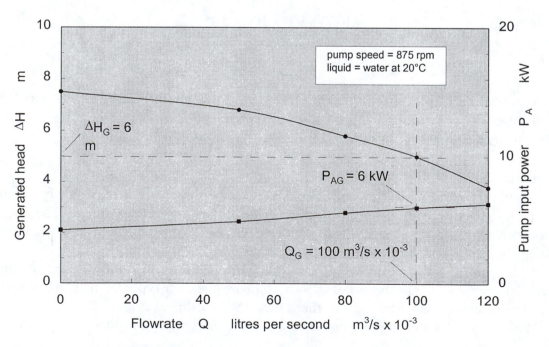

Fig. 12.6 Estimated performance curves — Option 2B

(iii) *Estimated annual energy consumption — Option* 2B

Table 12.5 Power consumed by flowrate band

%Q_p band $(Q_p = Q_G = Q_{bep})$	50 to 60	60 to 70	70 to 80	80 to 90	90 to 100	100 to 110	110 to 120
Annual hours run (Fig. 12.1)	500	100	500	1500	4000	0	0
Power absorbed (Fig. 12.6) kW	5.1	5.3	5.6	5.8	5.9	0	0
Yearly energy used in band kWh	2550	530	2800	8700	23,600	nil	nil

Estimated annual energy consumption = sum of energy used in each band

$$\approx 38,000 \text{ kWh} \quad \text{(to two significant figures)}$$

(iv) *Low flowrate protection requirements*

Pipichum $\varphi = \dfrac{P_A}{\rho_L\, Q_{bep}} \approx 0.06$ \qquad\qquad (Table 10.1, Category A)

$P_A = 6$ kW at Q_{bep}, $\rho_L = 1000$ kg/m^3, $Q_{bep} = 100$ m^3/s x 10^{-3}

(v) *Observations on cavitation risk to be taken into account when preparing a specification*

Ask for an Applications List showing where one or more plant similar to that offered is in operation.

An 8-pole (875 rpm) motor driven pump is likely to be a very safe option. It may be worth looking at whether an axial flow machine can offer advantages.

Low flowrate protection proposals require to be defined in the offer. A leak-off system is unnecessary.

Example 3

PART A — INPUT

Information provided by the pump user

Pumped liquid: Ammonia at 20°C (Ammoniacal liquor)

Plant flowrate Q_p: 10 litres per second

Plant generated head ΔH_p: 200 m (at 10 litres per second)

Estimated annualised plant running by flowrate band = see Fig. 12.1

NPSH(A): Not given; advise NPSH(R), i.e. minimum permissible NPSH(A).

Inlet is taken from a closed vessel that is at the vapour pressure of the liquid being pumped.

Prime mover: Electric motor running on 60 Hz supply

PART B — OUTPUT

(i) *The evaluation of pump design options*

The evaluation starts with the highest (and cheapest) prime mover speed of 3550 rpm. Progressive evaluation of an increasing number of stages is made until reasonable efficiencies (see Fig. 3.4) are encountered. Then the stage and speed combinations are examined until an acceptable NPSH(A) level is attained.

Table 12.6
Options considered and calculated data   (Single-entry options only)

Option	3A	3B	3C	3D	3E	**3F**	3G	3H
Pump speed n	3550	3550	3550	3550	3550	**3550**	1750	1750
No. of stages	1	2	3	4	5	**6**	6	7
n_s	216	355	450	600	705	**800**	346	395
Efficiency η	-	-	-	62	65	**67**	50	50
NPSH(R) m (n_{ss} = 2500)	-	-	-	7.4	7.4	**7.4**	2.9	2.9
NPSH(R) m (n_{ss} = 4200)	-	-	-	3.8	3.8	**3.8**	1.4	1.4

Other options considered and rejected include the following.

(i)   A double-entry impeller on the first stage would reduce NPSH(R) by 1.41. Such a small change is unlikely to justify the added mechanical complexity.

(ii)   Choosing Q_{bep} = 11 litres per second (presuming all 50%Q_p to 60%Q_p band running is low flowrate protection) gives only a 10% increase in Q_{bep} and a 5% increase in n_s and n_{ss}. These are not worth evaluating within the margins of error in the evaluation process.

(iii)   A pump in a pit or a vertical shaft pump down a hole would give a metre or so extra NPSH(A) relative to ground level. Dealing with noxious fumes and leakage rules this out.

NPSH(R) is, to a first approximation, the physical height of the liquid level when drawing from an inlet vessel where the liquid at the free surface is at vapour pressure. The choice depends upon the ability of the user to provide 7.4 m. This is assumed practical for the purposes of this example.

THE CHOSEN OPTION — Option 3F.   A six-stage pump with single-entry impellers running at 3550 rpm with NPSH(R) = 7.4 m.

(ii) *Estimated performance curves — Option* 3F

Contractual performance guarantees for the pump will almost certainly be the subject of tests at the pump manufacturer's works where the test liquid is water. A more powerful prime mover will have to be provided for pump testing to accommodate the additional power absorbed. It is useful to include a power curve for clean cold water so that any cost implications for the user (usually none) can be assessed. Estimated curves are presented in Fig. 12.7.

Table 12.7 Estimated performance curve data

%Q_{bep}		0	50	80	100	120
Q		0	5	8	**10**	12
ΔH		211	220	210	**200**	180
P_A	ammonia 20°C	6	11	16	18	20
P_A	water 20°C	10	19	25	30	33

The pump duty is given in **bold**.　　Data are presented to two significant figures.

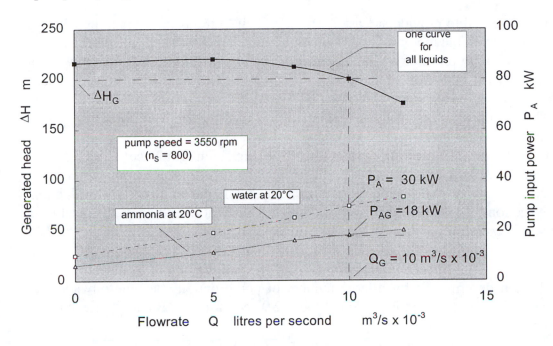

Fig. 12.7 Estimated performance curves — Option 3F

(iii) *Estimated annual energy consumption — Option* 3F

Table 12.8 Power consumed by flowrate band

% Q_P band ($Q_P = Q_G = Q_{bep}$)	50 to 60	60 to 70	70 to 80	80 to 90	90 to 100	100 to 110	110 to 120
Annual hours run (Fig. 12.1)	500	100	500	1500	4000	0	0
Power absorbed (Fig. 12.7) kW	12	14	16	17	18	0	0
Yearly energy used in band kWh	6000	1400	8000	25,500	72,000	nil	nil

Estimated annual energy consumption = sum of energy used in each band
$$\approx 113,000 \text{ kWh}$$

(iv) *Low flowrate protection requirements*

Pipichum　$\varphi = \dfrac{P_A}{\rho_L Q_{bep}}$　≈ 3　　　　(Table 10.1, Category C)

$P_A = 18.0$ kW at Q_{bep},　$\rho_L = 610$ kg/m³,　$Q_{bep} = 10$ m³/s x 10⁻³

(v) *Observations on cavitation risk to be taken into account when preparing a specification*

Ask for an Applications List showing where one or more plant similar to that offered is in operation. No allowance is made for any improvement in NPSH(R) on ammonia compared with water. Evidence of improvement on a near identical plant should be requested, particularly if the pump manufacturer claims a reduction in NPSH(3pc) is achieved when changing from water to ammonia for the pump being offered.

It is essential to provide at least 7.4 m in NPSH(A) if advantage is to be taken of the cheaper 2-pole motor drive. For anything less than 7.4 m a 2-pole offer must be backed up with proof of cavitation performance by an NPSH(3pc) test. Only proof of successful operation of the pump design offered in near identical conditions should permit consideration below an NPSH(A) of less than 3.8 m.

The argument is repeated for 4-pole motor driven pumps with NPSH(A) values of 2.9 m and 1.4 m, respectively.

Low flowrate protection is required. The pipichum guidance value of 3 is marginal. A leak-off system is probably required but this potential requirement needs to be confirmed when particular pump offers are available for evaluation. See section 13.3 for how this may be done. It is prudent to anticipate the need for a leak-off system when drawing up the pump specification.

Special abbreviations used in Examples 4 and 5

Written descriptions for a large number of pump options become unwieldy. A simple abbreviated form is used in Tables 12.9 and 12.11 (A, B and C). Each pump is described as

[impeller entry] **E**/[number of stages]**S**.

The impeller entry description is either single (1) or double (2) and relates only to the impeller at the pump inlet (the first stage impeller in a multistage pump). This should be the only location in the pump where there is a risk of cavitation.

e.g. 2E/5S denotes a **double-e**ntry first stage impeller in a **five-s**tage pump.

Example 4

PART A — INPUT

Information provided by the pump user

> Pumped liquid: Water at 150°C
> Plant flowrate Q_p: 100 litres per second
> Plant generated head ΔH_p: 2000 m (at 100 litres per second)
> Estimated annualised plant running by flowrate band = see Fig. 12.8
> NPSH(A): 50 m
> Inlet is taken from a closed vessel that is at the vapour pressure of the liquid being pumped.
> Prime mover: Electric motor running on 50 Hz supply

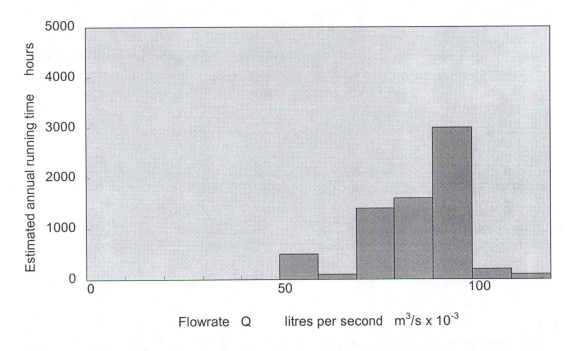

Fig. 12.8 Annual plant running estimate by flowrate band — Example 4

PART B — OUTPUT

(i) *The evaluation of pump design options*

The progressive evaluation of potential options is continued in this example with comments alongside each move to explain the logic applied in the rejection and selection process.

***Option* 4A**

Step 1. Assume $Q_G = Q_P$, — Spread just fits the range $50\%Q_{bep}$ to
(i.e. pump duty Q_G, $\Delta H_G = Q_P$, ΔH_P) $120\%Q_{bep}$

Step 2. Assume pump duty is at Q_{bep} — No options if cavitation risk is to be minimised

Step 3. Choose a pump speed
2950 rpm (4-pole) selected

Step 4. Choose to evaluate a single- — The basic design — see section 2.4.1
stage, single-entry impeller pump
Calculate specific speed

$$n_S = \frac{2950\,(100)^{1/2}}{(2000)^{3/4}}$$

$n_S \approx 96$ — Too low a specific speed (see Fig. 3.4)
Clearly a need for multistaging

***Option* 4B**

Choose a 10-stage pump

Step 4. Choose to evaluate single-entry impeller pump

Keep to single-entry for now. Leave the more complex double-entry option for "fine tuning" later.

Calculate specific speed

$$n_S = \frac{2950 \, (100)^{1/2}}{(200)^{3/4}}$$

n.b. 200 m per stage

$$n_S \approx 540$$

Estimate a best efficiency value

$$\eta \approx 75\%$$

Use Fig. 3.4

Step 5. Assess the adequacy of NPSH(A)

Calculate suction specific speed

$$n_{SS} = \frac{2950 \, (100)^{1/2}}{(50)^{3/4}}$$

$$n_{SS} \approx 1550$$

***Option* 4C**

The pumps considered so far have a low specific speed. A larger number of stages is mechanically possible but several factors mitigate against them, not the least of which is first cost. Indeed a 10-stage pump is not really favoured. Multistage pumps are not easy to maintain and increasing shaft length (to accommodate the stages) increases the risk of internal rubbing. With a fixed speed electric motor drive the next option to try is a pump driven through a speed step-up device (e.g. a gearbox). Such a pump will require a smaller number of stages.

Multistage pumps of up to five stages show good availability and return reasonable maintenance costs at speeds up to 5000 rpm. Above this speed a maximum of three stages has been shown in practice to be the reliable engineering choice (see Ref. 12.2).

Step 3. Choose a pump speed

5000 rpm selected

i.e. A 1.7:1 step up in speed

Step 4. Choose to evaluate a five-stage, single-entry impeller pump

Calculate specific speed

$$n_S = \frac{5000 \, (100)^{1/2}}{(400)^{3/4}}$$

$$n_S \approx 550$$

Estimate a best efficiency value

$$\eta \approx 75\%$$

Worse than Option 4B since gearbox losses have to be deducted.
Additional services — e.g. water cooling; oil system, etc. add to complexity and cost.

Step 5. Assess adequacy of NPSH(A)

Calculate suction specific speed

$$n_{SS} = \frac{5000 \, (100)^{1/2}}{(50)^{3/4}}$$

$$n_{SS} \approx 2570$$

***Option* 4D**

Step 3. Choose a pump speed
      8000 rpm selected   i.e. A 2.75:1 step up in speed

Step 4. Choose to evaluate a three-stage,
      single-entry impeller pump
      Calculate the specific speed

$$n_s = \frac{8000\,(100)^{1/2}}{(666)^{3/4}}$$

$$n_s \approx 610$$

      Estimate a best efficiency value   A probable 2% gain in efficiency for
      $\eta \approx 77\%$ choosing a higher speed and a simpler
                                      pump construction.

Step 5. Assess adequacy of NPSH(A)

$$n_{ss} = \frac{8000\,(100)^{1/2}}{(50)^{3/4}}$$

$$n_{ss} \approx 4300$$

    The use of a speed step up gives much greater flexibility. The numerical combinations this provides can be sampled to determine whether more viable options exist. It is only worth pursuing this until a specification can be produced that encompasses the likely "winner" and enough choice to generate commercial competition. When using generic data it is a mistake to try to optimise selection in detail; the basis of the methodology will simply not support this.

***Option* 4E**

Step 3. Choose (a) a ***booster*** pump speed
          of 2950 rpm   Direct drive from the electric motor
      (b) a ***main*** pump speed
          of 8000 rpm   A 2.75:1 step-up in speed

Choose (a) a three-stage, single-entry impeller main pump
      (b) a single-stage, single-entry impeller booster pump

Inlet conditions nearly always set the limit on acceptable designs that comprise a booster pump and a main pump mounted together as a pump set. When evaluating the performance of pump set combinations *always* start at the booster inlet.

Step 4.

      Calculate the suction specific speed
      for the booster pump

$$n_{ss} = \frac{2950\,(100)^{1/2}}{(50)^{3/4}}$$

$$n_{ss} \approx 1530$$

Choose a specific speed to give a good efficiency   ($n_s = 1500$; 85% — see Fig. 3.4)

      Calculate the generated head for the booster pump

$$\Delta H = \left\lceil \frac{2900\,(100)^{1/2}}{1500} \right\rceil^{4/3}$$

$$\Delta H = 52 \text{ m}$$

$$\text{Main pump NPSH(A)} = \text{Booster NPSH(A)} + \text{Booster } \Delta H$$
$$= 50 + 52$$
$$= 102 \text{ m}$$

$$\text{Main pump } n_{ss} = \frac{8000 \, (100)^{1/2}}{(102)^{3/4}}$$
$$\approx 6150 \qquad \text{Too high}$$

The results for all the options evaluated are summarised in Table 12.9.

Table 12.9 Summary of pump options

Option		4A	4B	**4C**	4D	4E booster	4E main
Pump speed n	rpm	2950	2950	**5000**	8000	2900	8000
Impeller configuration		1E/1S	1E/10S	**1E/5S**	1E/3S	1E/1S	1E/3S
Specific speed n_s		96*	540	**550**	610	1530	6150*
Suction specific speed n_{ss}		-	1550	**2570**	4300	1500	-
Best efficiency	%	-	75	**75**	77	85	-
Evaluation outcome		*	**	**	**	-	*

* = unacceptable   ** = basis for competitive tendering   chosen option in **bold**

For the continuation of this example Option 4C is chosen.

THE CHOSEN PUMP — Option 4C

A five-stage pump with a single-entry, first-stage impeller running at 5000 rpm with NPSH(A) = 50 m

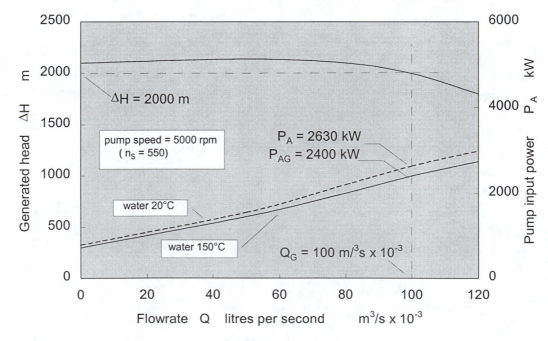

Fig. 12.9 Estimated performance curves — Option 4C

(ii) Estimated performance curves

$$P_A = \frac{917 \times 9.8 \times 100 \times 2000}{0.75}$$

$$\approx 2400 \text{ kW}$$

Using the same method as in earlier examples Fig. 12.9 is constructed.

(iii) Estimated annual energy consumption

Table 12.10 Power consumed by flowrate band — Option 4C

%Q_P band		50 to 60	60 to 70	70 to 80	80 to 90	90 to 100
($Q_P = Q_G = Q_{bep}$)						
Annual hours run		500	100	500	1500	4000
Power absorbed	kW	1500	1700	1900	2100	2300
Yearly energy used	MWh	750	170	950	3150	9200

Estimated annual energy consumption = 14,220,000 kWh

(iv) Low flowrate protection requirements

Pipichum $\varphi = \dfrac{P_A}{\rho_L Q_{bep}} \approx 26$ (Table 10.1, Category C)

$P_A = 2400$ kW at Q_{bep}, $\rho_L = 917$ kg/m^3, $Q_{bep} = 100$ m^3/s x 10^{-3}

(v) Observations on cavitation risk to be taken into account when preparing a specification

Ask for an Applications List showing where one or more plant similar to that offered is in operation.

A choice has to be made between extreme mechanical complexity (Option 4B: ten stage), limited mechanical complexity (Option 4C), and a simple construction with a potentially higher efficiency which has a significant risk from cavitation erosion (Option 4D: $n_{ss} = 4300$). Without particular plant design input it would seem prudent to produce a specification that allows any of the three to be offered and to make a final choice based on an assessment of data supplied with the tenders.

No allowance should be made for any improvement in NPSH(A) on 150°C water compared with 20°C water.

A leak-off system is essential. Consider duplicating the leak-off valves if the cost of pump seizure (pump repair cost plus consequential product loss) is very high.

Example 5

PART A — INPUT

Information provided by the pump user

Pumped liquid: Water at 20°C – 150°C

Plant flowrate Q_p: 500 litres per second

Plant generated head ΔH_p: 2000 m (at 500 litres per second)

Estimated annualised plant running by flowrate band = see Fig. 12.10

NPSH(A): Evaluate (a) 50 m, (b) 10 m

Inlet is taken from a closed vessel that is at the vapour pressure of the liquid being pumped.

Prime mover: Steam turbine — maximum speed 8000 rpm
— minimum speed 6000 rpm

System resistance equation: $h_{sys} = 1500 + 0.002\,Q^2$ (metres)

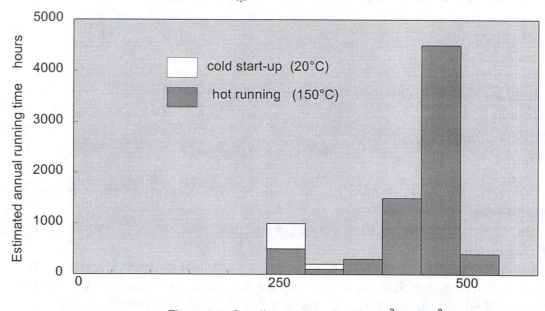

Fig. 12.10 Annual plant running estimate by flowrate band — Example 5

The calculation procedure is the same as that used for the previous examples. Only the tabulated results are presented. The data chosen to define each calculational option are <u>underlined</u>. The other numbers are calculated from them.

PART B — OUTPUT

(i) *The evaluation of pump design options*

Table 12.11A 1 x 100% pump options   NPSH(A) = 50 m

Option		5A	5B	5C	5D	5E
Pump speed n	rpm	<u>8000</u>	<u>8000</u>	7200	<u>8000</u>	<u>8000</u>
Impeller configuration		<u>1E/1S</u>	<u>2E/1S</u>	<u>2E/1S</u>	<u>1E/1S</u>	<u>2E/1S</u>
Specific speed n_s		1900	1900	<u>1700</u>	1350	1350
Suction specific speed n_{ss}		67,500*	47,900*	19,200*	6750	4800
Evaluation outcome		*	*	*	*	*

* = unacceptable

Table 12.11B 2 x 50%[booster + main] pump set options   NPSH(A) = 50 m

Option		5F	5G	5H	5J	5K	**5L**
BOOSTER PUMP							
Impeller configuration		1E/1S	**1E/1S**				
Suction specific speed n_{ss}		2400	**2400**	as	as	as	**as**
Pump speed n	rpm	2800	**2800**	5G	5G	5G	**5G**
Specific speed n_s		1500	**1200**				
Generated head ΔH	m	94	**122**				
Best efficiency η	%	85	**84**				
MAIN PUMP							
Impeller configuration		1E/1S	1E/1S	2E/2S	2E/3S	2E/4S	**2E/5S**
Generated head ΔH	m	1906	1878	**1878**			
NPSH(A)	m	144	172	**172**	as	as	**as**
Pump speed n	rpm	8000	8000	**8000**	5H	5H	**5H**
Suction specific speed n_{ss}		3040	2660	**1990**			
Specific speed n_s		440*	450*	540	735	905	**1120**
Best efficiency η	%			75*	80**	82**	**83****
Evaluation outcome		*	*	*	**	**	****

* = unacceptable   ** = basis for competitive tendering   chosen option in **bold**

Table 12.11C 2 x 50%[booster + main] pump set options   NPSH(A) = 10 m

Option		5M	5N	5P	**5Q**
BOOSTER PUMP					
Impeller configuration		1E/1S	*2E/3S*		
Suction specific speed n_{ss}		2400	*2400*	as	*as*
Pump speed n	rpm	860	*1230*		
Specific speed n_s		1000	*1000*	5N	*5N*
Generated head ΔH	m				
Best efficiency η	%				
MAIN PUMP					
Impeller configuration		-	1E/1S	2E/2S	*2E/3S*
Generated head ΔH	m	-	1960		
NPSH(A)	m	-	50	as 5N	*as 5N*
Pump speed n	rpm	-	8000		
Specific speed n_s		-	435	310*	*715*
Suction specific speed n_{ss}		-	6600*	4700	*4700+*
Best efficiency η	%	-			*77*
Evaluation outcome		-	*	*	*+*

* = unacceptable   *+ = best possibility — at the limit of acceptable industrial practice*

THE CHOSEN PUMP SETS — Option 5L: two 50%Q_p rated pump sets in parallel

A 2 x 50%Q_p pump set configuration is chosen from Table 12.11. Each of the two pump sets comprise a single-stage single-entry booster pump running at 2800 rpm, in series with a five-stage main pump running at 8000 rpm. The main pump first stage has a double-entry impeller. An NPSH(A) of 50 m will enable a design to be offered based upon normal industrial practice.

An estimated performance curve for one Option 5L pump set is shown in Fig. 12.11.

(ii) *Estimated performance curves — Option 5L*

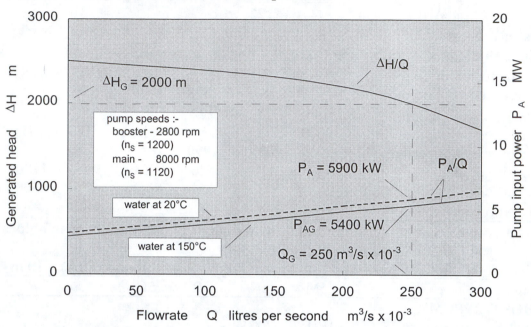

Fig. 12.11 Estimated maximum speed performance curves – Option 5L
(one pump set)   (n.b. a pump set = booster + main)

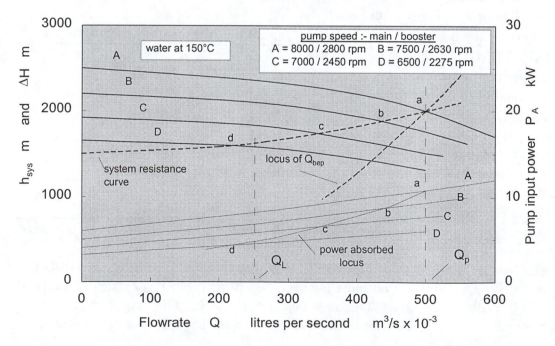

Fig. 12.12 Variable speed operation — Option 5L (two pump sets combined)

(iii) *Estimated annual energy consumption - Option 5L*

The variable speed control of a steam turbine enables it to provide pumping power over the required operating range as dictated by the system resistance curve. As the turbine speed is reduced the forward flowrate is correspondingly reduced. Operation is along the system resistance line a, b, c, d shown in Fig. 12.12. Pump speed and power absorbed reduce as performance is dictated by curves A, B, C and D. The power absorbed is given by the power absorbed locus a, b, c, d. For the particular example an estimate of the total energy consumption is calculated using Table 12.12.

Table 12.12 Power consumed by flowrate band
(Total for two pumps — Option 5L)

%Q_P band ($Q_P = 2Q_G = 2Q_{bep}$)	50 to 60	60 to 70	70 to 80	80 to 90	90 to 100	100 to 110
WATER AT 20°C						
Annual hours run	500	100	-	-	-	-
Input power P_A centre band values kW	4500	5100	-	-	-	-
Yearly energy in band MWh	2250	510	0	0	0	0
WATER AT 150°C						
Annual hours run	500	100	300	1500	4500	400
Input power P_A centre band values kW	4200	4400	4700	5000	5300	5700
Yearly energy in band MWh	2250	510	1410	7500	23,850	2280

Estimated annual energy consumption (double the single pump data in Fig. 12.11)

on water at 20°C = 2760 MWh,
on water at 150°C = 378,000 MWh.

Total estimated annual energy consumption ≈ 380,000,000 kWh
(to two significant figures)

(iv) *Low flowrate protection requirements (each pump set)*

Pipichum $\varphi = \dfrac{P_A}{\rho_L Q_{bep}}$ ≈ 24 (Table 10.1, Category C)

P_A = 5400 kW at Q_{bep}, ρ_L = 917 kg/m³, Q_{bep} = 250 m³/s x 10⁻³

(v) *Observations on cavitation risk to be taken into account when preparing a specification*

Ask for an Applications List showing where one or more plant similar to that offered is in operation.

Table 12.11A shows that a single 100%Q_P rated pump cannot meet the duty without serious risk of cavitation erosion attack. An n_{ss} value of 4800 (Option 5E) is well over the "best" figure for international practice. If such a pump were offered the basis for proof of performance — from operating data on very similar pumps and contractual proving tests — NPSH(4mm) essential — would need close examination.

Table 12.11B shows that a number of options exist, none of which involves risk from cavitation over the required operating range.

Conversely, as shown in Table 12.11C, reducing NPSH(A) from 50 m to 10 m means the only viable option identified requires machinery that is at the boundaries of pump technology. This exercise enables users to make an informed judgement as to whether the benefits in plant capital cost reduction are sufficient to justify taking the risk technological progress entails if a pump like this is offered.

No allowance should be made for any improvement in NPSH(R) on 150°C water compared with 20°C water.

A leak-off system is essential. Consider duplicating the leak-off valves if the cost of pump seizure (pump repair cost plus consequential product loss) is very high.

Note: The cost attributable to lost production is zero if a standby pump (making a 3 x 50% pump arrangement) is installed.

Example 6

One of the smallest pumps in common usage is the domestic central heating pump. Whilst it is inappropriate to produce a specification for such a mass-produced pump it is of interest to evaluate some aspects of performance.

Domestic central heating systems are either on or off. There is usually no attempt at flow control. The duty is fixed by a system resistance which changes as the heat sinks (radiators) are brought into and out of service.

PART A — INPUT

Information provided by the pump user

      Pumped liquid: Water at 80°C

      Plant flowrate Q_p: 0.5 litre per second

      Plant generated head ΔH_p: 3.5 m (at 0.5 litre per second)

      Estimated annualised plant running by flowrate band = 5000 hours at Q_p.

      NPSH(A): 0.5 m (a pessimistically low value)

      Prime mover: Electric motor running on 50 Hz supply

PART B — OUTPUT

  (i) ***The evaluation of pump design options***

Assume a single pump ($\Delta H_G = \Delta H_p$: $Q_G = Q_p$)

          a single-stage pump

          a single-entry impeller

Calculate specific speed and suction specific speed for likely pump speeds.

Table 12.13 Summary of design options

			Option 6A	Option 6B
Pump shaft speed	n	rpm	1900	1750
Specific speed	n_s	-	530	480
Pump best efficiency	η	%	30	25
Suction specific speed	n_{ss}	-	2200	2000

Cavitation is not a concern even when NPSH(A) is as low as 0.5 m (very hot water in the system, low header tank); most systems have a greater NPSH(A) at all times. The choice appears to be between noise (go slower = quieter) and motor cost (go faster = less copper in the motor windings). However, it is probably more important for this particular application to have a continuously falling $\Delta H/Q$ curve and an ability to run to much higher flowrates to cater not only to low system resistance (radiators closed) but for customers who grossly overestimate the system resistance. A further option is considered.

Option 6C   Choose $Q_p = 0.7$ litres per second and $\Delta H_p = 2.5$ m
i.e. Q_G (the duty flowrate) $= 0.5/0.7 = 0.66 Q_{bep}$
At 1900 rpm: $n_s = 800$ $n_{ss} = 2500$
$$P_{Abep} = \frac{\rho_L g \, \Delta H \, Q}{\eta} = \frac{970 \times 9.81 \times 3.5 \times (0.7 \times 10^{-3})}{0.25} \approx 100 \text{ watt}$$
At pump duty ($66\% Q_{bep}$) $P_A \approx 70\% P_{Abep}$ see Fig. 3.3(a)
i.e. pump absorbs ≈ 70 watt   (at $Q_{bep} = 2.5$ dm³/s at $\Delta H = 3.5$ m)

Allowing for start-up on cold water (higher density) and a low system resistance the power absorbed could be approximately $\dfrac{\rho_L \text{ (cold)} \times P_{Abep}}{\rho_L \text{ (hot)}} = \dfrac{1000 \times 100}{970} \approx 105$ watt

Selling to domestic consumers it would appear prudent to recommend provision for, say, 110 watt as the maximum power draw.

THE CHOSEN PUMP — Option 6C
A single-stage, single-entry 1900 rpm pump with NPSH(A) = 0.5 m
   (ii)   *Estimated performance curves*
Easily drawn — see Example 1 on how to do it. However, the pump manufacturer will probably opt to choose an area ratio **Y** to accentuate the falling $\Delta H/Q$ characteristic.
   (iii)   *Estimated annual energy consumption — Option 6C*
Annual energy consumption $= 5000 \times 70 \times 10^{-3} = 350$ kWh
   (iv)   *Low flowrate protection requirements*
Pipichum $\varphi = \dfrac{P_A}{\rho_L \, Q_{bep}} \approx 0.1$ (Table 10.1, Category A)
$P_A = 0.070$ kW at Q_{bep}, $\rho_L = 970$ kg/m³, $Q_{bep} = 0.7$ m³/s x 10^{-3}

   The cost of a leak-off system is likely to exceed the cost of a new pump. If the boiler reliably shuts down when water temperature is high using its own protection system then this is usually deemed adequate.

   (v)   *Observations on cavitation risk to be taken into account when purchasing or modifying a central heating system*
   Low flowrate conditions arise when the pump is on but demand is minimal. In these circumstances radiators, especially those fitted with thermostatic control valves, are closed down. If every radiator is switched off the mass of water in the small bore pipe is the churned mass. In many systems the water heats up until a thermostat trips the pump. If this fails the pump eventually becomes vapour locked and seizes. To minimise the possibility of this happening it is prudent to leave one radiator, preferably in a relatively cold room, without a radiator valve. The effect of the additional heat sink

and slight increase in "churned mass" is a significant delay in the time it takes to reach high temperatures. In systems with a low boiler rating the trip temperature may never be reached.

12.4.2 *A comment on the variety of options available*

It is evident from the examples evaluated that there is a huge variety of options available for each application. The variety is made up from the possible permutations of impeller design, impeller configuration, pump set design and choice of running speed(s). Choosing not to have the best efficiency flowrate at the "guarantee" conditions or reducing the NPSH required by seeking "best" practice pump designs backed by stringent test requirements rather than more "basic" technology adds further to the possibilities. Whilst the selection process soon "weeds out" pumps with inherent disadvantages it is clear that there will always be a wide choice of pumps for each duty.

12.5   The Benefits of Understanding the Selection Process

Working through the selection process before issuing a specification gives a guide to the value of NPSH(A) that pump manufacturers are likely to deem necessary. This enables the user to determine the ease or likely difficulty encountered in meeting this probable need.

An estimate of the annual energy consumption enables early consideration to be given to particular pump running pattern proposals. It also helps put a value on changes that could be made to meet other considerations; for example, plant design change options relating to the plant process that the pump maker may be unaware of or may not fully understand.

The examples presented are simplified to highlight how to deal with particular facets of performance. In pumping applications with complex running patterns or where the selection process has presented numerous pump design choices, the annualised power utilisation can be costed. A comparison can be made, for example, by evaluating the cost to improve performance by increasing NPSH(A). The data provide a basis for cost/benefit calculations *before* the specification is finalised. Choices that result in an unacceptable restriction on plant design can be identified. They can also be described as disadvantageous in the specification with particular confidence when more beneficial options are known to exist. The outcome can be a useful guide to the pump manufacturer without being overly prescriptive.

By appreciating the pump manufacturer's view a more appropriate specification can be produced. Serious dialogue between pump user and pump maker can be quickly initiated when an unacceptable risk is perceived. Appropriate confirmatory cavitation tests can be insisted upon.

Completing a full evaluation turns a pump user into an informed pump user.

References
12.1   Stepanoff, A. J., <u>Centrifugal and Axial Flow Pumps</u>, Wiley, New York, 1957.
12.2   Grist, E., <u>Advanced class boiler feed pumps — Operational experience</u>, BPMA Tech. Conf., pp. 1–21, 1981.

Chapter 13

New Centrifugal Pump Offers —
Technical Assessments and Cavitation

When the offers of pumps to meet the requirements of a technical specification are received a technical assessment should be carried out. The significance of differences between offer and specification has to be evaluated in terms of technical acceptability and commercial worth. It forms a technical underpinning for commercial decisions.

In moving to the assessment stage the basis of technical judgements changes. Generic evaluations are no longer appropriate. Particular data are always better than generic data. So only the hydraulic performance data included with the offer should be used.

Ideally an offer to supply a centrifugal pump should fully cover all aspects of cavitation performance. This should include (i) relevant Application Lists, (ii) comprehensive performance data, (iii) works test proposals and (iv) low flowrate protection proposals. It is often worth carrying out a preliminary assessment to check the completeness of the information supplied. A request for the information required "to enable the offer to be considered fully" usually meets with a prompt response. The following guidance for the technical assessment of cavitation performance assumes the required information has been received.

13.1 Application Lists

These are the most important part of the technical offer. Normally the lists are an edited extract from the manufacturer's order book. The key items to look for are

  (i)   pump model reference,
  (ii) the date pumps were supplied,
  (iii) the duty and operating speed (ΔH, Q and n),
  (iv) constructional arrangement (number of stages and impeller configuration),
  (v)   the pumped liquid (name, temperature and relevant properties),
  (vi) the customer.

It is not usual for the lists to include meaningful NPSH(A) data.

The first task is to categorise relevant information. It is worth remembering that the list is probably part of an on-going compilation. The last order is often the last on the list. Most manufacturers limit the data issued with a particular offer to a particular model. At first sight the lists appear to be over lengthy. This is understandable and is to be welcomed. It is time consuming for the manufacturer to segregate pumps for a particular application interest; it allows the user to appreciate the status of the model in the marketplace (i.e. well-proven old technology; recent design benefiting from latest ideas).

The date when particular pumps have been supplied is very relevant. It indicates whether the design deficiencies that occur in every new model have had time to be worked through. Note also that because many lists are based on entries at the "order confirmed" stage some of the pumps referred to may not yet have been manufactured.

By segregating the data in the Application Lists into a table of "near-identical" pump designs and a summary table of "relevant generic" pump applications a picture emerges of the likely proven experience of pump suppliers.

It should be appreciated that the information a pump manufacturer presents has had little chance of being updated after a pump has been delivered, installed, and commissioned. The exception is when there has been a need to visit a site to rectify deficiencies or to return a pump to the manufacturer's works to repair damage. Outside any guarantee period there is often perceived to be little commercial incentive for manufacturer/user dialogue.

It is not uncommon to conclude that no request for repairs or spare parts equates with proof of good performance. This might be the case, it usually is, but it is not always so. Instances are known to exist of pumps being so bad that they have been replaced by those of another manufacturer. More typical of the source of a "false-positive" message is the early decommissioning of a plant. The collapse of the market for the end product of a process plant or the rationalisation of plants following a company take-over can result in almost-new pumps being consigned to the scrap heap. There is no mechanism for this sort of information to filter back to the pump manufacturer, so whilst Application Lists are not presented with an attempt to mislead, the validity of the information in them cannot be assumed.

Questioning the validity of Application Lists with the pump manufacturer's previous customers should only be pursued when the adequacy of cavitation protection is marginal and the pumps are in the "near-identical" group. Contact should be direct, preferably with a person responsible for Operation and Maintenance activities. Prior permission from the pump manufacturer should be obtained. Remember that the fellow user may not wish to comment and will certainly not wish to be drawn into a lengthy probe: all users have a job to do and this is not part of it. However, most are willing to give an overall impression and confirm the extent of operation under the arduous conditions of interest. This is all that should be expected. A judgement has to be made on the level of objectivity and the competency of the person responding. Whatever this may be good manners dictate courtesy at all times.

13.2 Performance Data

An evaluation of the risk of unacceptable cavitation is achieved by considering in turn whether (i) cavitation is a risk, (ii) cavitation erosion is a risk, (iii) vapour locking during transient operation is a risk, and (iv) cavitation surging is a risk.

This requires the suction specific speed to be calculated from the data supplied as part of the offer. Knowing the NPSH(A) at the pump inlet enables the cavitation protection ratio **R** also to be calculated. Figure 13.1 presents graphically the essential pump performance data issues.

(i) *Is cavitation a risk?* All assessments are made assuming the operating range is limited to within the $50\%Q_{bep}$ to $120\%Q_{bep}$ range. Chapter 9 established that under such conditions a centrifugal pump is likely to be free from cavitation if the suction specific speed is 2500 or less (Guideline 1). A first check must therefore be to determine whether the pump offered passes this simple all-encompassing test. Information supporting the pump offered should include the pump speed at which the pump duty is met and the flowrate at which best efficiency point occurs together with the NPSH(3pc) value at this flowrate. From this data calculate the suction specific speed n_{SS} and compare it with 2500.

(ii) *Is cavitation erosion a risk?* Chapter 9 defined conditions for negligible risk as

(a) for speeds up to 3600 rpm

$$\mathbf{R}(3pc) \geq 2.5 \quad \text{at pump } Q_{bep} \qquad \text{(Guideline 2)},$$

(b) for speeds between 3600 rpm and 9000 rpm

$$\mathbf{R}(3pc) = 1.5 + \frac{n}{3600} \quad \text{at pump } Q_{bep} \qquad \text{(Guideline 3)},$$

$$\text{where NPSH(R)} = \text{R(3pc) x NPSH(3pc)} \qquad \text{(equation 4.3)}.$$

Notes: (i) the above calculations are carried out at the flowrate Q_{bep},
      (ii) the NPSH(A)/Q curve is calculated using equation 4.10.

The guaranteed duty of the pump(s) offered does not usually coincide with the best efficiency flowrate. Where NPSH is low enough to pose a risk of unacceptable cavitation the 50%Q_{bep} to 120%Q_{bep} limits for the operating range need to be borne in mind. This requires that the value of Q_{bep} be not less than 83% of Q_H (i.e. Q_H is not more than 120%Q_{bep}). In practice it is rare for Q_H to equal Q_G. Usually the uncertainties in calculating h_{sys} described in section 4.2.7 result in Q_H being much greater than Q_G.

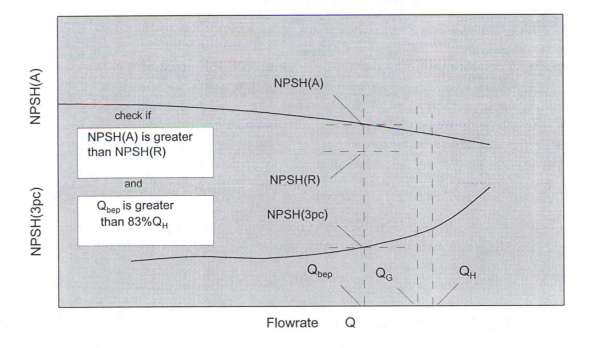

Fig. 13.1 Assessing the adequacy of NPSH(A) to prevent cavitation erosion

Check that NPSH(A) exceeds NPSH(R) at Q_{bep}. Failure to meet this requirement should lead to rejection of the offer unless either

(i) an Applications List pump has demonstrated proven performance in "near identical" conditions or

(ii) acceptable NPSH(4mm) data backed up by proof by test are offered.

An NPSH(3pc) test should be included in the offer where the NPSH(3pc) value is the basis for demonstrating cavitation performance adequacy. A cavitation test is unnecessary when NPSH(A) is at least twice the NPSH(3pc) at all flowrates in the operating range.

Where an NPSH(4mm) test is offered the pump is likely to be free from cavitation erosion if

$$\mathbf{R}(4mm) = 1.05 \text{ at all operational flowrates} \qquad \text{(Guideline 4)}$$
$$\text{and } NPSH(R) = \mathbf{R}(4mm) \text{ x } NPSH(4mm) \qquad \text{(equation 4.4)}.$$

Flowrates Q_L, Q_G and Q_H are the points where pump performance should be proven. Compare the calculated NPSH(R) with the value of NPSH(A) at these flowrates. Check that NPSH(A) always exceeds NPSH(R). Failure to meet this requirement should lead to rejection of the offer unless an Applications List pump has demonstrated proven performance in near identical conditions.

(iii) *Is vapour locking during transient operation a risk?* It is only at this stage that the full evaluation detailed in Chapter 6 can be carried out. The number of pumps and operating regimes are now definable. More importantly, the NPSH(3pc) curve is known. The minimum acceptable value that NPSH(A) should be allowed to fall to under any circumstances is given by

$$NPSH(A) = 1.3 \text{ x } NPSH(3pc) \text{ at all flowrates in the operating range.} \qquad \text{(equation 6.1)}$$

Failure to meet this requirement should lead to rejection of the offer. Either a different pump with a lower NPSH(3pc) should be chosen or restrictions on operating conditions to limit the transient fall in NPSH(A) must be imposed.

The NPSH(3pc) curve dictates the minimum level to which a transient depression in NPSH(A) may go. Its validity may have to be proven by test if calculations indicate conditions are very close to being unacceptable.

(iv) *Is cavitation surging a risk?* There is very little possibility of hydrodynamically induced cavitation surging occurring if

$$NPSH(A) \geq 1.3 \text{ x } NPSH(3pc) \text{ } \textbf{\textit{and}} \text{ } Q_L \geq 50\%Q_{bep} \qquad \text{(see Chapter 7)}.$$

If short periods of operation below $50\%Q_{bep}$ are inevitable try to evaluate the risk.

(i) Determine whether NPSH(A) is much in excess of 1.3 x NPSH(3pc) at the particular flowrates.

Note: Provision of NPSH(A) greater than that judged necessary to prevent cavitation erosion during continuous operation (normally 2.5 x NPSH(3pc) at Q_{bep}) is likely to ensure that pump operation is surge-free.

(ii) Determine whether the pump inlet design is one less conducive to the propagation of surging flows (e.g. one with a side entry into the impeller eye from the inlet pipe such as a vertical-in-line or a pump with a double-entry impeller).

(iii) If NPSH(A) is low and a long straight inlet pipe is to be used that leads flow directly into the eye of a single-entry impeller consider whether a backflow recirculation device (see section 7.2.5) can be fitted.

Be under no illusions. If hydrodynamically induced surging sets in, the consequences can be devastating. Do not take unnecessary risks.

13.3 Low-flow Protection

The proposals for low-flow protection need to be examined. The data supplied with the offer for the particular pump must be used. From a knowledge of the pump input power at zero flowrate P_{AO} and the volume of liquid enclosed within the pump casing V_C a worthwhile estimate of the maximum permissible time for action to be taken (pump shut-down or establishment of leak-off flow) can be calculated.

From equation 10.3,

time to reach limiting temperature $t_o = \dfrac{m\, C_P(T_2 - T_O)}{P_{AO}}$

$$\text{where } m = \rho_L\, V_C$$
$$T_0 = \text{liquid temperature in inlet pipe.}$$

The capability of the proposed equipment or operating procedures to meet this time needs to be considered. Where doubts exist a confirmatory site test requirement should be made a condition of placing an order for the pump.

To illustrate this numerically a continuing use is made of Chapter 12, Example 3. This assumes the following from the earlier work:

$$\text{NPSH(A)} = 3.8 \text{ m (a minimum requirement)}$$
$$T_0 = 20°C$$

from thermodynamic property $\rho_L = 610 \text{ kg/m}^3$
tables for liquid ammonia $C_p = 4.5 \text{ kJ/kg °C}$

Assume the offer includes the following:

$$P_{AO} = 18 \text{ kW} \qquad \text{(see Fig.12.7)}$$
$$\text{NPSH(3pc)} = 1.5 \text{ m} \qquad \text{(cold water data)}$$
$$V_C = 0.05\text{m}^3$$

From equation 10.2, $\Delta p_{vap} = \{\text{NPSH(A)} - 1.3\ \text{NPSH(3pc)}\}\ \rho_L\, g$
$$= \{3.8 - [1.3 \times 1.5]\}\ 610 \times 10^{-5} \times 9.81$$
$$= 0.108 \text{ bar}$$

From thermodynamic property tables for liquid ammonia,
vapour pressure at 20°C = 3.32 bara
saturation temperature at 3.43 bara ($\approx$ 3.32 + 0.108) = 21.5°C

From equation 10.3,

$$t_2 = \frac{m\ C_p\ (T_2 - T_0)}{P_{AO}}$$

$$t_2 = \frac{(610 \times 0.05)\ 4.5\ (21.5 - 20)}{18}$$

$$t_2 \approx 11 \text{ seconds}$$

The calculated time is likely to be a conservative estimate. The volume V_C does not take into account liquid migrating into and out of the pump casing and the NPSH(3pc) value offered is probably on the high side since the manufacturer must always be sure that it can be demonstrated by test. So for this particular example a time of less than about 15 seconds seems likely. This is too short to allow reliance to be placed upon human intervention effecting a shut-down. It confirms the earlier anticipation of the need for a leak-off system (Chapter 12, Example 3; pipichum = 3, Category C). A leak-off system that establishes a flow in the leak-off pipework in less than 10 seconds is needed.

It is instructive to examine the effect of the same requirement being imposed when pumping cold water.

The data required for calculation are

NPSH(A) = 3.8 m (a minimum requirement)
T_0 = 20°C

from thermodynamic property ρ_L = 998 kg/m³
tables for water C_p = 4.15 kJ/kg °C

Assume the offer includes the following:

P_{AO} = 30 kW   (see Fig. 12.7)
NPSH(3pc) = 1.5 m   (cold water data)
V_C = 0.05m³

From equation 10.2, Δp_{vap} = {3.8 - [1.3 x 1.5]} 998 x 10⁻⁵ x 9.81
= 0.181 bar

From steam tables,
vapour pressure at 20°C = 0.023 bara
saturation temperature at 0.204 bara ($\approx$ 0.023 + 0.181) = 60.5°C

From equation 10.3, $$t_2 = \frac{(998 \times 0.05)\ 4.15\ (60.5 - 20)}{30}$$

$$t_2 \approx 280 \text{ seconds} \quad (\approx 4.6 \text{ minutes})$$

As before the result is a conservative estimate and indicates a time which, for many practical situations, is too short for human intervention to take place. However, a much longer period for action to be initiated is evident even with a much larger power input. Examination of the calculations shows that this is primarily due to differences in the rate of change of vapour pressure with temperature. Cold water exhibits a particularly slow rate of change. A leak-off system designed for a cold water pump application will have a much longer time to respond than a hot water application or applications for most other liquids.

For the cold water application, action is required within, say, four minutes. Two possible approaches to low-flow protection are:

(i) programme the control system to stop the pump within four minutes of a reliable low flowrate signal being received — a period which allows some short transients in flowrate. The low flowrate value needs to be below the normal operating range but high enough for adequate measurement accuracy. In choosing the control system method care needs to be taken not to inhibit a pump start.

(ii) provide a leak-off system that responds within four minutes.

The leak-off option has the advantage of keeping the pump available for service following, say, reopening of the outlet control valve. The "pump stop" option is the cheapest.

The above shows the need for calculations to be undertaken when making a technical assessment of pump offers. It also shows that pipichum values are only indicative and suited to general guidance before detailed offers are received.

Chapter 14

Centrifugal Pumps in Service —
Resolving Cavitation Problems

14.1 Where to Start — Define an Acceptable Objective

Cavitation problems in an existing plant always present a test of ingenuity. Occasionally, a well-designed plant is of necessity run in a way that was never intended. In other instances the layout of the plant is poor, the pump selection is far from ideal, or the operating instructions are deficient. The opportunity for change is sometimes very limited, especially where operation of the pump is an essential part of a continuous and profitable process. A first step is to define the objective of remedial measures.

Existing constraints usually mean that a proposed solution is less than ideal. In these circumstances a more limited objective has to be accepted at the outset. Very often this comes down to the bringing about of a significant improvement with the least disruption. The criterion by which proposed actions are judged is nearly always cost effectiveness.

Plant design is sometimes so deficient that a programmed "outage" to facilitate essential changes has to be made. Where the cause of a cavitation problem is ill defined more than one approach has to be implemented simultaneously making it impossible to accurately attribute a subsequent improvement to a cause. In spite of this the bringing back of a pumping plant into reliable service is intellectually very rewarding and more than makes up for the "minor difficulties" encountered along the way. Making best use of the existing hardware and the pump location is a key element to achieving success whichever course of action is chosen.

14.2 The Identification of a Cavitation Problem

It is of course essential to make sure wherever possible that the problem being resolved is one caused by cavitation. There are a number of ways in which cavitation can be identified with some certainty.

14.2.1 *Listen for cavitation noise*

Well-developed cavitation in the stable zone (section 3.5) exhibits a characteristic crackling noise, especially in cold water. An industrial stethoscope confirms pump cavitation more clearly than the human ear. The crackling noise is heard when cavitation is well developed. This means that the net positive suction head available, NPSH(A), is much less than NPSH(3pc). Its presence indicates that serious erosion damage is a real possibility. The audible presence of cavitation is also a strong indication that serious deficiencies in pump hydraulic performance may be occurring or that the mechanical integrity of the pump is being threatened.

Cavitation in the unstable region produces a characteristic pulsating noise. The chuffing of a reciprocating steam engine exhaust is the best descriptor. The periodic pulsations usually have a frequency of between 2 Hz and 10 Hz. This form of surging almost always causes distress to the axial thrust bearing. In many cases the large changes in momentum caused by the oscillating cavitating flows make the pump pipework system shake violently.

14.2.2 *Look for a generated head-drop at a known non-cavitating Q/ΔH value*

At NPSH values less than NPSH(3pc) cavitation can cause the performance of a centrifugal pump to drop off noticeably. It is worthwhile seeing if a site test can be arranged in which NPSH is varied. This usually means working through to extremes in the inlet vessel level and the outlet vessel level. Where the inlet vessel is closed and is at vapour pressure it is sometimes possible to alter this sufficiently to make a noticeable change in NPSH(A). Occasionally in these circumstances a test where the liquid is not at full operating temperature can be carried out. The objective is to measure generated head at several flowrate values and see whether the effect of changing the NPSH(A) alters the pump hydraulic performance. If changing NPSH at a given flowrate has no effect on the generated head then it is probable that only a small amount of cavitation (if any) is present. It is at the planning stage for such tests that the benefit can be reaped from having NPSH(3pc) and ΔH/Q test data from the works tests carried out when the pump was manufactured. Knowing in advance where Q_{bep} is supposed to be and what value NPSH(3pc) should be at a given flowrate enables the value of testing within the constraints imposed by the plant to be determined at an early stage.

The above tests are usually easy to carry out although their value is sometimes marred by their dependence upon industrial grade measurement equipment, particularly that for the measurement of flowrate. It should go without saying that all instrument calibrations should be checked before testing commences.

14.2.3 *Examine operational records*

The frequency of maintenance activities, especially those that are unplanned, is a very good indication as to whether a pump is operating under duress. A record of hydraulic deterioration with time indicates an internal wear process. Cavitation erosion damage to blades can produce this effect when NPSH(A) is less than NPSH(3pc). As shown in Fig. 7.4 axial thrust bearings are subject to fluctuations in load when cavitation surging becomes fully developed. Buffeting of the shaft, particularly during cavitation surging, readily causes thrust bearing failure.

14.2.4 *Examine pump internals*

The evidence of cavitation in pump internals (pumps being repaired/maintained, photographs/reports of past damage, etc.) is very important. It is well worth the effort of trawling through maintenance records to find it. The significance of a change recorded in the maintenance log is sometimes only appreciated at a later date. Damage to pump components is not only a sign of the presence of cavitation but is a measure of the extent to which it occurs within the pump. The photographic evidence in Chapter 9 gives guidance on how to classify findings. Lengthy grooves or through penetration of impeller blades resulting from cavitation erosion damage indicate that the NPSH(A) is far from that which is sufficient to protect the pump.

14.2.5 *Confirm low-flow protection is functioning correctly*

The operability of leak-off protection equipment can be checked by simple functional tests. A failure to open promptly leads to large amounts of vapour in the pump followed by rubbing at narrow internal clearances. A test to crudely check operability can be made by (i) observing the time for movement to commence after the signal to initiate it is sent to the actuator (power line becomes live) and (ii) establishing the additional time to clear valve seating surfaces sufficient to establish flow (usually audible). This should be compared with the original design intent where this is known.

Instances are on record of the elements producing the pressure-drop within a pressure reducing device being incorrectly sized. Actuators are sometimes unreliable or too slow to act. The result of these deficiencies is intermittent vapour locking (leading to excessive internal wear and seizure) or hydrodynamically induced surging.

14.3 Examples from Experience

A small number of examples are now presented that show how cavitation related problems might be solved. These are mainly, though not exclusively, from the author's experience in the power generation industry. The details have been simplified to highlight the important points. The conclusions reached and lessons learned can be applied generally.

14.3.1 *Remedial measures — Make a change to operating procedures*

Section 12.3.2 identified conditions in which the unacceptable effects of cavitation during continuous service could be avoided with a high degree of certainty. For a particular pump these were achieved by limiting operation to a $50\%Q_{bep}$ to $120\%Q_{bep}$ flowrate band and providing NPSH(A) to give a suction specific speed n_{ss} of lower than 2500. Remedial measures will be effective if they move existing plant operation closer toward these conditions. This implies being able to ascertain the value of Q_{bep} at a given pump speed and then measure the generated head at this point on the hydraulic performance characteristic. The calculated value of NPSH(R) — the minimum permissible value of NPSH(A) — should then be compared with the actual value of NPSH(A) at Q_{bep}. The shortfall in NPSH determines the measures needed to rectify a deficiency caused by cavitation. For some pumps — mainly those with shaft speeds in excess of 3600 rpm — the shortfall in NPSH(A) is numerically too great to be made up by minor adjustments to the pump or to the pumping plant. Conversely, for many pumps running at lower speeds some minor changes can improve the pump operating regime considerably. A judgement has to be made in each case as to whether the improvements being considered represent practical options and whether it is worth pursuing them — perhaps for a trial period — where the shortfall in NPSH(A) after applying them remains large.

(i) *Increasing NPSH(A).*   It is often worth reviewing the factors that determine the "low" liquid level in the inlet vessel. Dealing with an actual pump rather than a design study means that operating performance requirements can be quantified with more certainty. The need to operate down to the lowest levels can be questioned. Very often the pump and system have been oversized and this allows the low level to be increased. Remember that a gain of as little as 0.5 m may be a substantial part of the NPSH(A) shortfall on a slow speed pump.

(ii) *Optimising the operating flowrate range.*   In a plant where pumps run periodically to fill or empty vessels a small amount of choice in running pattern is often available. Filling the inlet vessel higher — a greater NPSH(A) — reduces the generated head required so pump flowrate is higher. In systems where pumps are manually switched on and off the running of the pumps more frequently from higher levels might be advantageous, especially where cavitation surging at low flowrates is suspected.

(iii) *Controlling the performance of pumps running in parallel.*   In many power and process plants 3 x 50% pumps are installed. The design intent is to achieve the plant design flowrate Q_p with two pumps running. In Fig. 14.1 a single pump performance curve is shown as A-B. Assuming the pumps are identical and therefore share load equally then in the two pump operating mode the maximum possible flowrate is X1. Halving the two pump flowrate at the intersection of curve A-C with the system resistance curve gives the flowrate through each pump X2. As explained in section 4.2.3 this is where the pumps will operate unless surplus energy is destroyed by some means. Usually this is by an outlet control valve. As a consequence the flowrate at X1 is somewhat greater than Q_P.

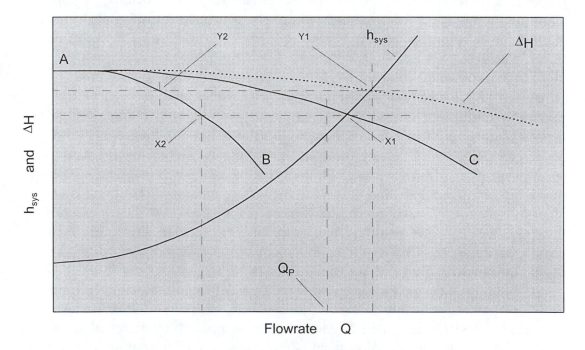

Fig. 14.1   Potential problems with pumps running in parallel

The third pump is an installed spare to enable full load to be maintained should one of the two running pumps have to be taken out of service. It is not a design intent to run all three together. Running all three pumps together can be very inefficient, particularly when a greater part of the system resistance is made up from friction losses (e.g. pumping through extensive pipe networks) as shown in Fig. 14.1. To increase total flowrate to Y1, the maximum possible value permitted by the system resistance, each pump passes a flowrate of Y2. Y2 is one-third of Y1 if load is shared equally.

An example of how problems arise from unplanned three pump operation was observed in a power plant that was unexpectedly required to produce maximum output as an essential part of maintaining the daytime security of the electrical grid. Efficiency considerations were temporarily suspended and a three condensate extraction pump operation was introduced. In normal two pump running, each pump ran at about 90%Q_{bep}. With minimal flow control valve resistance the two pumps ran out to about 100%Q_{bep} each. Running three pumps added a further 10%Q_P — a valuable contribution in the particular circumstances. This meant that each pump was, in theory, working on average at 37%Q_{bep}. In fact small differences in the slopes of the individual pump generated head/flowrate curves probably gave a flowrate per pump between 35%Q_{bep} and 40%Q_{bep}. So even when running "flat out" the minimum continuous flowrate guideline was being breached. In running the extra flow through a common inlet pipe the NPSH(A) was also reduced. These practices alone would perhaps have been tolerable. The feature of operation that produced cavitation damage was the swing in demand overnight. As demand fell back the three pumps were throttled back by the outlet control valve so that individual pump flowrates were very much less than 30%Q_{bep}. Failure to switch off a pump as soon as maximum demand fell by more than about 10%Q_P meant cavitation erosion damage and an unnecessarily high maintenance bill.

It is well worth examining in detail all the advantages and disadvantages of running fewer or more pumps than originally intended. Avoidance of unnecessary operation outside the 50%Q_{bep} to 120%Q_{bep} range should be the aim.

14.3.2 *Remedial measures — Make changes to plant component design or layout*

Example 1 — Cavitation caused by major plant derating

(i) *A description of the problem.* Pumps form an integral part of many process plant designs. The selection of a suitable pump is usually the result of lengthy deliberation. A great deal of effort is put into getting it right. On occasion however the "big picture" turns out to be not as originally planned. The plant sometimes has to be downrated for reasons unconnected with the pump. Often the aim is to bring the plant up to its original design specification at a later date. Sometimes this can take a long time — measured in years. Sometimes the aim is abandoned and the plant is permanently derated. With all the attention focused on plant output many a pump is left with permanent part-load operating conditions. The increased risk of unacceptable cavitation arising at the new low %Q_{bep} conditions often turns a marginal cavitation erosion attack and a mild propensity to surge situation into a very serious pump-life threatening one.

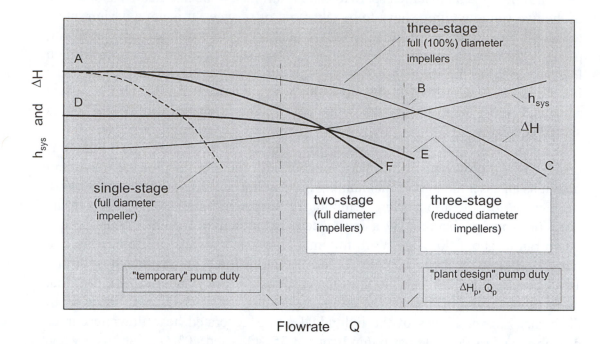

Fig.14. 2 Potential remedies to extensive low %Q_{bep} operation

(ii) *A typical example — multistage pumps.* In the early days of nuclear power plant technology long delays were experienced in getting reactors to full power. This meant that the pumps serving the reactor had to run continuously at part plant design load. The plant design was based on a philosophy that assumed the reactor and its pumps would run continuously at full power, the swings in demand being taken up by fossil fuel stations. The NPSH(A) was reduced to give a narrow flowrate band protection. A value of less than 2 x NPSH(3pc) at Q_{bep} was provided. The significant reduction in Q_p brought about by plant operating restrictions meant that the pumps were subject to cavitation erosion and cavitation surging. The damage recorded was significant.

The solution to be applied in these circumstances is obvious. The pump has to be changed to match its "temporary" duty. Additionally, the change must be carried out in a way which permits the pump to return to its original design duty later.

There are two ways in which this might be achieved; either (i) reduce an impeller diameter or (ii) remove an impeller stage. The choice depends on particular circumstances and practical considerations. A reduction to below 80%Q_p usually makes (i) inappropriate. The outcome from each of the two is shown in Fig. 14.2. The original duty ΔH_p, Q_p is shown as point B on curve ABC. Cutting the impeller diameter gives curve DE. Removing one of three stages gives curve AF.

Option A — Cutting the impeller diameter. Impeller cuts are a first choice. In a multistage pump it may be possible to secure the required reduction by cutting the first impeller only. The effect of an impeller cut of less than 20% can be considered identical to a change in speed of the same amount. New impeller(s) replacement is the only component change needed to bring the pump back to the original duty.

Trimming an impeller diameter reduces the generated head approximately in direct proportion to the pump speed. This is because only the peripheral velocity of the impeller is reduced in proportion to the impeller diameter cut. The flow velocity pattern within the impeller hydraulic passages is unchanged; the width of the impeller is changed very little. For the affinity laws derived in section 3.2.1 to apply the impeller hydraulic passages (i.e. the width as well as diameter) as well as the flow velocities would have to change by the same proportion. This is clearly not the case when the impeller diameter is reduced.

Option B — Removal of an impeller stage. Removal of an impeller enables a more substantial reduction to take place. The increased losses arising from drawing liquid through the empty passages of a first stage casing go unnoticed in such a large change. The impeller taken out of the pump can be put into a store ready for immediate restoration of the original duty when the need arises.

(iii) *Commentary*. The changes to be made are simple. The decision to do this is sometimes a long time being made. Plant designers and plant users must be alert to the need to consider a change. The increased absorbed power cost and the accumulating consequential damage to a pump are almost always sufficient to justify early action.

Example 2 — Gross inadequacy of NPSH(A) in original plant

(i) *A description of the problem*. The overall efficiency of a steam turbine/generator unit in a power station benefits from the water which is being pumped to the boiler being preheated by turbine exhaust steam. On the steam side of the heater, condensate is formed which falls to the bottom of the heater. This has to be pumped away and returned to another point in the closed water/steam cycle. If the system design is poor the liquid level in the heater bottom becomes unstable and is regularly drawn down into the pump inlet pipe. The detectors producing signals for the level control equipment which actuates the pump outlet control valve are unable to respond in an accurate or timely way. Pump vapour locking ensues.

(ii) *A typical example*. The initial design of a power station heater drains system is shown diagrammatically in Fig. 14.3. In this arrangement level control is achieved by means of a pump outlet valve that responds to a level signal in the bottom of the heater vessel. In operation it soon becomes clear that the condensate level in the bottom of the heater is readily drawn into the pump inlet pipe. When this occurs the NPSH(A) falls to practically zero and the pump is vapour locked. The continuing flow of condensate soon backs up into the heater and this mass of liquid effectively reprimes the pump. The cycle is then repeated with a cycle time dependent upon the amount of steam being exhausted to the heater.

With a pump having internal clearances and mechanical seals that depend upon a pumped liquid film being present at all times gross rubbing and seizure were frequent events — almost daily.

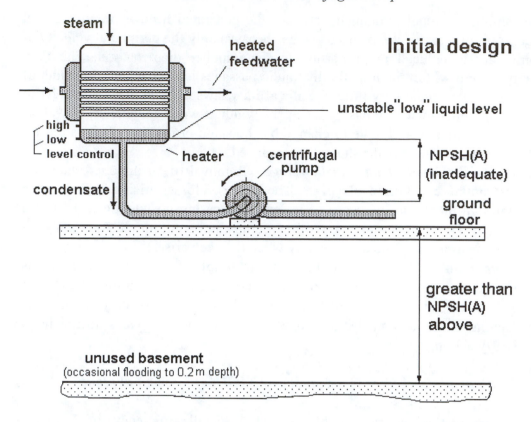

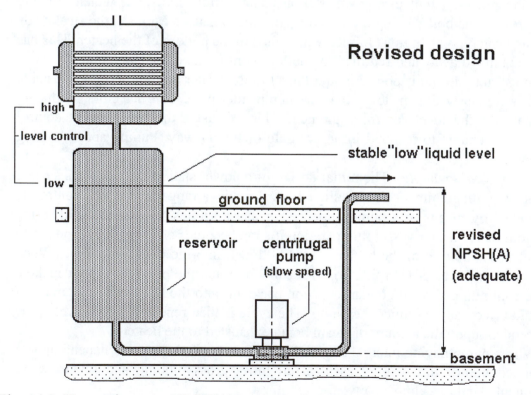

Fig. 14.3 How to increase NPSH(A) — A new basement mounted pump
and improved level control

The solution in such circumstances is to review the features thought to be causing failure and to determine the likely contribution from each. It is soon evident that marginal changes to the initial design are unlikely to remove the causes of pump failure. A change of pump and related plant is required. Having decided on such a costly and time consuming remedial step, it is necessary to consider inclusion of more than one of the measures that will improve performance. The marginal cost increase for additions is sometimes very small. In the particular example the beneficial objectives were listed as:

(i) to provide more NPSH(A),
(ii) to reduce NPSH(R),
(iii) to improve level control.

To meet them the following package of measures was decided upon:

(a) utilise the basement below the pump,
(b) choose a new and more reliable pump that could be shown by test to withstand a limited period of vapour locking and that would be suitable for the unfavourable operating environment of the basement.

The basement below the original pump gave a substantially greater NPSH(A). To enable the pump to function during occasional flooding of the basement and to improve mechanical reliability (see Ref. 14.1) a low speed vertical-in-line pump was chosen. A large diameter reservoir tank was interposed between the heater and the centrifugal pump so that when the level fell from the heater a new slowly falling (and very stable) free surface was established.

The way in which the above was interpreted is illustrated diagrammatically in the revised design shown in Fig. 14.3.

(iii) *Commentary* It may well be that a change to the pump alone, moving the existing pump to the basement, or just improving the level control by a reservoir vessel on the ground floor would have produced acceptable performance. However, once committed to an "outage" all the listed advantages could be included. The risk of failure after implementing the minimum change was unacceptable. The payback, the power station efficiency being raised by nearly 0.2% (a restoration of what should have been there in the first instance, not an "improvement"), clearly made the certainty of success associated with implementing all of the recommendations worthwhile.

The opportunity to increase NPSH(A) by moving to an otherwise unused basement does not often arise. An alternative worth investigating when this is the case is the use of a vertical centrifugal pump mounted in a caisson. Again the NPSH(A) is substantially increased and the liquid volume in the caisson prevents the liquid level falling rapidly into the impeller eye. An example of this pumping arrangement (as used in a condenser extraction duty where a high generated head was also required) is shown in Fig. 14.4. Tests to demonstrate the operating capability of the caisson mounted pump during liquid level draw down are easily carried out at the pump manufacturer's works.

Note that the inlet and outlet branches can be at any level on the caisson wall. It is often a convenient expedient on new plant designs to have this pipework run in trenches in the ground floor.

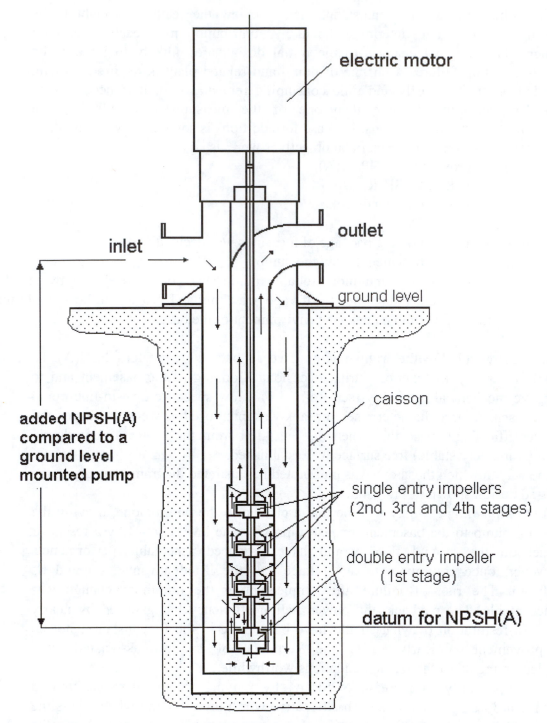

Fig.14.4 How to increase NPSH(A) — A pump inside a caisson

Reference

14.1   Grist, E., <u>Standard vertical-in-line pumps</u>, Pumps-pompes-pumpen,
         Trade & Technical Press, Morden, Surrey, U.K., pp.550 – 551, No.100, 1975.

Chapter 15

Centrifugal Pumps and Cavitation —
A View of the Past

15.1 Engineering History — A Personal View

An interest in pumping machinery is greatly enhanced by an understanding of where, when and why ideas originated. This chapter is based in part on work undertaken by the author which led to an engineering degree. As a young student already with considerable experience in centrifugal pump construction, design, and testing it was necessary to choose the subject of a dissertation — a major exercise in report writing for which a whole academic year was allowed. The title, "The Development of the Centrifugal Pump", was chosen and the dissertation (Ref. 15.1) duly produced.

The perspective of engineering development gained by such a personally driven review of history is well worth the considerable effort it involves. Reading through

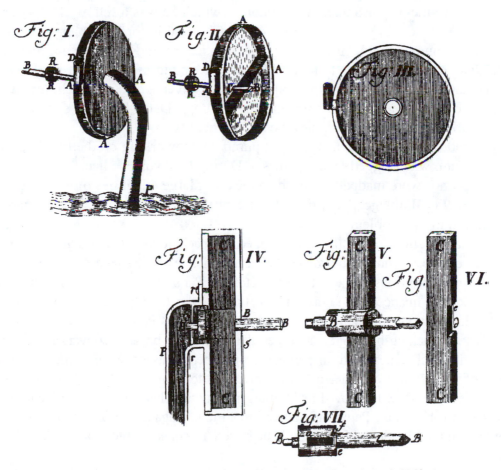

Fig. 15.1   A forerunner of the centrifugal pump   (Papin, circa 1689)

books of the 17ᵗʰ, 18ᵗʰ and 19ᵗʰ centuries as part of a search for information not only allows pump development to be explored but sets in place the contemporary discoveries in medicine, astronomy and other forms of engineering together with the influence of political history. A good engineering education can bring the words and drawings of the early engineers to life. The reader becomes acutely aware of understanding the intellectual isolation some of these early inventors faced.

This chapter does not pretend to result from the rigorous endeavours of a seasoned historian; it is the personal view of a practising engineer. Access to the earliest books was secured by visits to the Central Reference Library, Manchester and the British Museum Library, London.

15.2   The Centrifugal Pump

15.2.1   *Denis Papin — Intuitive engineering design of the highest quality*

The first true centrifugal pump was made in about 1689 by Denis Papin (Ref. 15.2). Whilst his earliest design (Fig. 15.1) was somewhat primitive in nature, a later design (Fig. 15.2) circa 1705 illustrates his appreciation that for efficient pumping to take place the pump should comprise a spiral-shaped tympanum (a true volute casing), a tangential discharge pipe and multiple wings (a multi-bladed impeller). The 1705 design is clearly described by Fig. 15.2 and a letter from Papin (Ref. 15.3) to the Royal Society of London of which he was a fellow. The letter is copied in Appendix A*. The essentials of Papin's description could apply in most respects to pumps made today. Yet the politics of the time meant that his contemporaries were unable to take advantage of his ideas. It was not until over a century later that the centrifugal pump was re-invented in a much less well engineered form. The first name in centrifugal pumps, Denis Papin, is synonymous with intuitive engineering design of the highest quality.

Papin had originally designed his first pump whilst working on the building of a canal through waterlogged land in Prussia. During this work he had found that the existing pumps were inadequate for the removal of large quantities of water needed for satisfactory drainage. The principle of centrifugal pumping evolved as part of his solution to the problem.

Papin had a strong scientific and engineering background. He designed a direct displacement steam pump and evolved the basic principle of the steam engine. His early designs sprang from his work in the Laboratory in Paris under the physicist Huygens. This work was financed by Louis XIV of France who through taxation was able to allocate large sums to scientific experimentation. When Papin became assistant to Huygens, after taking his degree at Angers University, he became conversant with the experimental techniques used, in particular, on experiments concerning the expansive power of gunpowder and steam. After travelling to London to avoid the religious persecution in France, he moved to Marburg in Prussia where his work on centrifugal pumps was carried out. Papin then attempted to return again to London but due to a dispute with the Munden Shippers Guild he suffered a considerable financial loss from

---

* Papin's letter of 1705 refers to publication of his invention in 1699. By Ref. 15.2 it is known that it was presented in 1689.

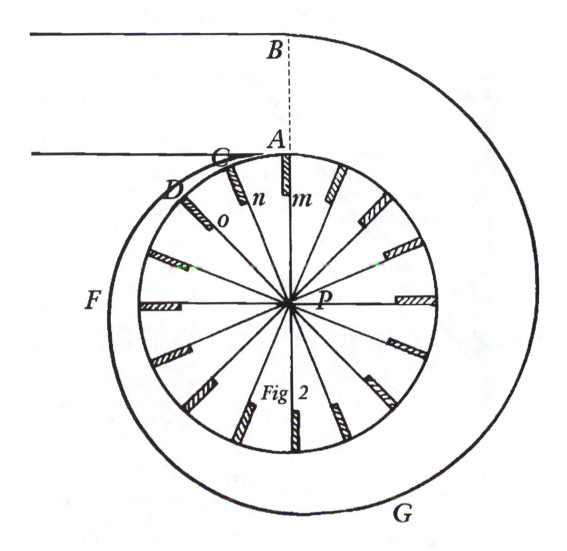

Fig. 15.2   The first centrifugal pump (Papin, circa 1705)

which he never recovered, his ship and possessions being smashed. This event is of great significance in the development of the centrifugal pump for although Papin eventually arrived in England he was unable to arouse interest in his work and died in poverty in about 1714. This lack of an active following meant that his invention was not developed further and eventually was forgotten.

Many later historical accounts of pumping equipment make no reference to Papin's work. It was only in the mid-twentieth century that engineers and historians together realised the significance of his work on the centrifugal pump. The great potential of the centrifugal pump remained unexploited. There is a sad irony in the lack of appreciation and support. The King of England and his nobles had witnessed Papin's pump at work during tests on the banks of the River Thames. That the demonstrated potential of the centrifugal pump should then be so readily forgotten is today difficult to comprehend.

15.2.2 *The "dark ages" of centrifugal pumping*

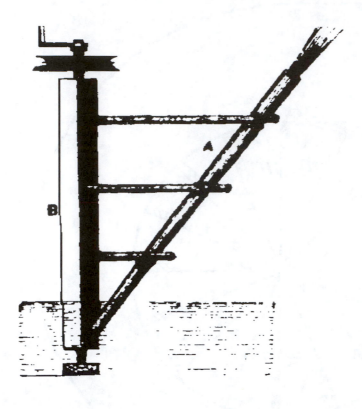

Fig. 15.3 The rotating tube pump (Le Demour, 1732)

After Papin's death primitive pumps using centrifugal force were produced by Le Demour in 1732 (Fig. 15.3) and D. G. Fahrenheit in 1736. Another form of centrifugal device (Fig. 15.4) appeared shortly afterward, this again being on the same principle as Le Demour's machine, fashioned in the way of a reversed Barker's Mill (an early water turbine). This was subsequently improved by the French engineer Jorge in 1816.

Meanwhile, in 1750, the mathematician Euler had conducted the first mathematical analysis of centrifugal pumps and turbines. Using Le Demour's pump as a basis he proposed more advanced forms of the machine, recommending parabolic form between inner and outer walls of the "diaphragm" (impeller). He was however concerned only with the mathematics of the device and his work, although representing the first analytical approach to the problem, had no practical significance in aiding pumping machinery development.

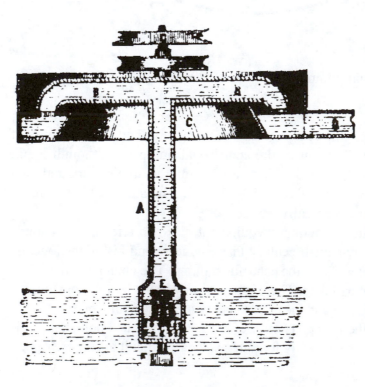

Fig.15.4 The inverted Barkers Mill (pre-1816)

The idea of applying centrifugal force to pumping machinery became neglected in spite of the great need in Western Europe for the machine which could have been developed from these early ideas. Indeed, it appears that for the centrifugal pump the right people had not, so far, been in evidence at the right time.

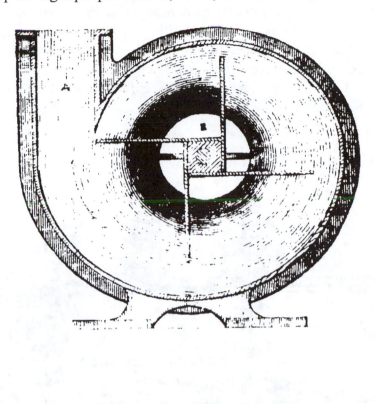

15.2.3
The evolution of the modern centrifugal pump

In 1818 the centrifugal pump re-appeared in a practical form in Boston, Massachusetts. The exact origin of the pump and the name of its inventor are unknown, and because of this the pump became referred to as the "Massachusetts" or "Boston" pump. The design of the Massachusetts pump (Fig. 15.5) possessed many hydraulic faults, by far the most significant being the radial vanes. Although its beginnings were obscure this pump became the basis of a chain of developments from which the modern centrifugal pump evolved.

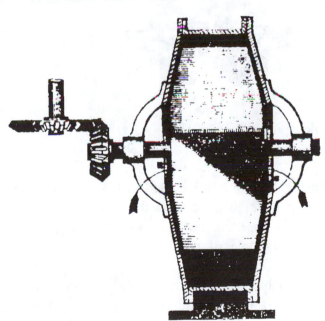

Fig. 15.5   The Massachusetts pump (1818).

The Massachusetts pump was soon improved by other American engineers as the potential of the design became apparent. In 1831, Blake at the New Steam Mills,

Connecticut, brought out a pump (Fig. 15.6) in which the impeller consisted of radial vanes mounted on a single disc. The advantages of this modification were, however, counteracted by the delivery columns being positioned in an axial direction toward the periphery of the casing thus introducing unnecessary losses due to turbulence. Whilst this last point is clearly evident to an hydraulic engineer today, it must be recalled that during this period very little was known in the way of theories of fluid flow. The only available information in this direction was probably limited to the works of European engineers of the eighteenth century such as Pitot, Bernoulli, Chezy, Borda and Venturi and these were perhaps not wholly understood by the practical engineers of the day.

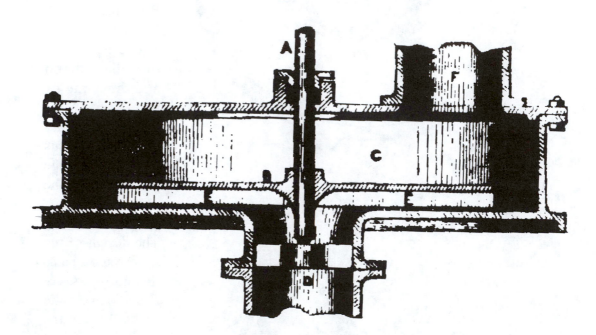

Fig. 15.6   The centrifugal disc pump   (Blake,1831)
(pump incorporating an impeller with a back shroud)

The first patent for a centrifugal pump in the United States was taken out in 1839 by W. D. Andrews. His pump differed little from the Massachusetts pump, the main feature being the rapidly tapering radial vanes and elliptically shaped collecting ring (the volute). This type of pump was used later in New York harbour.

So far, an inherent feature of each design has been the radial vanes. This fault may have been recognised by James Whitelaw of Johnstone who published sketches of pumps working on the inverted Barker's Mill principle and showed the relative effect of straight and curved vanes. Whitelaw's pump was reputed to have attained an efficiency of 77%. This invention does not appear to have been developed much further than the laboratory experiment stage, however, the pump referred to lifting water to a height of less than 1 m. No arrangement was shown for collecting the discharged water in a practical form (it was probably collected in an open circular trough during the experiment) and, unlike centrifugal pumps of the time, was not in a form easily adaptable to a variety of applications.

Working in the United States, James Gwynne of England then conducted a series of experiments in Pittsburgh in 1844 during research into the uses of centrifugal forces. From the results he developed his "direct acting balanced pressure centrifugal pump". This pump was first exhibited publicly at the Passaic Copper Mine in 1849 and in the following year a U.S. patent was taken out for the invention. Like the previous American designs Gwynne's pump had an impeller with radial vanes and a large diameter (about 4 m). The large dimensions of the pump were necessary because of the lack of high speed prime movers to generate centrifugal forces. At that time centrifugal pumps, to be of any practical use, had either to be driven by belts from the main engine of a factory or be driven directly by a steam turbine. The steam turbine was found to be most suitable, but the financing of this combination proved restrictive in many cases. American engineers continued to lead the field in centrifugal pumping and with increasing mechanisation the popularity of this pump continued to grow enabling it to replace the cumbersome reciprocating pump for the higher flowrate duties.

In England Henry Bessemer took out a patent in 1845 for a centrifugal pump. This was very similar to Blake's earlier design and it is possible that Bessemer's idea originated from this since the activity in America on the centrifugal pump could hardly have passed unnoticed in England. Bessemer did, however, claim with this pump to have produced "the first recorded invention for impelling fluids by centrifugal force generated in a revolving disc". This false statement, even when forgetting the earlier work of Papin, caused a great controversy as to the true inventor of the centrifugal pump. This controversy was to culminate later in London at the Great Exhibition of 1851 where Bessemer, Gwynne and Appold all exhibited centrifugal pumps.

Andrews improved on his earlier pump by introducing in 1846 an impeller shrouded both on back and front sides but unfortunately reverted to a 90-degree turn at the end of his volute. This must have adversely affected the performance of the pump and reduced the advantage gained by the double shrouded impeller.

All the pumps developed from the Massachusetts pump had, so far, an inherent fault in that each had used the radial type blade. Many articles appeared denouncing attempts to use curved blades. Typical of these is one that appeared describing the advantages of radial blades in Gwynne's pump. This article states, "in attempting to construct a centrifugal pump, a very large number of inventors have exhausted their skill in making arrangements of spiral or curved arms on an axle, or in endeavours to find the supposed angle of curve, being in utter ignorance of the fact that in obedience of the law of central forces, the escaping fluid takes the shortest line to reach the circumference; or in other words, that each particle of matter in a state of rotation when free to escape, moves directly in the line of the radius until it reaches the circumference, and thence follows the tangent line until influenced by gravity or some other disturbing force". This false argument was finally exposed by Appold's centrifugal pumps and the results of independently adjudicated tests at the Great Exhibition of 1851.

In 1848 Appold constructed a model centrifugal pump for use in draining marshes (Fig. 15.7). In this design he set a number of straight vanes at 45 degrees to the diametrical axis of the impeller. This proved a major improvement, giving as it does a good approximation to the shape of the curved vane ideal. Appold must have realised this from the results of his experiments for in later pumps he introduced fully curved vanes.

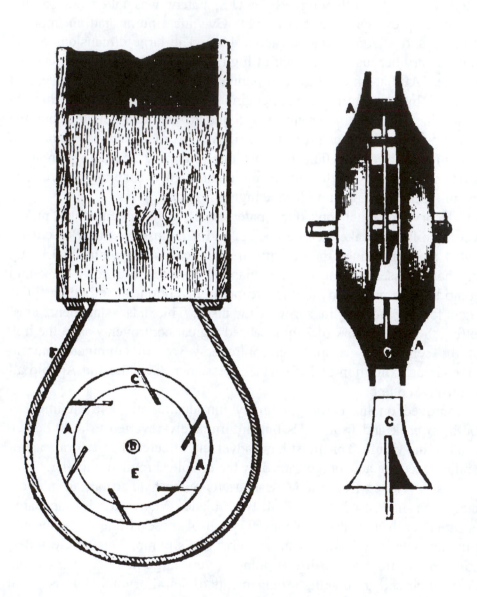

Fig. 15.7   Centrifugal pump with straight blades at 45 degrees (Appold, circa 1848)

The introduction of curved impeller blades (Fig. 15.8) had a great effect on the performance of the centrifugal pump, and the importance of Appold's invention of this form cannot be overestimated. This was, perhaps, best illustrated by Appold's performance figures obtained during his experiments, which he subsequently published. These results are summarised in Table 15.1.

Table 15.1 The effect of impeller blade shape — Test data by Appold (see Ref. 15.4).

Vane shape	Flowrate (imperial gallons per minute)	Generated head (feet)	Speed (rpm)	Efficiency (%)
Curved arms	1236	19.4	788	68.0
Straight arms inclined at 45 degrees	736	18.0	690	43.4
Radial arms	474	18.0	720	24.3

The results of his experiments, in which he had also determined the effect of the number of vanes, attracted much attention both from the British Association in Birmingham and the Institution of Mechanical Engineers in London. All these improvements were derived from experimental results alone, no attempt being made to analyse the problem theoretically.

Bessemer, in the following year, developed his radial bladed "disc, lift or suction pump" and, in a patent that was later issued relating to this, many variations of constructional arrangement.

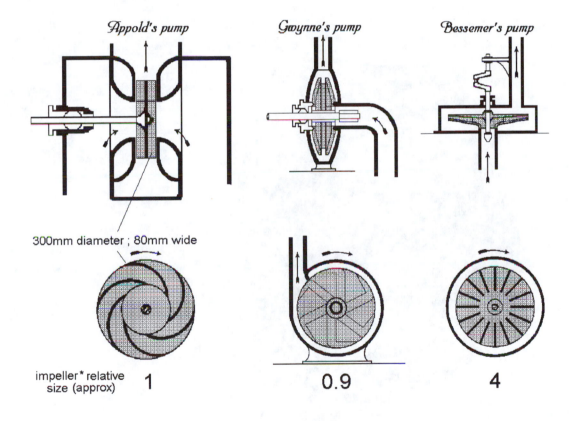

Fig. 15.8 Appold, Gwynne and Bessemer centrifugal pumps at the time of
The Great Exhibition, London, 1851
(Composite constructed from contemporary illustrations and drawings)

As mentioned earlier, claim to the origin of the centrifugal pump had started a great controversy. This situation became even more inflamed when *The Times* newspaper in London (Ref. 15.5) in reviewing the hydraulic machinery to be shown at the Great Exhibition implied that Gwynne, whose pumps were at the Exhibition, had no right to the claims of certain improvements made. The editor of the paper was later made to retract his statement publicly but unfortunately this did not stop the circulation of injurious statements to Gwynne's reputation. The tense rivalry of the three engineers Bessemer, Gwynne and Appold increased as the date of the Great Exhibition grew nearer. It was evident that whoever produced the pump judged by the jurors to be the best would gain world wide acclaim. Gwynne, as head of the company formed in 1849 in London, issued a challenge to Bessemer and Appold to "try the relative merits of the three designs by a year's duty" with the losers to pay £1000 to the London Mechanics Institution. In 1851 £1000 was a very considerable sum of money. The challenge remained unaccepted. This had the effect of restoring, to some extent, the general belief in Gwynne's right to his claim.

At the 1851 Great Exhibition the three centrifugal pumps were examined for scientific merit and commercial value. Tests showed conclusively that Appold's design was by far the best. This is illustrated by the efficiency figures recorded by jurors at the exhibition. These figures are summarised in Table 15.2.

Fig. 15.9 A 4.5 ft. (1.4 m) diameter impeller — Appold, 1852
A double-entry impeller with spiral blades that was in service for twenty-five years.

Photograph courtesy of
Science Museum / Science and Society Picture Library, London

Table 15.2 Test results from the Great Exhibition, London 1851 (see Ref. 15.6).

Inventor	Flowrate		Generated head		Speed	Efficiency
	gpm	dm³/s	feet	m	rpm	%
Appold	1236	93.0	19.4	5.9	788	68
Gwynne	290	21.9	13.8	4.2	670	19
Bessemer	846	63.7	3.28	1.0	60	22.5

gpm = imperial gallons per minute

The results of these tests, which were readily available to all engineers, proved once and for all the beneficial effects obtained by the use of curved vanes. Individual improvements had eventually become present in one design. The curved blade, the volute, the shrouded impeller, axial suction and tangential discharge all became embodied in a basic design format from which the development of the centrifugal pump was subsequently able to proceed in a logical manner.

In reviewing the contemporary literature it should be noted that although the word "impeller" was used in its current sense at the beginning of the twentieth century, different terms were used in the early years of centrifugal pump development. In 1851 it was described by Appold as "the fan". Bessemer described it as "the disc". The word impeller was used by Gwynne to describe "the division or filling-up piece between the sides of the disc...for the purpose of effectively imparting a circular motion to the water...". By 1905 the term was being used (Ref. 15.7), as today, to describe the complete rotating component as shown in Fig. 15.9.

An Appold impeller from this period is shown in Fig. 15.9. It was reported to have been used "constantly" from 1852 to 1877 at Wittlesea Mere, U.K (Refs. 15.8 and 15.9B)). The pump duty was 15,000 imperial gallons per minute (1140 m³/s x 10^{-3}) at a generated head of between 4 and 5 feet (1.2 to 1.5 m). The 4.5 feet (1.4 m) diameter impeller ran at a speed of 90 rpm. The impeller was subsequently presented to and is now kept by the Science Museum, London.

15.2.4 *Early uses and limitations of the centrifugal pump*

Following the establishment of a basic design for the centrifugal pump attention was turned to extending its limited range of application. Although many companies commenced manufacture of the pump in Europe and the United States it still did not seriously threaten the reciprocating pump as first choice for the majority of applications. Steam was the accepted method of power generation and the techniques involved in its economic use were known by all the engineers of the day. Being carried forward on a wave of industrial expansion in Western Europe and America steam driven reciprocating pumping engines appeared to have reached an unassailable position, especially with the introduction and rapid development of the direct-acting form by Henry R. Worthington.

The centrifugal pump, which at that time necessitated a steam engine and its accompanying crankshafts for producing rotary motion, became restricted to a small range of applications. The limitations on rotating speed imposed by the action of the steam engine effectively reduced its use to low lifts and it became used mainly in land drainage and irrigation.

John S. Gwynne had, as early as 1851, realised the generated head limitation of the centrifugal pump and had also seen that even if a relatively higher speed was available, the pump would still not be able to generate a head in excess of about 30 m in its present form due to the undesirable effects arising from impellers of large diameter and mass revolving at such speeds. This led to his invention of the multistage pump (Fig. 15.10). The specification in his patent shows his understanding of the practical uses of the design. In this specification he says of the drawing shown, "It is a vertical longitudinal section of the compound centrifugal apparatus which I employ for raising fluids from great depths or to great heights, as in mining operations, also for obtaining a powerful blast for pumping air or gases or for any other purpose where considerable pressure is to be obtained with a low rate of revolution of the pump." Gwynne's multistage design does not, however, appear to have been used to any extent in practice and this was probably due to his desire to perfect the impeller in his earlier designs before pursuing developments further following his experiences at the Great Exhibition.

Spiral passages, to give rotary motion to the fluid before it entered the impeller blades, were introduced by Bourne in 1853 and patented later by Thomson, but apart from this doubtful improvement hardly any further developments occurred over the next 15 years.

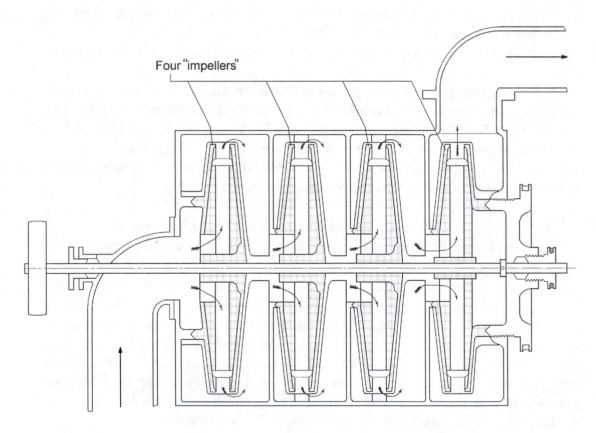

Four "impellers"

Fig. 15.10 Multistage pump patent — Gwynne, 1851   (Ref. 15.10)
(Composite constructed from contemporary illustrations and drawings)

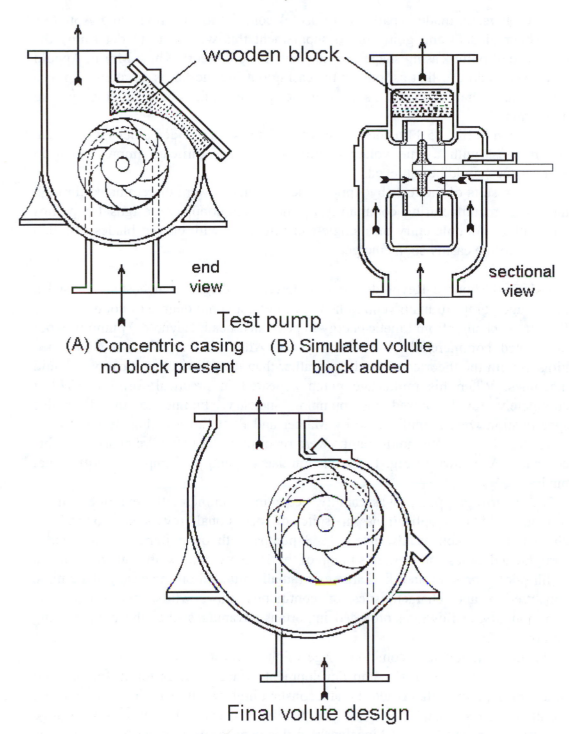

wooden block

end
view

sectional
view

Test pump

(A) Concentric casing
no block present

(B) Simulated volute
block added

Final volute design

Fig. 15.11 The pump casing shapes tested by R. C. Parsons, — 1876 (Ref. 15.11)
(Constructed from contemporary drawings)

A theoretical analysis of centrifugal pump performance was now sought. With the inadequacies of the Robertson theory (Ref. 15.9A) exposed by Appold's results proposals were presented by R. C. Parsons (Ref. 15.11) and by Unwin (Ref. 15.12). Both used the work of Euler (Ref. 15.13) as a base.

R. C. Parsons made a particularly useful contribution by reviewing Appold's experimental data and going on to supplement this with tests to determine the influence of pump casing shape. He saw that "the old theory which Morin, Appold and others held......that as long as the casing outside the fan is large enough it is immaterial what shape it is......can be proved false both in theory and experiment".

As shown in Fig. 15.11 a concentric casing fitted with wooden blocks was tested. A final test with a true volute confirmed that a significant improvement in performance could be achieved.

As the results of the tests by Appold and by Parsons were disseminated through the pump manufacturing community the now commonplace designs for pumps with either a double-entry or a single-entry impeller with curved blades installed inside a volute casing soon emerged.

In 1875 Osbourne Reynolds, head of Owens College, Manchester, devised a multistage pump. In his invention he tried to effect some degree of control on the fluid in its change from kinetic energy to generated head. Reynolds' pump was not introduced commercially for several years. After his invention he concerned himself with the theories of fluid flow rather than the practical application of fluid machines. When his multistage pump appeared in practical form in 1878 it completely revolutionised the pumping industry. Pumps to the Reynolds specification were manufactured by Mather and Platt Ltd. of Manchester and by 1893 the Mather – Reynolds pumps were reported to be made "regularly" by this company. As shown in Fig. 15.12 this multistage pump had impellers with radial blades.

The centrifugal pump was rapidly becoming commercially exploited in its limited field of applications in spite of occasional reversions to inferior characteristics such as Gwynne's 1868 pumps with open impellers and partly straightened vanes and Bernay's pump of the 1873 Vienna Exhibition which had a "whirlpool" chamber and discharge exit radially situated on the casing. Its simple construction and characteristics of continuous flow, whilst permitting easy variation of the flowrate, provided important advantages over the reciprocating pumps.

The introduction into common usage of the electric motor was of very great significance in centrifugal pump development. Manufacturers found for the first time a prime mover that could run at a constant high speed, was simple to operate, required modest foundations, and was relatively cheap. The formation of large industrial electricity undertakings enabled this new form of power to be exploited to the full. With the standardisation of the frequency of alternating current systems, pumps were able to be designed to run below the synchronous speed of the two-pole motor or submultiples of it. By 1899 Sulzer, Switzerland, made a notable step in the application of multistage pumps by installing one in the Horcajo Silver Mine, Spain. This was possibly the first mine drainage pump with an electric drive.

At the turn of the century the multistage pump with an electric motor had become firmly established. The development of the new prime mover, the electric motor, had occurred at the right time and pump manufacturers were able to consolidate and then capitalise on the work previously done using steam engines.

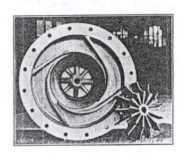

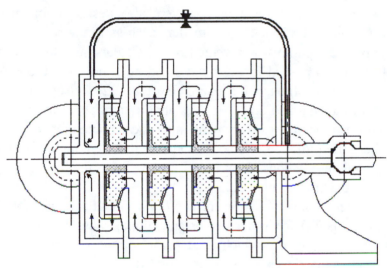

Fig. 15.12 An 1887 Mather — Reynolds pump (Taken from Refs. 15.8 and 15.14)
(Composite constructed from contemporary illustrations and drawings)
Photograph of the assembled pump courtesy of
Science Museum/Science and Society Picture Library, London

15.2.5 *Centrifugal pumps in the twentieth century*

The application of centrifugal pump designs proved almost limitless. The strong growth in the engineering industries enabled many new centrifugal pump manufacturers to start trading.

The multistage centrifugal pump was not yet sufficiently advanced technically to completely oust the reciprocating pump from favour. Efforts were made to remedy this and in 1900 Mather and Platt Ltd. and Sulzer Brothers Ltd., who virtually monopolised the high-pressure field of centrifugal pumping in Europe, came to an agreement providing for an interchange of the "future improvements" made by the two companies. The agreement ended by mutual consent in 1904.

The main defect of the multistage design was an inherent large axial thrust from pump impellers mounted in series and facing in the same direction. Sulzer tried the balancing disc then later moved to mounting impellers back to back in pairs. A. C. E. Rateau of Paris and Byron Jackson of San Francisco soon followed Sulzer with further improvements in axial thrust balancing devices. Vanes on the back shrouds of impellers were tried but with little commercial success. With the rapid commercial development of high-pressure centrifugal pumps efforts were made to simplify the design. A big step in this direction was made by John Richards who appears to be the first to have used guide vanes in a removable annular ring. In 1911 Mather and Platt Ltd. centrifugal pumps were being made by the unit stage assembly method using axial bolts to hold the casing segments together. This subsequently became common practice in Europe.

The American market, led by Byron Jackson, preferred multistage pumps based on complex split-casing castings with external interconnecting piping. Apart from this "national" preference for a particular design, by 1912 the multistage pump had reached a form suitable for rapid commercial manufacture.

High pressure fire pumps to the Rees–Roturbo patent of 1906 were much in evidence in the United Kingdom. Centrifugal pumps were also designed for such arduous applications as pumping fibrous and solid matters such as in the sewage industry. In 1910 the "stereophagus" pump was invented by R. C. Parsons (Ref. 15.15). In this pump the rotating blades pushed forward and shredded solid and fibrous matter that would otherwise not pass through the pump. By the 1920s electricity was readily available for most industrial and domestic uses and brought about the first noticeable trends away from steam power. It was proving much cleaner and more economical than steam power where the initial cost of boilers and ancillary equipment was becoming prohibitive for small installations. Centrifugal pumps for office and factory heating (Ref. 15.16) provided a thriving market.

Large electricity generating stations were another important market for centrifugal pump manufacturers. Boiler feedwater pumps and cooling water pumps were in great demand. Hydro-electric schemes which used pump-turbine combinations for pump storage to meet peak demand were also proving cost effective.

The borehole pump was chosen for use by many waterworks' engineers as it needed a bore of only a few inches diameter to enable a pump to be inserted at great depths. Sulzer in 1923 supplied a 38-stage pump 115 m long of maximum diameter less than 200 mm. In 1932 wet rotor borehole pumps were being sold by Isler and Co. Ltd. in London. This latter design avoided the necessity for a long multi-bearing shaft down the well since the electric motor was permanently under the water level. Canned pumps (wet rotor/dry stator) appeared using a squirrel cage induction motor suitable for single- or three-phase alternating current with a non-corrodable rotor running in water and a thin alloy cylinder being used to protect the stator winding. A "Mobila" canned pump was produced in the 1930s in Switzerland suitable for use in a domestic central heating system; it was compact enough to be carried in the hand.

Drysdale and Co. Ltd. of Glasgow in 1930 patented a centrifugal oil circulating pump for electric transformers. Centrifugal pumps came to be used in the chemical and process industries being constructed in almost every conceivable material including stoneware, rubber and even glass. Beer, milk, fruit juices and soups, all were successfully pumped by special adaptations of the centrifugal pump. Volatile liquids in the chemical and process industries created a need for even more research into the effects of cavitation.

By 1950 demand for electricity was increasing rapidly world wide. Power generation was, almost without exception, achieved by moving to ever larger steam turbine-generator units. In the United Kingdom the adoption of a 660 MW size in 1965 (ten times that of a decade earlier) necessitated a radical rethink in the design of centrifugal pumps for this industry (Refs. 15.17 and 15.18). The "Advanced Class" boiler feedwater pump which emerged represented the culmination of the many years of pump development for these special needs.

In the United States the developments in the space industry, culminating in the Saturn rocket pump, necessitated entry into a similarly demanding technology. Also at this time in the United States and in France power generation technology based upon the pressurised water reactor (PWR) led to the development of very large reactor coolant pumps.

Table 15.3 High-technology pumps of the 1980s

Pump configuration	multistage single-entry impellers Fig. 15.13	single-stage single-entry impeller Fig. 15.14
Application	high-pressure steam-turbine driven boiler feedwater	reactor coolant circulation
Pump design duty (approx.)	*	**
flowrate litres per second	500	65,000
generated head ΔH metres	1980	90
pumped liquid	feedwater at up to 185°C	borated water up to 325°C
Other operating and design data		
pump speed rpm	7500 max.	1500 (synchronous)
power kW	(variable)	6000 (motor rating)
weight tonne	11,000 (absorbed)	99 (pump+motor)
principle dimension (approx.) m	4 (pump only) 1.5 (length)	6 (height)

*   Ferrybridge C power station (1 prototype pump for a 500 MW unit).
**   Sizewell B power station parameters (4 pumps for the two 660 MW units).

Paralleling the development of centrifugal pump hardware the affinity laws (section 3.2.1) emerged semi-empirically. The following quote (Ref. 15.19) illustrates how this result is easily obtained from simple reasoning.

> "If then the velocity were that of 32 ft. in a second, the water resistance would rise to a vertical height of 16 ft., and would rise to a greater or smaller height according to the square of the velocities (as can be shown by the theory of Galileo). It will thus be easy to find by calculation what speeds will be needed for raising water to any desired height; and as speed can be infinitely increased there is no height which cannot be attained by means of this machine provided that sufficient power be applied."

This statement was made by Denis Papin in 1689!

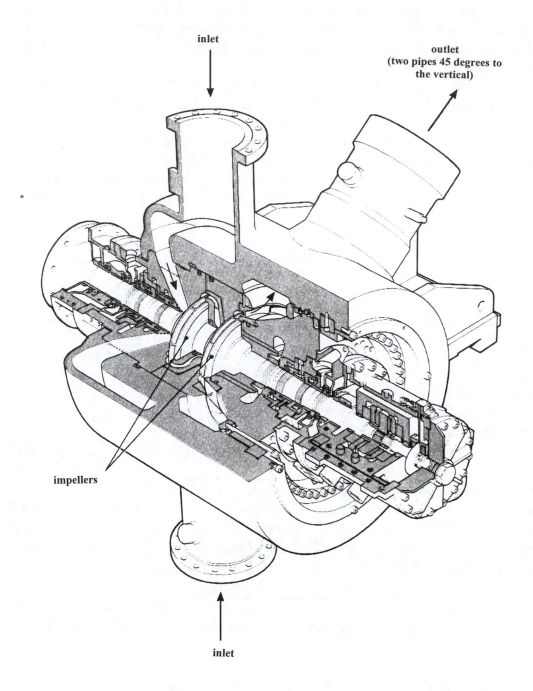

inlet

outlet
(two pipes 45 degrees to
the vertical)

impellers

inlet

Fig. 15.13 A high-technology multistage pump of the 1980s
An "Advanced Class" boiler feedwater pump by Weir Pumps Ltd., UK
Courtesy of Weir Pumps Ltd

The pump illustrated in Fig. 15.13 is the prototype pump which, after service at Ferrybridge C power station between 1969 and 1980, now resides in the Museum of Science and Technology, Manchester, U.K. Numerous pumps based on this design were subsequently used in power station service.

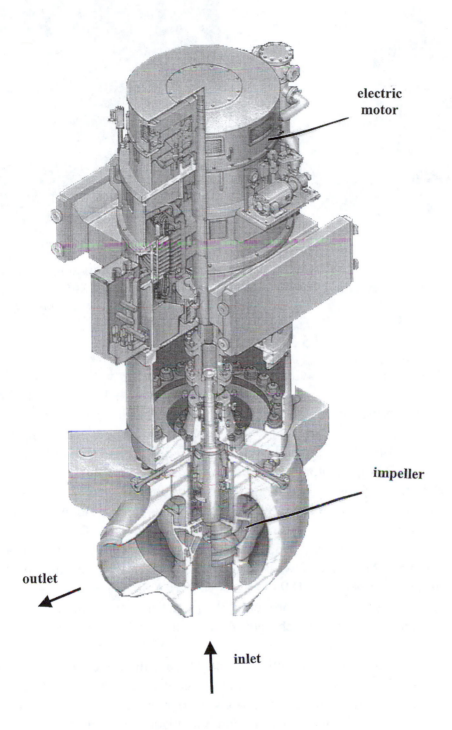

electric
motor

impeller

outlet

inlet

Fig. 15.14   A high-technology single-stage pump of the 1980s
A reactor coolant pump by Westinghouse Electric Corporation, USA
Courtesy of Westinghouse Electric Corporation

Fig. 15.14 shows a section view of the pumps at Sizewell B, the only UK pressurised water reactor (PWR) power station. Table 15.3 gives more detail.

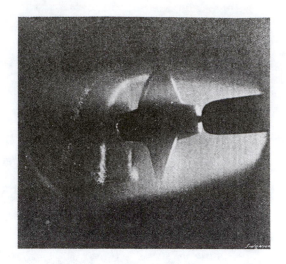

(a) The rig with viewing facility   (b) Example of photographs obtained

Fig. 15.15   The first cavitation test rig — photographs taken circa 1895

see Richardson Ref. 15.36

15.3 Cavitation

The first use of the word cavitation to describe vapour within a liquid is attributed (Ref. 15.20) to a statement made by R. E. Froude in 1895. At that time the bringing into service of very large ocean liners and ever more powerful warships was accompanied by problems with propellers. These were twofold (i) damage which could not be explained by the then frequent "bad metal" from the foundry and (ii) high "slip" factors which began to limit the maximum speed (and usable power) of these ships. The issue was famously brought to a head by C. A. Parsons (Ref. 15.21) who, starting in 1893, had sought to extend the use of steam turbines into ship propulsion. In the ship built to demonstrate this, the *Turbinia* (a 30 m-long, 3 m-beam, 44-tonne displacement vessel permanently preserved in the Newcastle Discovery Museum, Newcastle, U.K.), sea trials revealed unexpected propulsion problems. By April 3rd 1895 Parsons wrote,

> "In further reference to the propeller, the matter is now, I think, quite cleared up by a paper, proof of which I got this morning, by Thornycroft.
>
> He makes out from trials of the *Daring* and two other very high-speed boats that if the mean pressure exceeds 11¾ lb. then vacuum, or as Froude has termed it 'cavitation' is set up, the slip goes up enormously as well as the power required for a given revolutions. In our present screw we have some 60 lb. per sq. in. mean pressure on the blades and therefore enormous 'cavitation' set up."

With the mark of true engineering genius Parsons set about understanding the new phenomenon by building what was to be the world's first cavitation test rig. This rig, shown in Fig. 15.15(a), is also on display in the Newcastle Discovery Museum. This test rig, referred to as a cavitation tunnel, consists of an oval closed circuit of uniform rectangular cross section. The test propeller was initially driven by a steam turbine but later by the electric motor shown. Photographs, an example of which is shown in Fig. 15.15(b), were taken through windows at either side. When in use a plain mirror was fixed to an extension of the shaft. This reflected light from an arc lamp onto a parabolic mirror, which in turn lit the propeller for a fixed period during each revolution. An illumination period of 1/3000 second is said to have been achieved.

Showing great insight Parsons anticipated the interaction between local (static) pressure and propeller speed. The rig was designed so that the static pressure could be changed. In Parsons' words

> "To enable the propeller to cause cavitation more easily the tank is closed and the atmospheric pressure removed from the surface of the water above the propeller by an air pump. Under these conditions the only forces tending to hold the water together and resist cavitation are the small head of water above the propeller, and capillarity. The propeller is 2-inch diameter and 3-inch pitch; cavitation commences at about 1200 revolutions and becomes very pronounced at 1500 revolutions. Had the atmospheric pressure not been removed, speeds of 12,000 and 15,000 revolutions per minute would have been necessary, rendering observation more difficult. The shape, form and growth of the cavities about the blades could be clearly seen and traced. It appeared that a cavity or blister first formed a little behind the leading edge, and near the tip of the blade: then as the speed of revolution was increased, it enlarged in all directions until at a speed corresponding to that in the Turbinia's propeller, it had grown so as to cover a sector of the screw disk of 90 deg."

The unacceptable problems arising from cavitation in centrifugal pumps were referred to in 1915 in a book by Daugherty (Ref. 15.22) who opined that the reason for the generated head reduction observed at high flowrates in tests arose because

> "...the absolute pressure at entrance to the impeller becomes so low, due to the high velocity in the suction pipe with a consequent loss of head, that 'cavitation' is set up in the eye of the impeller. That is, the eye of the impeller, and consequently the impeller passages is not completely filled with water. Thus the actual rate of discharge of the pump is decreased below the expected amount"

Meanwhile the problems arising from cavitation erosion had been addressed by Rayleigh who, by 1917, had presented his paper (Ref. 15.23 — referred to in Chapter 5) on the conditions governing the collapse of a cavity. Parsons and Cook (Ref. 15.24) in 1919 published detailed metallurgical studies of the erosion of propellers.

Appendix II of Ref. 15.22 included the outline mathematics of cavity collapse controlled by inertia forces alone. The paper was reprinted in the wide circulation magazine *Engineering* (Ref. 15.25) so that by 1920 most engineers, including those engaged in centrifugal pump manufacturing, were aware of cavitation, its effect on hydraulic performance and its ability to cause rapid and damaging erosion. The 1895 cavitation test rig with its visual observation facility had demonstrated a practical way of addressing such problems although there is little evidence to show that this was effectively pursued.

A major step forward in the development of a theoretical analysis applied to hydraulic machines such as centrifugal pumps came with the publication by Taylor and Moody (Ref. 15.26) of the simplest form of similarity law for the cavitation performance of centrifugal pumps and water turbines. This became referred to as the Thoma – Moody similarity law. It provided a basis for comparing performance data for a given machine running at different speeds; viz.,

Thoma cavitation parameter σ = $\dfrac{\text{inlet head above vapour pressure of liquid}}{\text{generated head}}$

$$\sigma = \frac{NPSH}{\Delta H} \qquad \text{---------------------------15.1}$$

In 1932 Pfleiderer (Ref. 15.27), aware of the earlier work of Thoma (Ref. 15.28) and Busemann (Ref. 15.29), tried to explain centrifugal pump erosion damage caused by the presence of cavitation. Conclusions were only qualitative but made a valuable start to driving forward basic research into this aspect of centrifugal pump performance.

The value of extending similarity considerations to cavitating conditions at a centrifugal pump inlet was identified by Wislicenus, Watson and Karassick (Ref. 15.30) in 1939. By simple logic it was shown that a constant for a given cavitation condition is:

$$\frac{n\ Q^{1/2}}{NPSH^{3/4}}$$

Wislicenus, Watson and Karassick identified that the parameter "has, in regard to the cavitation characteristics, a significance very similar to the specific speed with respect to general operating conditions of centrifugal pumps". The parameter was named suction specific speed by Bergeron (see the addendum to Ref. 15.30). Lazarkiewicz and Troskolanski (Ref. 15.31) claim that this same mathematical grouping was used in Russia by Rudnieff in 1934. Whatever the true historical origin the reality is that it was the widespread publication of the work of Wislicenus and his colleagues that made the term one of common usage in centrifugal pumping.

Wislicenus also showed that specific speed and suction specific speed are related to each other by the Thoma parameter. This is easily demonstrated from the work presented in Chapter 3.

From equation 3.11 $\quad n_{SS} = \dfrac{n\ Q^{1/2}}{NPSH^{3/4}}$ $\qquad$ from equation 3.4 $\quad n_S = \dfrac{n\ Q^{1/2}}{\Delta H^{3/4}}$

so that $\qquad \dfrac{n_{SS}}{n_S} = \sigma^{3/4}$ $\qquad$ ----------------------15.2

Stepanoff (Ref. 15.32) in 1957 used Wislicenus, Watson and Karassick (Ref. 15.30) to show that for a series of pumps of consistent design, a continuous relationship between σ and n_s should be obtained since all factors governing cavitation (eye area, number of impeller blades etc.) are a continuous function of n_S. He demonstrated that this gives a curve of the form

$$\sigma = k_4 \, n_S^{4/3} \qquad \text{-----------------------------------} 15.3$$

From this it is evident, as deduced already in section 3.8 and demonstrated by experience in Table 9.1, that a single value of suction specific speed can be used to describe the cavitation performance of centrifugal pumps.

The use of NPSH(3pc) as a basis for comparing cavitation performance has been common since the 1950s. The references in Chapter 3 have been chosen to demonstrate this fact. The use of NPSH(4mm), as described in Chapter 9, has been included in contractual requirements describing cavitation performance requirements since 1974.

Hydrodynamically induced cavitation surging in centrifugal pumps is often thought of as a recent phenomenon. This is partly true in that the severest form manifests itself only in pumps that have a pump input power to impeller churned mass ratio that is very high, and these machines only became commonplace after about 1950. Chapter 7 includes examples of the visual records made from 1968 onward. Indeed, observation of the extraordinary "reverse-direction" spiralling flows has, in terms of a history of two or three centuries, occurred only recently. Reading carefully the earlier descriptions of pump performance, however, it is clear that the problem of cavitation surging has been around a long time before being formally identified. Daugherty (Ref. 15.22) in a 1915 description of centrifugal pump operation on suction lift stated,

> "As vapour is liberated and fills part of the space, the rate of discharge decreases. This causes the pressure to rise, the vapor again becomes liquid, the *pipe* is filled with water and the discharge increases. But as it does so the pressure drops again and thus a pulsation of flow is set up. If there is just about balance this pulsation will be perceptible to the ear only. If the pressure is further decreased the pulsation is sufficient to cause the suction line and pump to vibrate".

Applying the description to the pump impeller passages instead of the adjacent inlet *pipe* an accurate description of cavitation surging presents itself.

15.4 In Summary

Historically the path of centrifugal pump development has been a tortuous one. The almost forgotten work of Papin contrasts vividly with the excitement surrounding the rapid progress made by Appold leading up to the 1851 Great Exhibition in London.

C. A. Parsons' early work on propellers gave extraordinary impetus to cavitation studies and identified the opportunity, which was not immediately taken up, offered by pursuing solutions using visual observation. Understanding the fundamental processes that result in cavitation surging and cavitation erosion have since that time progressed slowly. Application to centrifugal pumps has often been piecemeal.

Perhaps the lesson of such a history is to be forever vigilant in reviewing the merits of old and new ideas.

References

In the Text - Centrifugal pumps

15.1   Grist, E., The Development of the Centrifugal Pump,
        Royal Technical College, Salford,1960, (unpublished).
15.2   Acta Eruditorium, 1689, —
        Leibnizens und Huygens briefweshel mit Papin, Berlin,1881.
15.3   A letter from Mr. D. Papin to Dr. F. Slare re improvement to Hessian
        Bellows, Philosophical Transactions of the Royal Society of London,
        Vol. XXIV, No.100, pp.1990–1993, 1705, copied in Appendix A.
15.4   Glyn, J., Power of water, pp. 146-148, 1853.
15.5   The Times, London, p. 5, col. 6, 15 April 1851.
15.6   The Great Exhibition, London,
        Report of the Jurors, Vol.1, class V, p. 384, 1851.
15.7   Webber, W. O., Some types of centrifugal pump,
        Trans. ASME, Vol. 24, pp. 764–800, 1905.
15.8   Westcott, G. F., Pumping Machinery,
        Science Museum, London, pp. 105, 106 and plate VII, 1932.
15.9A   Robertson, A. J., On the mathematical principles involved in the centrifugal
        pump, Proc. IMechE, pp. 99 – 109, 1852.
15.9B   Appold, J. G., Discussion following 15.9A,
        IMechE Proc., pp. 153 – 163 and plates 68 and 69, 1852.
15.10   British Patent 13577, 1851.
15.11   Parsons, Hon. R. C., The theory of centrifugal pumps, as supported by
        experiment, Proc. Institution of Civil Engineers, Vol. 47, pp. 267 – 282, 1876.
15.12   Unwin, W. C., The centrifugal pump,
        Proc. Institution of Civil Engineers, Vol. 53, pp. 249 – 281, 1877.
15.13   Euler, Recherche sur l'effet d'une machine hydraulique proposee par
        M. de Mour, Academie des Sciences et Belles Lettres a Berlin, p. 303, 1751.
15.14   Hopkinson, E. and Chorlton, A. E. L.,
        The evolution and present development of the turbine pump,
        The Engineer, p.100, 101, 130 and 131, 1912.
15.15   British Patent 26913, 1910.
15.16   Anon., Geneology of the central heating pump,
        Heating and Air Treatment Engineer, Dec. 1958.
15.17   Advanced-Class Boiler Feed Pumps, IMechE Conf., Vol. 184, Part 3N, 1970.
15.18   Glandless Pumps for Power Plant, IMechE Conf.,. Vol. 184, Part 3K, 1970.
15.19   Hodgeson, J. E., The inventor of the centrifugal pump,
        Engineering, pp. 670 and 671, 1890.

In the Text - Cavitation

15.20   Thornycroft, J. I. and Barnaby, S. W., <u>Torpedo-boat destroyers</u>,
         Proc. Institution of Civil Engineers, p. 67, 1895.
         Statement attributed to R. E. Froude.

15.21   Sir Charles Parsons, <u>Applications of the compound steam turbine to the
         purposes of marine propulsion</u>,
         Trans. Inst. Naval Architects, Vol. 38, pp. 232, 1897.

15.22   Daugherty, R. L., <u>Centrifugal Pumps</u>, McGraw-Hill, New York, 1915.
         (see pp. 40 – 42 for remarks on cavitation induced surging;
         see pp.92 –94 for general reference to cavitation).

15.23   Rayleigh, Lord, <u>On the pressure developed in a liquid during the collapse
         of a spherical cavity</u>, Phil. Mag., Vol. 34, pp. 94 – 98, 1917.

15.24   Parsons, C. A. and Cook, S. S., <u>Investigations into the causes of corrosion
         or erosion of propellers</u>, Inst. of Naval Architects, Vol. 61, pp. 223 – 240,
         1919.

15.25   Parsons, C. A. and Cook, S. S., <u>Investigations into the causes of corrosion
         or erosion of propellers</u>, Engineering, pp. 515 – 519, 1919.

15.26   Taylor, H. B. and Moody, L. F., <u>The hydraulic turbine in evolution</u>,
         Engineers and Engineering, Vol.39, No.7, pp. 241 – 259, 1922.

15.27   Pfleiderer, C., <u>Die Kreiselpumpen</u>,
         Chapter J Kavitation, p.235, Springer, Berlin, 1932.

15.28   Thoma, D., <u>Experimental research in the field of water power</u>,
         Proc. First World Power Conference, pp. 536 – 551, London, 1924.

15.29   Busemann, A., <u>Das forderhohenverhaltenis radialer kreiselpumpen mit
         logarithmisch spiraligen schaufeln</u>,
         Z. angew Math Mech., Vol. 18, p. 372, 1928.

15.30   Wislicenus, G. F., Watson, R. M. and Karassik, I. J., <u>Cavitation
         characteristics of centrifugal pump as described by similarity
         considerations</u>, Trans. ASME, Vol. 61, pp. 17 – 24, 1939.

15.31   Lazarkiewicz, S. and Troskolanski, A. T., <u>Impeller Pumps</u>,
         Pergamon, Elmsford, New York, 1965.

15.32   Stepanoff, A. J., <u>Centrifugal and Axial Flow Pumps</u>, Wiley, New York, 1957.

Further Reading

(i) *Important contributors to centrifugal pumps and cavitation*

15.33   Denis Papin FRS, (1647 – 1714);
         see H. P. Spratt, <u>The prenatal history of the steamboat</u>,
         Trans. Newcomen Society, London, pp. 16 –17, Vol. 30, 1955.

15.34   John George Appold FRS, (1800 – 1865);
         see <u>Obituary notices of fellows deceased</u>, Proc. Royal Society of London,
         insert p. (i) – (vi) (between p. 518 and 519), Vol. XV, 1866.

15.35   Charles A Parsons FRS, (1854 – 1931);
         see <u>Sir Charles Parsons and cavitation</u>,
         Proc. Inst. Marine Engineers, Vol. LXIII, No. 8, pp. 149 – 167, 1951.

(ii) *Historical reviews*

15.36 Anon., The centrifugal pump, The Practical Mechanics Journal,
       pp. 121 – 126 and plate 77, 1 Sept 1851.

15.37 Harris, L. E., Some factors in the early development of the centrifugal
       pump, 1689 – 1851, Trans. Newcomen Society, London,
       pp. 187 – 202 and plates XX to XXIV, Vol.28, 1953.

15.38 Appleyard, R., Charles Parsons and his Life Work, Constable, 1933.

15.39 Richardson, A., The Evolution of the Parsons Steam Turbine,
       Offices of Engineering, 1911.

15.40 Parsons, G. L., Ed., Scientific papers and addresses of
       the Hon. Sir Charles Parsons with memoir by Lord Rayleigh,
       Cambridge University Press, 1934.

15.41 Smith, K., The Story of Charles Parsons and his Ocean Greyhound,
       Newcastle Libraries and Tyne & Wear Museums, 1996.

Chapter 16

Centrifugal Pumps and Cavitation —
A View on the Future

16.1   Engineering Progress — A Personal View

In 1960 the author expressed the view (Ref. 16.1) that the future for centrifugal pumps would be dominated by two things: (i) a move to high speed pumps and (ii) the corporate amalgamation of the many pump manufacturers. Space rocket engines and power station feedwater pumps did indeed grow to include very high speed pumps. Large international pump firms also emerged from a lengthy period of pump industry restructuring. Whilst high-speed pump technology became widespread it did so using electric motor/gearbox and steam turbines as the prime mover. The anticipated use of variable-frequency high-speed electric motors did not become commonplace. Now, with space (and defence) related objectives largely met and the general move away from large high-pressure steam-cycle based power stations the source of financial support has, since about 1990, been removed. Projects requiring new high-energy high-speed pumps have almost disappeared. It might appear from this that, as in the seventeenth and eighteenth centuries, centrifugal pump development has entered a quiescent period.

Being older, wiser, and less imbued with the certainties of youth discourages an enthusiastic supporter of technological progress from engaging in the optimism that forecasting the future can bring. Pragmatism now prevails. However, it is possible to identify pressing needs and commercial opportunities.

Changes in engineered products like the centrifugal pump are more than ever driven by short-term realities. The commercial advantage pump manufacturers now seek in this respect may well be provided by new design variation opportunities afforded by utilising pump speeds up to and beyond 10,000 rpm. Allied to the use of automated machining and assembly techniques a significant drive forward looks probable for pumps up to 100 kW. High-quality engineering is required in designing much smaller pumps; the real opportunity lies much more in successfully combining the best in hydraulic and mechanical design with the best of production engineering skills.

Commercial investment in research and development is more than ever driven by the effect of pump failures in service. Failures have always been seen as requiring prompt action to avoid besmirching a pump manufacturer's good name. It now seems only a matter of time before litigation arising from consequential damage or personal injury assumes greater importance and starts to drive the technical "need to know" forward.

There are two causes of serious damage to centrifugal pumps that are of concern to users. They are (i) sub-synchronous shaft vibrations (not dealt with in this book — for

311

further information see Ref. 16.2) and (ii) hydrodynamically induced cavitation surging. The response on both these topics from too many of the pump manufacturers is reactive; that is to say when the problem occurs the only remedy offered is recourse to empirically derived corrective measures which are then implemented on the operating plant. The failures produced by each of the two phenomena, though infrequent, are usually extremely costly both in the consequential damage caused and the extent of remedial pump/plant design changes necessary to secure safe operation. Pump users, quite rightly, insist that the problem be addressed at the pump and pump/plant design stages. If restrictions are to be imposed on a new plant then these should be quantified at the outset. Preferably such restrictions should be based upon a general theoretical understanding. A restatement of the major engineering research and development issues relating to surging that need to be addressed are described in section 16.2.

The descriptions and photographs of the unacceptable forms of cavitation surging and the methodology of preparing a specification based upon generic data described in earlier chapters provide users with a commercially relevant route to choosing an acceptable pump. The background to the methodology provides a high degree of confidence in the outcome. The emphasis is on acceptability borne of reduced risk.

The data supporting the route to avoidance of unacceptable cavitation erosion are also well documented. However, in regard to proof of performance it could be improved. Indeed for high-energy pumps it must be improved. Review of the earlier chapters shows where improvements should be made. Some possibilities are described in section 16.3.

There must be a place for new ideas and for revitalised old ones. As a better understanding of basic mechanisms emerge it is worth looking away from the conventional designs and considering what rewards innovation might offer. Section 16.4 offers a little stimulus for those interested.

16.2 Cavitation Surging

16.2.1 *Hydrodynamically induced surging*

The spiralling flows of cavitation that characterise this form of surging are impressive in their violence and apparent repeatability. In spite of the certainties of the phenomenon, the prediction of its effect in pumping plant appears limited to empirical methods based on test observations. Given the phenomenon was adequately described in 1915 by Daugherty (Ref. 16.3), and again in 1976 by the fully documented visual test results of Massey and Palgrave (Ref. 16.4), progress has indeed been slow.

Test results are readily obtained that allow a surge characteristic to be drawn up. An idealised characteristic is presented in Fig. 16.1. Such qualitative results, welcome though they are, are a poor substitute for a comprehensive analysis by which the frequency and surge peak pressure of the phenomenon can be predicted and the limits of stability of the twisting columns of cavities calculated and demonstrated by test. The

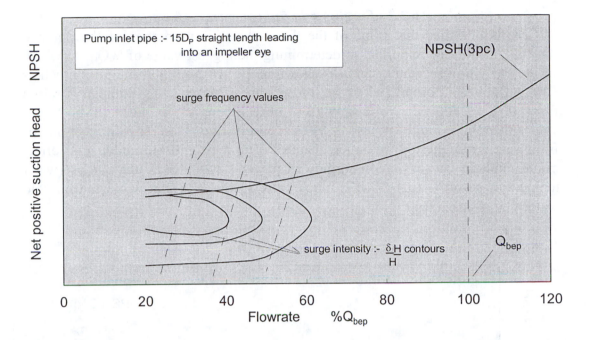

Fig. 16.1 Idealised hydrodynamically induced cavitation surge characteristic

"cut and try" approach often employed when violent surging occurs is not good enough for pump users. What needs to be pursued with vigour is research into the underlying mechanism of hydrodynamically induced cavitation surging.

The aim should be a quantitative method which enables pump manufacturers to take pump impeller geometry, inlet pipe geometry and pump speed and from them predict for given NPSH(A) values (i) surge frequency with flowrate, (ii) surge length with flowrate and (iii) surge pressure pulsation intensity with flowrate. This work should then allow numerical forecasting of the effect of the pulsations as they affect (a) shaft torsional load and (b) axial thrust. The forecast should cover at least the normal pump operating range of $50\%Q_{bep}$ to $120\%Q_{bep}$. The results should be achievable to an uncertainty less than +/-10%.

A means of describing the onset of surges with an unacceptable magnitude needs to be devised. Preferably this should be accompanied by a commercially acceptable way of obtaining proof of this performance boundary by test so that it can be demonstrated that particular designs are not at risk. A test method which can be applied as a site test as well as a manufacturer's works test is the ideal since most problems are encountered on high-energy pumps, many of which cannot be tested in a manufacturer's works because of absorbed power limitations.

16.2.2 *Surging and fluid property effects*

There is evidence that some of the thermodynamic properties of the pumped fluid might be very important in determining the highest value of $\%Q_{bep}$ at which severe cavitation surging occurs. As shown in Fig. 16.2 an increase of as much as $10\%Q_{bep}$ appears possible when the temperature of water being pumped is raised from 38°C to 100°C. Fortunately these observations are at much lower NPSH values than those judged necessary by the conservative measures to protect pumps from surging suggested in earlier chapters. The apparent thermodynamic effect might explain why cavitation surging in power station feedwater pumps (water temperatures 150°C to 180°C) is much more severe than that witnessed on cold water works-based tests.

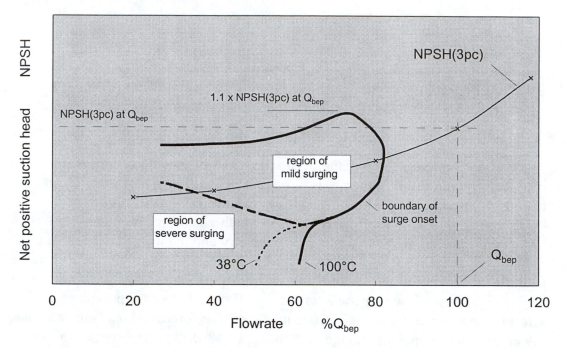

Fig. 16.2 Boundary of severe cavitation surge onset for water at 38°C and 100°C (500 mm diameter straight inlet pipe into inducer/impeller combination)

16.2.3 *Thermodynamically induced surging*

In practice the conditions that are conducive to severe surge are limited to high-energy pumps that are rapidly brought down to "leak-off only" conditions just as the leak-off valve is opening. The heat losses from this churning low flowrate operation rapidly lowers the NPSH(A). These conditions often result from a failure to implement correctly the pump "shut-down" procedures. The unquantified presence of forces sufficient to break a pump shaft or to destroy an axial thrust bearing that this produces is of concern to pump users. The inability to predict the magnitude of the axial thrust that arises when, inevitably, a low-flow protection system fails needs to be addressed. In the meantime the cost of improving reliability by duplicating the leak-off system falls on the pump user.

16.3 Cavitation Erosion

Cavitation erosion risk is currently inferred from a combination of past experience obtained at similar NPSH(A) and flowrate levels together with evidence gathered during visual cavitation tests. It is not a direct measurement of damage or of damage rate. Cavitation erosion is best avoided, but this is not always possible. The concept of "acceptable" damage was described in section 9.2.1. Its application was dependent upon being able to identify "progressive" damage. What is needed is a test that meets the requirements listed in Table 16.1.

Table 16.1

Test requirement	Comment
(i) Records actual cavitation damage by (a) location and (b) rate of attack	
(ii) Uses actual "contract" impeller	Small but sometimes significant geometry differences negate the use of a prototype
(iii) Uses the actual "contract" pump	Not dependent upon visual observation — this permits high temperature operation where appropriate
(iv) Has a cost which allows it to be carried out on all impellers at risk	On an order for several multistage pumps this entails testing each first-stage impeller

The measures outlined in the table take "proof by test" one step further than the paint erosion test. They describe a test that shows not only (i) cavitation erosion is taking place at a particular location, but also (ii) quantifies the rate at which it is occurring. Unlike the paint erosion test, the test objective described is intended to provide proof of performance in a contractual setting.

An attempt to devise such a test was made in work carried out at Nottingham University, UK, in the 1980s. The results did not provide a commercial way forward. However, as the methodology might give a stimulus to more successful efforts it is described in outline here.

The method consisted of

    (i) the contract impeller in the contract pump being the basis of tests,

    (ii) a multi-layered coating to provide a surface record of cavitation damage.

By choosing alternating layers which are of visually contrasting materials a "contour map" of damage from which the area and depth of penetration is obtainable. Layers of the order of 0.2 mm were used. The overlay materials chosen were capable of being laid down and removed at a temperature that was low enough to prevent any unacceptable changes to the heat treatment properties of the impeller material on which it was laid. This enables the contract impeller to be the subject of a cavitation erosion test. By making the first layer of a relatively low melting point material the consequence of heating the impeller after testing is to bring off all the layers and return the impeller to a "ready for contract use" state.

The idea was good. The implementation proved too difficult. The choice of cadmium, with successive layers of electroplated copper and cadmium, was too difficult to control.

16.4   The Influence of Pump Design on Cavitation Performance

16.4.1   *Constructional arrangements*

A number of pump constructional arrangements are usually offered for a particular application. Pump users often have a preference for particular designs (end-suction, vertical-in-line, double-entry), which are strongly influenced by ease of maintenance, space saving and, of course, first cost. None of these preferences relate directly to hydraulic performance. This is partly because pump users are generally unaware of how strong the risk of unacceptable cavitation is in low-NPSH(A)/low-$\%Q_{bep}$ conditions. The evidence available shows impellers taking liquid into their eye from a long straight length of inlet piping (for many years promoted as the "ideal" pump inlet arrangement) are the most susceptible to severe forms of hydrodynamically induced cavitation surging. Installation of a back flow recirculation device has been shown to prevent such surging in an elegantly practical way by presenting the flow to the impeller at a desirable angle down to low $\%Q_{bep}$ values. Pump users need to be presented with qualitative data that shows how significant in reducing the risk of surging the use of an asymmetrical flow in "side-inlet" constructions is. Inlets of this type are usually found on double-entry, vertical-in-line and multistage pumps. If the presence of inlet guide vanes is an advantage this benefit also needs to be quantified.

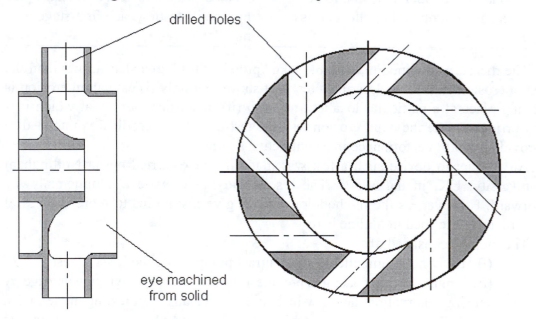

Fig. 16.3   The solid disc* impeller
   (*i.e. not cast — machined from solid metal for hygienic applications)

16.4.2   *Impeller design*

This is not an area for the pump user other than to point out that it is very difficult to weigh the balance of advantage and disadvantage of design innovations. As surging is dominated by impeller recirculation caused by the rectangular cross-sectional shape of passages between the blades it is right to question how novel ideas to disrupt recirculating flow patterns and prevent or minimise surging could be assessed.

Unconventional impellers using multi shrouded (Ref. 16.5) impeller passages could have distinct advantages. Even the author's range of solid disc impellers (Ref. 16.6 and Fig. 16.3) with a circular cross-sectional passage shape might prevent formation of pulsating flows. History has shown that dismissing novel solutions out of hand can be a commercial risk for pump manufacturers.

Standard test configuration and standard test conditions need to be devised for the hydrodynamically induced surge testing of conventional and unconventional pumps. The criteria for acceptability also need to be established.

16.5 Finally

It is hoped that these few comments, added to the detail included in earlier chapters, are of interest and may even stimulate new thinking. The partnership between pump user and pump manufacturer needs to be understood and nurtured so that progress is maximised. Students of engineering should learn from the views presented throughout this book that although a tenable position for selecting and operating pumps has been established more problems await solutions. The opportunity for innovation still presents itself.

References

In the text
16.1   Grist, E., The Development of the Centrifugal Pump, Royal Technical College, Salford, 1960 (unpublished).
16.2   IMechE Conf., Vibrations in Centrifugal Pumps, 1990.
16.3   Daugherty, R. L., Centrifugal Pumps, p. 42, McGraw-Hill, New York, 1915.
16.4   Massey, I. C. and Palgrave, R., The suction instability problem in rotodynamic pumps, NEL Conf. Pumps and Turbines, Vol.1, Paper 4-1, Sept. 1976.
        (The name of coauthor R. Palgrave was omitted from the published paper.)
16.5   Barske, U. M., Development of some unconventional centrifugal pumps, IMechE, 1960.
16.6   Grist, E., Investigation into the performance characteristics of a centrifugal pump of unconventional design, Royal Technical College, Salford, 1961 (unpublished).

Important sources of further reading on cavitation in centrifugal pumps
16.7   IMechE Cavitation Conference,
        Institution of Mechanical Engineers, London, 1974, 1983, 1992.
16.8   ASME Cavitation Forum,
        American Society of Mechanical Engineers, New York, Annual.
16.9   IHRA Cavitation Conference,
        Kluwer Academic Publishers, Dordrecht, The Netherlands, 1996.

Appendix A

The First Centrifugal Pump —
As Described by Denis Papin in 1705

The following is an extract from *Philosophical Transactions of the Royal Society of London*, 1705 (see Ref. 15.3). It is recorded as "Part of a letter from Mr D Papin … concerning an improvement to the hessian bellows". Notice the reference to "His Serene Highness", a description given only to the King of England. A portrait of Papin is presented in Fig. App-1. Copies of the drawings to which the text refers are presented at the end of this appendix as Fig. App-2.

The wording is a copy of the original text but, because page ageing stains have been removed by computer, is not an identical copy.

I Am bufie at prefent for a Coal-mine, which hath been left off becaufe of the impurity of the Air : I have therefore improved the *Heffian* Bellows : I don't queftion but you have feen that new contrivance, printed *Lipfiæ in Actis Eruditorum anno* 1699. with this Title, *Rotatilis Suctor & Preffor Haffiacus :* And it may be apply'd for Wind as well as for Water. At that time the fhape of the *Tympanum* was Cylindric, as may be feen Fig. 1, where D A F C is the circumference : C P, D P, A P, are the *Radii* which bear the Wings C m, D n, A o : C E is the aperture through which the Wind muft be driven in the direction of the Tangent C B : And it may be obferv'd that when the Engine is working, every Wing from the end of the aperture E, till it comes to the beginning of the fame aperture C, drive always the fame Air, with the fame fwiftnefs, and at the fame diftance from the Center : So that in perufing all that circumference, the Air doth find refiftance by friction, and gets nothing at all. I do therefore now make the circumference of the *Tympanum* in a Spiral fhape, which is to be feen *Fig.* 2. where the Spiral circumference is A F G B, the *Radii* are A P, C P, D P, *&c.* The Wings are A M, C N, D O, *&c.* The aperture is A B. And it is to be obferved, that every Wing in going round drives new Air, becaufe the Air which is firft in motion finds place to recede from the Center towards the Spiral circumference ; and fo it gives room to new Air to come

come to the Wing: And when the Wings come near to the Aperture, they drive their new Air into the Aperture without any friction ; and the Air which hath been first driven and removed from the Wing, cannot lose its swiftness, because the Wings which continually follow do continually drive new Air, which keeps that which is before always in the same swiftness. This new shape of the *Hessian* Bellows affords also another advantage ; because the Air in going round follows the Spiral line, which is nearer to the straight line than a Circular circumference ; and when the Air comes to the Aperture, it gets into it without any loss of substance ; but in the Cylindrical Machine, *Fig.* 2. the Air doth always go round in a Circular circumference ; and when it comes to the Aperture, the Wind is driven directly in the direction of the Tangent but just in the beginning at C ; and afterwards the Impulsion is oblique : And this obliquity is always increasing until the Wing comes to the *punctum* A : Now it is known how much diminution such an obliquity can make to the strength. I believe therefore that this Spiral figure is a good improvement to this Engine. And indeed I have made such Bellows where the *Radius* A P is but 10½ inches, the Wing A m 2 inches broad and 9 inches high ; because the *Tympanum* is also so high, or little more ; the Aperture A B is also 9 inches, or a little more, so that it makes a square hole. When I work this Engine with my Foot, it makes such a Wind that it may raise up two pounds weight ; and without doubt a stronger Man could do much more : But this is more than sufficient for our purpose, since we must but drive Air enough for the respiration of such Men that can work in the Mine ; and we may easily with Boards make Wooden Pipes, to carry the Wind to the very bottom : So that the Air within will be continually renewed as well as without.

His Serene Highness being gone to *Shualbach*, I must expect his return to apply the Engine to the Mine ; and I
hope:

hope then I fhall be able to impart the fuccefs to the Royal Society.

About the Engine, proved before a Committee of Parliament, to demonftrate the power of Water expanded by Fire, I will tell you that we have here made very good Experiments of that matter before Winter. We have raifed Water to the heighth of 70 foot, by a very commodious way, which may be yet very much improv'd ; and becaufe his Serene Highnefs was defirous to fee fomewhat more, the Engine was left too long in the River, fo that the Ice broke it, and carried away part of the fame ; and fince that time other defigns have been undertaken, fo that this Water Engine is not yet repaired : I hope in time we fhall again work about that as well as about a Furnace, to which the *Heffian* Bellows will be very ufeful. I have already made a little tryal of it, and I had a very ftrong Fire in a Furnace, to melt Glafs, Iron, or any other hard Mettal ; and yet could open the Furnace above the matter to be wrought upon, and yet no Flame would get out through the Aperture ; nor cold Air from without get into the Furnace So that it is very like this will be a great conveniency for feveral forts of Work, fince Men may work the Matters when they are moft foftned in the Fire ; and they may be drawn up Perpendicularly, that they may not be bent, as they are when we draw them Horizontally. I believe that would be good, efpecially to make eafily Glafs Pipes and Looking-glaffes of an extraordinary bignefs. It would be too long to give now the defcription of thefe Inventions and I have made mention of thefe but by the by, to fhew that the *Heffian* Bellows are an Invention that may be apply'd to feveral good ufes, and fo that deferve very much to be improved.

Fig. App-1

Denis Papin, 1647 – 1714

(a) A tangential off-take

(b) A perfect volute

Fig. App-2 Replicas of the original drawings by Papin published in 1705
(The lower drawing is shown earlier as Fig. 15.2)

INDEX